Lecture Notes in Physics

For information about Vols. 1–99, please contact your bookseller or Springer-Verlag.

Vol. 100: Einstein Symposion Berlin. Proceedings 1979. Edited by H. Nelkowski et al. VIII, 550 pages. 1979.

Vol. 101: A. Martin-Löf, Statistical Mechanics and the Foundations of Thermodynamics. V, 120 pages. 1979.

Vol. 102: H. Hora, Nonlinear Plasma Dynamics at Laser Irradiation. VIII, 242 pages. 1979.

Vol. 103: P. A. Martin, Modèles en Mécanique Statistique des Processus Irréversibles. IV, 134 pages. 1979.

Vol. 104: Dynamical Critical Phenomena and Related Topics. Proceedings, 1979. Edited by Ch. P. Enz. XII, 390 pages. 1979.

Vol. 105: Dynamics and Instability of Fluid Interfaces. Proceedings, 1978. Edited by T. S. Sørensen. V, 315 pages. 1979.

Vol. 106: Feynman Path Integrals, Proceedings, 1978. Edited by S. Albeverio et al. XI, 451 pages. 1979.

Vol. 107: J. Kijowski, W. M. Tulczyjew, A Symplectic Framework for Field Theories. IV, 257 pages. 1979.

Vol. 108: Nuclear Physics with Electromagnetic Interactions. Proceedings, 1979. Edited by H. Arenhövel and D. Drechsel. IX, 509 pages. 1979.

Vol. 109: Physics of the Expanding Universe. Proceedings, 1978. Edited by M. Demiański. V, 210 pages. 1979.

Vol. 110: D. A. Park, Classical Dynamics and Its Quantum Analogues. VIII, 339 pages. 1979.

Vol. 111: H.-J. Schmidt, Axiomatic Characterization of Physical Geometry. V, 163 pages. 1979.

Vol. 112: Imaging Processes and Coherence in Physics. Proceedings, 1979. Edited by M. Schlenker et al. XIX, 577 pages. 1980.

Vol. 113: Recent Advances in the Quantum Theory of Polymers. Proceedings 1979. Edited by J.-M. André et al. V, 306 pages. 1980.

Vol. 114: Stellar Turbulence. Proceedings, 1979. Edited by D. F. Gray and J. L. Linsky. IX, 308 pages. 1980.

Vol. 115: Modern Trends in the Theory of Condensed Matter. Proceedings, 1979. Edited by A. Pekalski and J. A. Przystawa. IX, 597 pages. 1980.

Vol. 116: Mathematical Problems in Theoretical Physics. Proceedings, 1979. Edited by K. Osterwalder. VIII, 412 pages. 1980.

Vol. 117: Deep-Inelastic and Fusion Reactions with Heavy Ions. Proceedings, 1979. Edited by W. von Oertzen. XIII, 394 pages. 1980.

Vol. 118: Quantum Chromodynamics. Proceedings, 1979. Edited by J. L. Alonso and R. Tarrach. IX, 424 pages. 1980.

Vol. 119: Nuclear Spectroscopy. Proceedings, 1979. Edited by G. F. Bertsch and D. Kurath. VII, 250 pages. 1980.

Vol. 120: Nonlinear Evolution Equations and Dynamical Systems. Proceedings, 1979. Edited by M. Boiti, F. Pempinelli and G. Soliani. VI, 368 pages. 1980.

Vol. 121: F. W. Wiegel, Fluid Flow Through Porous Macromolecular Systems. V, 102 pages. 1980.

Vol. 122: New Developments in Semiconductor Physics. Proceedings, 1979. Edited by F. Beleznay et al. V, 276 pages. 1980.

Vol. 123: D. H. Mayer, The Ruelle-Araki Transfer Operator in Classical Statistical Mechanics. VIII, 154 pages. 1980.

Vol. 124: Gravitational Radiation, Collapsed Objects and Exact Solutions. Proceedings, 1979. Edited by C. Edwards. VI, 487 pages. 1980.

Vol. 125: Nonradial and Nonlinear Stellar Pulsation. Proceedings, 1980. Edited by H. A. Hill and W. A. Dziembowski. VIII, 497 pages. 1980.

Vol. 126: Complex Analysis, Microlocal Calculus and Relativistic Quantum Theory. Proceedings, 1979. Edited by D. Iagolnitzer. VIII, 502 pages. 1980.

Vol. 127: E. Sanchez-Palencia, Non-Homogeneous Media and Vibration Theory. IX, 398 pages. 1980.

Vol. 128: Neutron Spin Echo. Proceedings, 1979. Edited by F. Mezei. VI, 253 pages. 1980.

Vol. 129: Geometrical and Topological Methods in Gauge Theories. Proceedings, 1979. Edited by J. Harnad and S. Shnider. VIII, 155 pages. 1980.

Vol. 130: Mathematical Methods and Applications of Scattering Theory. Proceedings, 1979. Edited by J. A. DeSanto, A. W. Sáenz and W. W. Zachary. XIII, 331 pages. 1980.

Vol. 131: H. C. Fogedby, Theoretical Aspects of Mainly Low Dimensional Magnetic Systems. XI, 163 pages. 1980.

Vol. 132: Systems Far from Equilibrium. Proceedings, 1980. Edited by L. Garrido. XV, 403 pages. 1980.

Vol. 133: Narrow Gap Semiconductors Physics and Applications. Proceedings, 1979. Edited by W. Zawadzki. X, 572 pages. 1980.

Vol. 134: γγ Collisions. Proceedings, 1980. Edited by G. Cochard and P. Kessler. XIII, 400 pages. 1980.

Vol. 135: Group Theoretical Methods in Physics. Proceedings, 1980. Edited by K. B. Wolf. XXVI, 629 pages. 1980.

Vol. 136: The Role of Coherent Structures in Modelling Turbulence and Mixing. Proceedings 1980. Edited by J. Jimenez. XIII, 393 pages. 1981.

Vol. 137: From Collective States to Quarks in Nuclei. Edited by H. Arenhövel and A. M. Saruis. VII, 414 pages. 1981.

Vol. 138: The Many-Body Problem. Proceedings 1980. Edited by R. Guardiola and J. Ros. V, 374 pages. 1981.

Vol. 139: H. D. Doebner, Differential Geometric Methods in Mathematical Physics. Proceedings 1981. VII, 329 pages. 1981.

Vol. 140: P. Kramer, M. Saraceno, Geometry of the Time-Dependent Variational Principle in Quantum Mechanics. IV, 98 pages. 1981.

Vol. 141: Seventh International Conference on Numerical Methods in Fluid Dynamics. Proceedings. Edited by W. C. Reynolds and R. W. MacCormack. VIII, 485 pages. 1981.

Vol. 142: Recent Progress in Many-Body Theories. Proceedings. Edited by J. G. Zabolitzky, M. de Llano, M. Fortes and J. W. Clark. VIII, 479 pages. 1981.

Vol. 143: Present Status and Aims of Quantum Electrodynamics. Proceedings, 1980. Edited by G. Gräff, E. Klempt and G. Werth. VI, 302 pages. 1981.

Lecture Notes in Physics

185

Hampton N. Shirer
Robert Wells

Mathematical Structure of the Singularities at the Transitions Between Steady States in Hydrodynamic Systems

Springer-Verlag
Berlin Heidelberg GmbH 1983

Authors

Hampton N. Shirer
Department of Meteorology
The Pennsylvania State University
University Park, PA 16802, USA

Robert Wells
Department of Mathematics
The Pennsylvania State University
University Park, PA 16802, USA

AMS Subject Classifications (1980): 58C27, 58C28, 76E30

ISBN 978-3-540-12333-0 ISBN 978-3-540-40963-2 (eBook)
DOI 10.1007/978-3-540-40963-2

Originally published by Springer-Verlag Berlin Heidelberg New York Tokyo in 1983

2153/3140-543210

Dedicated to our wives, Becky Shirer and Valerie Wells, without whose encouragement this monograph would not have been completed.

PREFACE

Since its introduction by René Thom, catastrophe theory has been a potentially valuable instrument for discovering the nature of transitional behavior in physical systems. In the excitement generated by his new view of the world, a central technical obstacle was considerably underrated, with the result that now an army of critics has replaced the original multitude of proponents.

The great promise of catastrophe theory is that with it, out of a vast number of influences on an evolving system, we may select a small number through which the rest will act to control the transitional behavior of that system. From this situation, we may obtain the classical, canonical pictures of surfaces of steady solutions and sets of bifurcation points, parameterized by a few numbers quantifying the few controlling influences. However, to apply catastrophe theory as it was originally formulated, we must have a potential or Lyapunov function for our evolutionary system. This requirement is the central technical obstacle to the rigorous application of catastrophe theory, the obstacle which was not overcome adequately in the early applications. Unfortunately it is in general, extremely difficult, if not impossible, to show that such a function exists, let alone to produce it. Consequently, most attempts at realization of the full promise of the theory cannot even get started.

Yet the canonical surfaces and singularity sets of catastrophe theory have appeared, independently of that theory, in the description of the behavior of a wide variety of physical systems. This fact is closely related to a singularity theory originated by John Mather during his work to establish the mathematical foundations of catastrophe theory. Besides enjoying the inestimable advantage of being mathematically rigorous, this generalization completely by-passes the central technical difficulty of catastrophe theory: Mather's Theory requires no Lyapunov function, and yet it can do everything that catastrophe theory, in the presence of a Lyapunov function, can do. In fact, now the appearance of the canonical surfaces and singularity sets in systems not regulated by a Lyapunov function is explained completely by Mather's Theory.

Unfortunately, Mather's Theory also includes an obstacle; it is as inaccessible to an applied physicist as anything in mathematics can be. Accordingly, to fill the gap between theory and utilization, in this monograph we first describe Mather's Theory operationally using examples instead of proofs, and then we develop a procedure for its application to physical problems whose dynamics are governed by systems of ordinary differential equations. We demonstrate the utility of our procedure by applying it to three different hydrodynamic systems. We show first how to identify the crucial parameters in the equations and then how to associate them with the corresponding physical effects. Consequently, by finding these parameters, we obtain systems that no longer must be unrealistically ideal because certain of their crucial parameters need not remain identically zero. The strength of our application of singularity theory is that we obtain a theoretical model whose solutions are directly comparable with experimental observations.

An apparent defect of Mather's Theory is that it does not, as it stands, describe the stability characteristics of the stationary solutions of a dynamical system. In particular, it does not respect Hopf bifurcations. However, it is readily extendable to a theory which does describe the stability characteristics, and we describe this extension in the final chapter.

We are deeply grateful to Professor John A. Dutton for the encouragement and advice freely given us during the lengthy evolution of this monograph from a jumble of ideas to six chapters of organized material. We also thank him for his many constructive criticisms of earlier versions of this manuscript that allowed better presentation of its contents.

We greatly appreciate the interest and useful comments given us by our colleagues. In particular, we thank Mr. David A. Yost for his help in unraveling the subtleties of horizontally and vertically heated convection, Dr. Kenneth E. Mitchell for his advice concerning quasi-geostrophic flow in a channel, and Dr. Peter Kloeden for directing us to the appropriate low-order model of rotating convection.

Finally, we are indebted to Mrs. Lori Weaver for her patient and meticulous efforts in typing the nearly unending stream of revisions of this manuscript, and to Mr. Victor King for his excellent drafting of the figures.

The research reported here was sponsored by the National Science Foundation through grants ATM 78-02699, ATM 79-08354, and ATM 81-13223 and by the National Aeronautic and Space Administration through grants NSG-5347 and NAS8-33794.

May 1983

Hampton N. Shirer

Robert Wells

TABLE OF CONTENTS

1. INTRODUCTION 1

1.1 Transitions in Hydrodynamics 1

1.2 Modeling Observed Transitions 3

2. INTRODUCTION TO CONTACT CATASTROPHE THEORY 7

2.1 The Stationary Phase Portrait 7

Example 1. The cusp and hysteresis 8

2.2 The Definitions of Mather's Theory 12

Example 2. A contact map to the cusp: embedding and hysteresis 14

Example 3. A contact map to the cusp: embedding and bifurcation 14

Example 4. A contact map to the cusp: extension 15

Example 5. A contact map to the cusp: transformation of coordinates 17

Example 6. Destruction of information: loss of periodic solutions 18

Example 7. Versal unfolding of $f(x) = x$ 20

Example 8. A versal unfolding of the Lorenz (1963) model: a preview 21

2.3 Mather's Theorems 24

Example 9. The cusp and Mather's Theorem I 25

Example 10. A versal unfolding of the Lorenz model: Mather's Theorem II 28

2.4 Altering Versal Unfoldings 30

Example 11. Codimension and the cusp 32

Example 12. Versal unfoldings of the Lorenz model: elementary alterations 33

Example 13. Versal unfoldings of the Lorenz model: alterations 36

2.5 The Lyapunov-Schmidt Splitting Procedure 38

Example 14. A versal unfolding of the Lorenz model: splitting and reducing lemmas 45

2.6 Vector Spaces and Contact Computations 47

Example 15. Codimension: Propositions 2.2 and 2.3 48

Example 16. The dimension of $\xi(n)/\xi_2(n)$: quotient spaces . . . 49

Example 17. Codimension of x^3: versal unfoldings 51

Example 18. Unfoldings of $\pm x^k$, $k \geq 2$: minimal versal forms in codimension 1 52

TABLE OF CONTENTS (Con't)

Example 19. The hyperbolic umbilic: minimal versal unfoldings 53

Example 20. The elliptic umbilic: minimal versal unfoldings 56

2.7 Classification of Singularities 57

Example 21. A versal unfolding of a nonpolynomial function: contact equivalence to a polynomial 58

Table 2.1 Corank 1 unfoldings 61

Table 2.2 Corank 2 unfoldings 61

2.8 Summary 61

3. RAYLEIGH-BÉNARD CONVECTION 67

3.1 Classification of the Singularity 69

3.2 Physical Interpretation of the Unfolding 73

4. QUASI-GEOSTROPHIC FLOW IN A CHANNEL 82

4.1 Heating at the Middle Wavenumber Only 83

4.2 Singularities in the Vickroy and Dutton Model 91

4.3 Butterfly Points in the Rossby Regime 94

5. ROTATING AXISYMMETRIC FLOW 114

5.1 The Butterfly Points 116

5.2 Unfolding about the Butterfly Point: The Hadley Problem 121

5.3 Unfolding about the Butterfly Point: The Rotating Rayleigh-Bénard Problem 123

5.4 Dynamic Similarity 125

5.4.1 Horizontal heating 131

5.4.2 Tilting domain 135

5.4.3 Other candidates 137

5.4.4 Final comments 144

6. STABILITY AND UNFOLDINGS 145

6.1 Invariant Sets of Matrices 146

Example 1. Some invariant subsets of M_2 148

6.2 Smooth Submanifolds of R^n 153

Example 2. The sphere: a 2-submanifold of R^3 154

Example 3. The double cone: a subset which is not a submanifold of R^3 155

Example 4. The cone: a subset which is not a smooth submanifold of R^3 157

Example 5. Invariant submanifolds 159

Example 6. The orbit of a matrix 159

Example 7. Some orbits in M_2 164

TABLE OF CONTENTS (Con't)

6.3 Transversality and Tangent Space 167

Example 8. Transversal curves and surfaces 167

Example 9. Transversality of two circles in the plane 169

Example 10. The tangent space at the fold on a cusp surface 176

Example 11. The tangent space of Orb(Γ) 177

Example 12. The spaces associated with transversality of a map on the cusp surface 180

Example 13. Computational verification of transversality of a map on the cusp surface 183

Example 14. Transversality of maps associated with the hyperbolic umbilic 184

6.4 Versal Unfoldings and Contact Transformations of the First Order 188

Example 15. An extended hyperbolic umbilic 191

Example 16. First-order contact transformations of the extended hyperbolic umbilic 196

Example 17. First-order contact transformation of the hyperbolic umbilic 196

6.5 Stability and First-Order Versal Unfoldings and Contact Transformations 201

Example 18. The modified Lorenz system unfolded further 203

Example 19. The stability phase portrait of a first-order versal unfolding of the Lorenz system 213

Example 20. The stability phase portrait of the original unfolding of the modified Lorenz system 217

6.6 First-Order Mather Theory 223

Example 21. The first-order Versal unfolding of x^n 233

Example 22. First-Order Versal unfolding of a fold 235

Example 23. The stability phase portrait of a general first-order versal unfolding of $g(x) = [x_2, - x_1, x_3^2]^T$ 247

6.7 Conclusion 253

APPENDIX SUMMARY OF SPECTRAL MODELS 256

A.1 The Lorenz Model 256

Table A.1 Dimensional Variables: Lorenz Model 257

Table A.2 Nondimensional Variables & Parameters: Lorenz Model 258

TABLE OF CONTENTS (Con't)

A.2 The Vickroy and Dutton Model 259
Table A.3 Nondimensional Variables & Parameters: Vickroy and Dutton Model 261
A.3 The Charney and DeVore Model 265
Table A.4 Dimensional Variables: Charney and DeVore Model 265
Table A.5 Nondimensional Variables & Parameters: Charney and DeVore Model 266
A.4 The Veronis Model 268
Table A.6 Dimensional Variables: Veronis Model 269
Table A.7 Nondimensional Variables & Parameters: Veronis Model 270
Table A.8 Butterfly Points in the Veronis Model 273
REFERENCES 274

"There is an almost forgotten branch of mathematics, called catastrophe theory, which could make meteorology a really precise science."

--from a conversation between the Venerable Parakarma and Mahnayake Thero in The Foundation of Paradise by Arthur C. Clarke, 1978.

CHAPTER 1

INTRODUCTION

This monograph presents a new method for determining the crucial parameters controlling transitions from one steady state to another in nonlinear systems of ordinary differential equations. These dynamical systems arise naturally in hydrodynamics and are called spectral models by atmospheric physicists; the systems are obtained via Fourier transformation of the governing nonlinear partial differential equations, which are usually a form of the Navier-Stokes equation together with forms of an equation of state, the First Law of Thermodynamics, and the continuity equation. Parameters in these systems represent physical effects such as externally imposed thermal gradients or rotation rates. Accordingly, the procedure includes a means by which knowledge of the crucial parameters can be used for identification of the corresponding crucial physical effects. In this monograph, we describe our technique in detail and then by way of examples, show how to use our method for gaining significant insights into the properties of the modeled physical system itself. A brief summary of the results discussed here is given in Shirer and Wells (1982).

1.1 Transitions in Hydrodynamics

A basic characteristic of nonlinear physical systems is that they exhibit transitions of one type of flow to another as the magnitude of the external forcing is varied slowly. Rayleigh-Bénard convection (Krishnamurti, 1970a, b, 1973) and heated flow in a rotating annulus (Fultz et al., 1959) are two such laboratory systems that serve as prototypes for two different scales of atmospheric motion. Both these systems exhibit sequences of flow transitions from time-independent states to increasingly more complex temporal structures as the magnitude of the thermal forcing (measured by a Rayleigh number) is increased. Moreover, most of the energy of these states is contained in a few spatial harmonics, and the number of significant harmonics increases discretely after each transition.

For example, in her Rayleigh-Bénard convection experiments, Krishnamurti (1970a, b, 1973) found the following hierarchy of transitions (Fig. 1.1). First, a motionless conductive state was observed that was replaced by a temporally-independent, two-dimensional roll dominated by one horizontal and one vertical wavenumber. For some values of the Prandtl number, which is the ratio of viscosity and thermometric conductivity, this solution exchanged stability with a stationary three-dimensional one, called bimodal convection, that is composed of two orthogonal rolls (Krishnamurti, 1970a). Eventually, the steady states are replaced by recurrent ones that are first characterized by one temporal period, and then by two periods after another transition has occurred. For large enough magnitudes of the vertical temperature gradient, the flow becomes turbulent (Krishnamurti, 1970b,

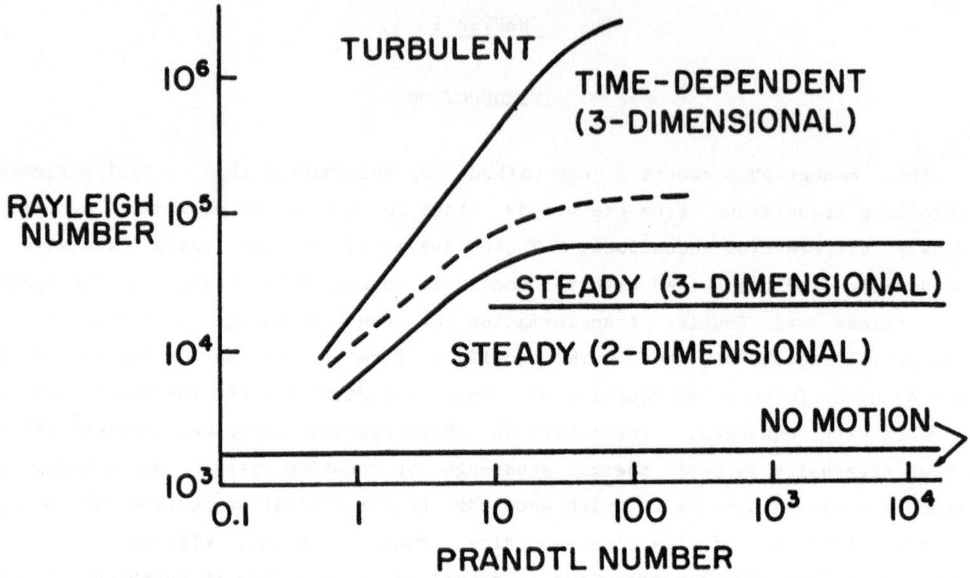

Fig. 1.1 Regimes of flow observed in Rayleigh-Bénard convection experiments, shown as functions of the Prandtl number and of the Rayleigh number, which is proportional to the vertical temperature difference (after Krishnamurti, 1970b).

1973). In their studies of Rayleigh-Bénard convection Ahlers and Behringer (1978a, b), Swinney (1978), Gollub and Benson (1978), Maurer and Libchaber (1979) and Fenstermacher et al., (1979) have reported flow transition sequences similar to those cited above. For a recent review of the subject, see the collection of articles compiled by Swinney and Gollub (1981).

Transitions between steady states are found to occur in two different ways. First, as the magnitude of the forcing is varied slowly, the original flow may be replaced smoothly by a new one. For example, such gradual transitions characterize Rayleigh-Bénard convection when the conductive state is replaced by a two-dimensional roll (Krishnamurti, 1970a). Second, the original flow may be replaced suddenly by a dramatically different one. In many of these cases, hysteresis, which appears when the values of the controlling parameters at which the transition occurs depend on the history of the system, is also observed. These types of transitions occur between convective states in Rayleigh-Bénard convection (Krishnamurti, 1970b, 1973) and between different waves in the Rossby regime of the rotating annulus (Fultz et al., 1959).

The atmosphere provides many examples of flows dominated by a few spatial harmonics in the horizontal and the vertical. This is commonly observed in large-scale mid-tropospheric flow fields, for example, but it is seen also in some cases in the planetary boundary layer as two-dimensional rolls (Sommeria and LeMone,

1978). As the values of certain external parameters are varied, each of these flows can change form suddenly as well.

In order to gain knowledge about the basic physical causes of the two types of transitions seen in hydrodynamic systems, we must develop nonlinear mathematical tools simple enough to utilize analytically but also complex enough to capture the complete range of possibilities in the neighborhood of the transitions. To accomplish this we must consider both the spatial resolution and the physical mechanisms driving and shaping the states of the system. The physical characteristics of a system are incorporated in part in the nondimensional parameters in the governing partial differential equations. In order to determine the fundamental causes of the transitions for a given spatial resolution, we must distinguish between those parameters that are crucial to describing all possible ways by which the transitions between the observed states may occur and those parameters that are important only to specifying accurately the amplitude of the solutions. In the following section, we outline how we use recent advances in nonlinear mathematics to locate these crucial parameters and their corresponding physical origins.

1.2 Modeling Observed Transitions

Once a set of partial differential equations has been converted to a system of ordinary differential equations via a Fourier transformation, it is logical to use the observed fact that laboratory and atmospheric systems often exhibit flows dominated by a few spatial harmonics to truncate the system as far as possible. The primary advantage of creating these truncated spectral models is that their steady states and temporally periodic solutions can be found analytically in many cases. Transitions in the physical system are determined by varying the values of parameters in these models and finding at which values one state exchanges linear stability with another. This can be determined relatively easily, and regions of sudden and smooth changes in flow type can be found. Because parameters in the spectral model represent physical effects, the fundamental causes of the observed transitions are revealed by such an analysis.

Although the relationship between the solutions of the governing partial differential equations and the corresponding severely truncated ordinary differential system has not been established, some spectral models of convection contain a hierarchy of transitions whose resemblence to observations of laboratory systems is striking (McLaughlin and Martin, 1975; Curry, 1978; Shirer and Dutton, 1979; Shirer, 1980). Moreover, the topological form of the solution surfaces of these and other spectral models of either shallow convection or large-scale atmospheric flow provide simple explanations for the cases of smooth transitions (bifurcation) and the cases of sudden transitions (catastrophe) and hysteresis (Lorenz, 1962, 1963; Veronis, 1966; Ogura and Yagihashi, 1970; McLaughlin and Martin, 1975; Curry, 1978, 1979; Vickroy and Dutton, 1979; Shirer and Dutton, 1979;

Wiin-Nielsen, 1979; Boldrighini and Franceschini, 1979; Charney and DeVore, 1979; Shirer, 1980; Lorenz, 1980; Mitchell and Dutton, 1981; Shirer, 1982; Yost and Shirer, 1982). However, applicability to the complete system of the temporal states in some of these ad hoc models is in question; Marcus (1981) studied numerically different truncations of the shallow Boussinesq equations and found that temporal solutions in the severely truncated models of Lorenz (1963), McLaughlin and Martin (1975) and Curry (1978) disappeared and were replaced by steady states as the modal truncation was relaxed.

If we accept the hypothesis that low-order spectral models can reproduce qualitatively the transitions observed in hydrodynamic systems, then mathematical techniques must be developed by which truncated spectral models can be made efficacious. If a model is designed properly, then the form of its solutions would be insensitive to the absence of neglected terms; mathematical tools are being developed that allow testing for such sensitivity.

For example, we seek the smallest spectral model able to represent each observed transition in such systems as Rayleigh-Bénard convection; once these are developed, relatively simple spectral models that can describe adequately the entire transition to turbulence might even be possible. Certainly, the determination of the fundamental ingredients of the transition to turbulence would advance our understanding of the dynamics of the atmosphere tremendously. For example, the development of a turbulent tail in the energy spectrum of small-scale features provides important dissipative effects for the large-scale laminar flows (Dutton, 1982).

Transitions in the mathematical model and the physical system occur at critical values of the parameters where two or more solutions meet. These critical values are called bifurcation points, or singular points, at which linear stability is also exchanged between the two states. Although the values of the singular points can be determined from a linear analysis of the model, the degree of complexity of the branching solutions can be found only from a nonlinear analysis. This degree of complexity depends on both the number of and the form of the solutions that can emanate simultaneously from the singularity on the known solution. According to the catastrophe theory of Thom (1976) (which is reviewed excellently by Poston and Stewart, 1978) or the singularity theory of Mather (1968), in some cases, the singularity type is one of only a small number of possibilities and the nonlinear structure of the steady states in the neighborhood of the singularity can be described by low-degree polynomials of a few independent variables. The low-order terms of these polynomials contain independent coefficients whose changes in values lead to fundamentally different branching behavior. When a steady-state polynomial has been written in its canonical general form, we say that the singularity has been unfolded. The coefficients of the required low-order terms are parameters that correspond in the modeled system to the crucial physical effects that govern the transitions among the observed steady states. This is a natural approach for analysis of truncated spectral models because their steady states are controlled by low-degree polynomials.

The procedure we develop in the next chapter is an application of Mather-Thom theory which can be applied to any finite dimensional system; the version we use we refer to as contact catastrophe theory. The advantage of this theory is that the Lyapunov or potential function required in catastrophe theory is not needed so that contact catastrophe theory can be applied more generally. (In catastrophe theory, the potential function allows determination of the stability of the steady states; but a suitable extension of contact catastrophe theory still allows recovery of stability information). To apply our theory, we find the singularity type, determine the required general polynomial form for the branching steady solutions and finally identify the resulting parameters with physical effects in the governing equations. This identification is accomplished by comparing the terms of the unfolded spectral model with those of the original system. Thus our approach is similar in philosophy to earlier, largely ad hoc, applications of catastrophe theory, which was used to suggest theoretically new terms in the equations governing a physical system (e.g. Chillingworth and Holmes, 1980). Our approach is more objective, however, because not only do we discover the number of critical parameters that are missing from the original model, but we determine the number of possible positions for each of them within the spectral system. In many cases, this information will lead to the discovery of a missing parameter in the governing partial differential equations themselves. More than one location for a parameter is possible, and these different implied terms in some cases can be related to vastly different physical effects. Thus we can find when different external conditions lead to qualitatively similar internal dynamics, and this can be a useful tool in finding simple ways to model in the laboratory complex forcing effects found in the atmosphere.

Moreover, there are significant improvements in the description of the observed flows that can be gained simply by adding a parameter to the equations but which eludes the addition of more harmonics to the truncation. Physical evidence of this is supplied for Rayleigh-Bénard convection by Tavantzis et al., (1978). When only a vertical heating parameter is included, they noted that some important aspects of the branching convective solution found in Rayleigh-Bénard experiments cannot be modeled adequately. The proper branching picture for Rayleigh-Bénard convection was obtained by Yost and Shirer (1982) who added a horizontal heating term to the three-component model of Lorenz (1963). Thus, horizontally varying heating rates, although weak, can change markedly the qualitative behavior of the first bifurcation; but these must be considered, because small non-zero values of this horizontal temperature gradient are always present. In Chapter 3 we show how our procedure leads to this conclusion.

Although the methods discussed in Chapter 2 apply only to transitions between steady states, their use surely will lead to improved description of temporally periodic solutions. Sometimes the coordinate system can be moved at a speed that causes the periodic states to appear stationary; in this situation, contact catastrophe theory can be applied directly. In other cases, periodic solutions branch from stationary ones, and the development of the temporal states can be reproduced well only after an adequate description of the steady ones is obtained. Finally, spectral models containing periodic solutions that depend on a parameter μ may have the steady states of other models as solutions in the limit $\mu = 0$; for example, the Shirer (1980) model reduces to the Lorenz (1963) system and the Mitchell and Dutton (1981) model reduces to a special case of the Vickroy and Dutton (1979) one. In these cases it seems reasonable to expect that the periodic solutions for $\mu \neq 0$ are represented adequately only when the solutions in the limiting steady models are fully general. In Chapter 6, we discuss an extension of contact catastrophe theory that preserves information about stability of the steady states. In this application, Hopf bifurcation points are preserved, although the stability of the branching temporally periodic solution may not be. Additional parameters are found to be required in this case, but application to specific systems is postponed to a future article.

We illustrate the technique of contact catastrophe theory with three spectral models based on two-dimensional flows. We do this for two important reasons. The first observed nonlinear states of many physical systems are quasi-two-dimensional. In order that the three-dimensional ones be described correctly, the two-dimensional flows must be represented completely. This is determined most easily from two-dimensional systems because a correctly designed three-dimensional model still will contain the two-dimensional flows as formal solutions. However, the method discussed here is not limited to two-dimensional models; it can be applied to analyze the steady states of any spectral model of any order. The ones based on two-dimensional dynamics contain fewer equations, so presentation of the technique is easier.

In the next chapter we give an overview of the contact singularity theory of Mather (1968) and then we develop our application of contact catastrophe theory for designing truncated spectral models. We work many examples to illustrate terminology and emphasize application of the therorems; the end of each example is denoted by the symbol # . The same convention is used in Chapter 6.

To show how we apply our procedure, we discuss in the remainder of the monograph spectral models of three hydrodynamic systems: the Lorenz (1963) model of Rayleigh-Bénard convection (Chapter 3), the Vickroy and Dutton (1979) model of quasi-geostrophic flow in a channel (Chapter 4) and the Veronis (1966) model of rotating axisymmetric flow (Chapter 5). These models and the principal results of our analyses of them are summarized in the Appendix.

CHAPTER 2

INTRODUCTION TO CONTACT CATASTROPHE THEORY

Truncated spectral models that we use to study nonlinear hydrodynamic systems are systems of ordinary differential equations. Thus there is a large body of mathematical theory available, but in order to apply some of it we must organize it into a suitable form. In this chapter we accomplish this organization by giving the requisite definitions and theorems. As we introduce new words and concepts we clarify their necessity and power with examples. For ease in application to the spectral systems of Chapters 3-5, we summarize the entire method in Section 2.8. Comments on extensions to Hopf bifurcations in ordinary differential systems is given in Chapter 6.

2.1 The Stationary Phase Portrait

Suppose that

$$\dot{x} = F(x,\lambda) \quad \text{with} \begin{cases} x = (x_1,\ldots,x_n) \\ F = \left(F_1(x,\lambda),\ldots,F_n(x,\lambda)\right) \\ \lambda = (\lambda_1,\ldots,\lambda_p) \end{cases} \tag{2.1}$$

is a family of smooth differential equations that are parameterized smoothly by the p-dimensional parameter λ. A smooth differential equation is one whose right side F is composed of C^∞ functions, which are functions for which all the partial derivatives are continuous. We seek information about the stationary phase portrait of each of the differential equations in the family, as well as the phase portrait of the family as a whole. That is, for each differential equation, we would like to know the location, number and type of its stationary points $F(x,\lambda) = 0$ as the value of λ varies. For the family as a whole we would like to know the location and type of its bifurcation points, of its hysteresis loops and of its exchanges of stability. Bifurcation points or singular points are the values of λ at which two or more solutions of $F(x,\lambda) = 0$ meet; a hysteresis loop occurs when the exchange of stability between two states is different depending on whether the magnitude of the parameter is increasing or decreasing. A closed hysteresis loop is one in which the initial and final values of x are equal when the initial and final values of λ are equal. We use the term stationary phase portrait to denote loosely all the above types of information relating to stationary behavior that a single differential equation or a parameterized family of differential equations might contain.

A fundamental characterization of this stationary phase portrait is the number of stationary points of (2.1) for a given value of λ. We learn significant qualitative information about the stationary phase portrait by answering such questions as: How does this number change as the value of λ is varied? At which values of λ does the number change? Thus, for the corresponding differential equation, we may divide λ-space (or parameter space) into regions, each characterized by a fixed number of stationary points, and then we may try to perceive the spatial arrangement of these regions.

A more sophisticated aspect of the stationary phase portrait is the way in which the corresponding steady states change. Transitions between steady states can occur in either a smooth or sudden manner; an example of sudden transitions is to be found in hysteresis. Here we seek to answer such questions as: What is a sufficient condition for the occurrence of sudden transitions? What are the ranges of values of the parameter λ that compose the hysteresis loop? To clarify the above questions, let us consider a simple form of (2.1)

Example 1. The cusp and hysteresis

We choose the one-dimensional system

$$\dot{x} = -x^3 + \lambda_1 x + \lambda_2 \tag{2.2}$$

For a given value of $\lambda = (\lambda_1, \lambda_2)$, the stationary points of (2.2) are the real solutions of

$$0 = x^3 - \lambda_1 x - \lambda_2 \tag{2.3}$$

The transition from three simple real roots to a single simple real root occurs at those values of (λ_1, λ_2) for which some solution of (2.3) also satisfies

$$0 = 3x^2 - \lambda_1 \tag{2.4}$$

Eliminating x from (2.3) and (2.4), we see that those values of λ are the ones satisfying

$$\lambda_1^3/27 - \lambda_2^2/4 = 0 \tag{2.5}$$

whose locus is the cusp pictured in Fig. 2.1. For the magnitudes of λ within the cusp, (2.3) has three simple real roots; for the values outside, (2.3) has a single simple real root. At the vertex $\lambda_1 = \lambda_2 = 0$, (2.3) has a triple real root, and the vertex is called a cusp point; on the rest of the cusp, (2.3) has a single simple real root as well as a double one, and these points are called fold points. Moreover, because the points that compose the cusp are points at which two or more solutions meet, we call these points singular points, and they are given by (2.5).

The situation may be clarified further by plotting the locus of roots of (2.3) in $(x, \lambda_1, \lambda_2)$-space; the resulting figure, called a cusp surface, is shown in Fig.

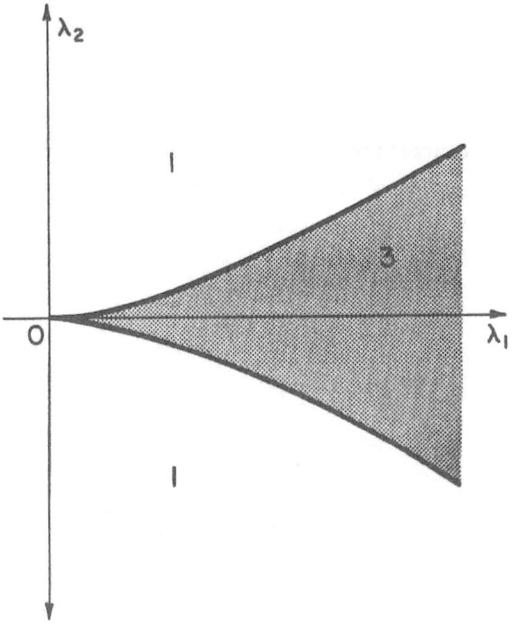

Fig. 2.1 Fold points at which two or more steady states of (2.2) meet; the locus of points is given by (2.5) and forms a cusp. For values of λ_1 and λ_2 inside the cusp $(\lambda_1{}^3/27 - \lambda_2{}^2/4 > 0)$, 3 real equilibria exist, and for values outside the cusp, only 1 exists.

2.2. When $\lambda_1 > 0$ and $\lambda_2 = 0$, the three roots x_s of (2.3) are

$$x_s = \begin{cases} \sqrt{\lambda_1} \\ 0 \\ -\sqrt{\lambda_1} \end{cases} \tag{2.6}$$

which obey the relation $-\sqrt{\lambda_1} < 0 < \sqrt{\lambda_1}$. We obtain the stability of the solutions x_s by writing

$$x = x' + x_s \tag{2.7}$$

substituting the result into (2.2), and then neglecting the terms involving x'^2 and x'^3. Thus we obtain

$$\dot{x}' = -(3\,x_s{}^2 - \lambda_1)x' \tag{2.8}$$

The solution of (2.8) is

$$x' = \hat{x}\exp(\omega t) \tag{2.9}$$

in which

$$\omega = -(3x_s^2 - \lambda_1) \tag{2.10}$$

If $\omega > 0$, then the perturbation x' grows and the steady state is unstable or unobservable; if $\omega < 0$, then x' decays and the steady state is stable or observable. We find the stable and unstable solutions in (2.6) by substituting them into (2.10) to obtain respectively

$$\omega = \begin{cases} -2\lambda_1 \\ \lambda_1 \\ -2\lambda_1 \end{cases} \tag{2.11}$$

so that $-\sqrt{\lambda_1}$ and $\sqrt{\lambda_1}$ are stable stationary points of (2.2) while the other, 0, is unstable. Because $\omega = -3x_s^2 + \lambda_1$ vanishes on the surface of Fig. 2.2 exactly along the two curves separating the middle pleat from the others, we conclude that the middle pleat consists of unstable stationary points while the other two consist of stable stationary points.

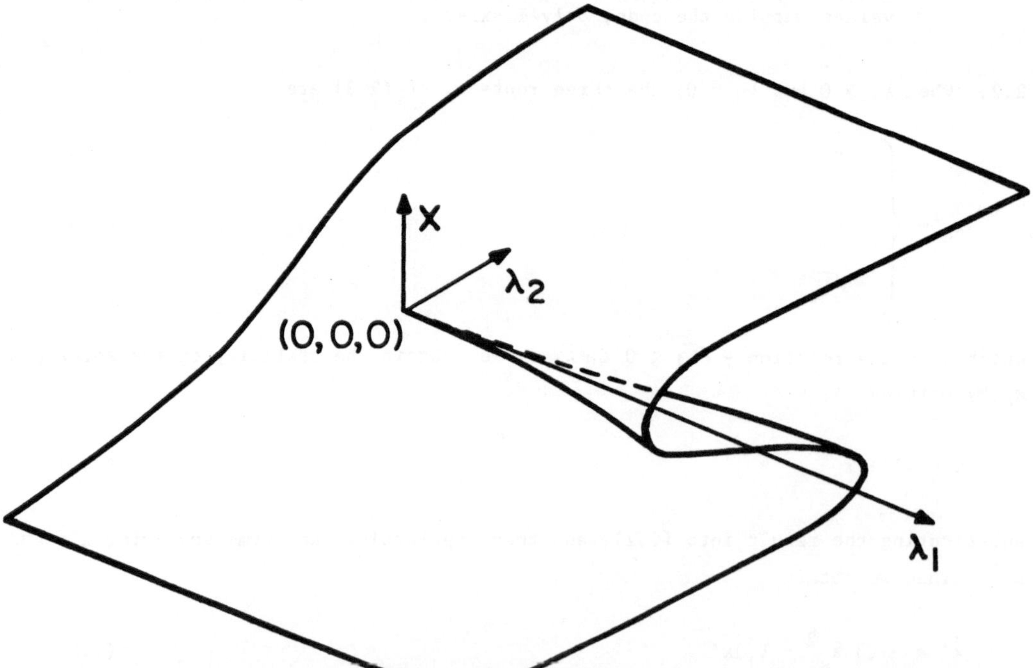

Fig. 2.2 The magnitude x of the steady states of (2.2) as functions of the parameters λ_1 and λ_2; the resulting surface is called a cusp surface.

We remark that when $\lambda_1 = 0$ and $\lambda_2 = 0$, which is a bifurcation point of cusp type, the three solutions of (2.3) meet at $x = 0$. At this bifurcation or singular point, we see from (2.10) that $\omega = 0$ so that the trivial solution is neutrally stable there. Thus, we see that singular points and stability exchange are intimately linked. In practice we find singular points by first performing linear stability analyses to find the parameter values at which stability is lost (i.e. at which $\mathrm{Re}(\omega) = 0$).

Suppose now that we hold λ_1 in (2.2) fixed at some positive value and for some value $\varepsilon > 0$ we vary λ_2 about a fold point between the values $-2(\lambda_1^3/27)^{1/2} - \varepsilon$ and $2(\lambda_1^3/27)^{1/2} + \varepsilon$. Then the solution to (2.2) will tend asymptotically to, say, the lower stable stationary point (Fig. 2.3). But as the value of λ_2 is increased past $2(\lambda_1^3/27)^{1/2}$, this stationary point ceases to exist and the solution of (2.2) will tend to the upper stable stationary point. Now as we decrease the value of λ_2, the solution to (2.2) will be on the upper stable stationary point, even as the value of λ_2 is decreased past $2(\lambda_1^3/27)^{1/2}$. This situation will continue until the value of λ_2 is decreased further past $-2(\lambda_1^3/27)^{1/2}$, at which value the upper stable stationary point ceases to exist and the solution of (2.2) must snap back to the lower stable stationary point. Thus, as we vary the values of λ_2 back and forth between $-2(\lambda_1^3/27)^{1/2} - \varepsilon$ and $2(\lambda_1^3/27)^{1/2} + \varepsilon$, the upper and lower stable stationary solutions trade places as we increase the value of λ_2 past

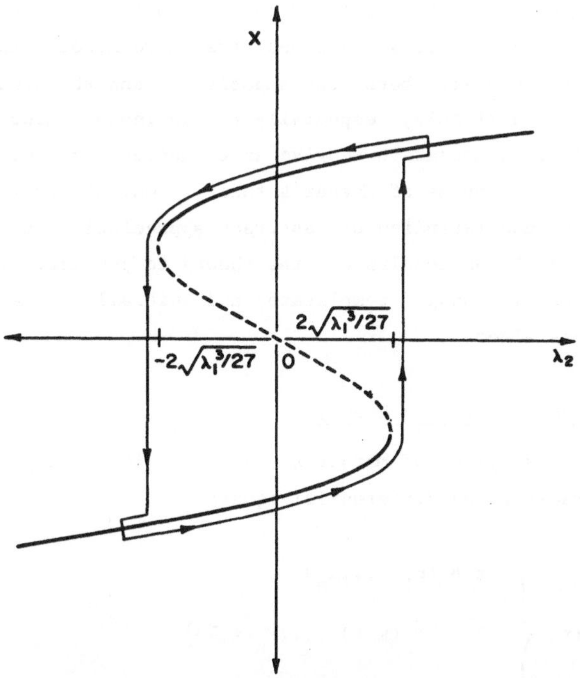

Fig. 2.3 A closed hysteresis loop of steady states obtained by varying the value of only λ_2 when $\lambda_1 > 0$ in (2.2). Stable, or observable, branches are denoted with solid lines and unstable, or unobservable, ones with dashed lines.

$\lambda_2 = -2(\lambda_1{}^3/27)^{1/2}$ or past $\lambda_2 = 2(\lambda_1{}^3/27)^{1/2}$. We find a delay in the exchange because we must pass $-2(\lambda_1{}^3/27)^{1/2}$ from right to left and $2(\lambda_1{}^3/27)^{1/2}$ from left to right in Fig. 2.3 for the snap-back to occur. Thus, Example 1 displays a closed hysteresis loop as well as simple transitions in the number of stationary points as the value of λ is varied. We note that if we had chosen our limits on λ_2 to be $-2(\lambda_1{}^3/27)^{1/2} + \varepsilon$ and $2(\lambda_1{}^3/27)^{1/2} + \varepsilon$ then the hysteresis loop would not have been closed because the initial and final values of x would have been different. #

Returning to our general parameterized differential equation (2.1), $\dot{x} = F(x,\lambda)$, we localize the problem by considering only stationary points near the origin $x = 0$ and by only considering values of λ near the singular point $\lambda = 0$. If the stationary solution of our original differential system (2.1) is not $x_s = 0$, then we may homogenize our system by writing $y = x - x_s$ and $\mu = \lambda - \lambda_s$, substituting these into (2.1) and then obtaining a new system $\dot{y} = G(y,\mu)$ which has stationary points near $(y,\mu) = (0,0)$. Moreover, we have that $G(0,0) = 0$. Now we may regard $\dot{x} = F(x,\lambda)$ as a parameterized perturbation of

$$\dot{x} = f(x) = F(x,0) \tag{2.12}$$

and we assume that $f(0) = F(0,0) = 0$.

With the assumptions above, we now face the nontrivial problem of describing the stationary phase portrait of (2.1) near the origin, with values of λ small. Mather's Theory of Contact Transformations gives a remarkably thorough solution to this problem. Unfortunately, both the exposition and the proofs of this theory (Mather, 1968) are formidable, especially to an investigator not familiar with module theory. In this monograph we give an elementary account--without proofs--of the basic notions and theorems of Mather's theory. Our objective is not to supply a general theoretical understanding or abstract appreciation of the theory, but to expedite application of the results of the theory to physical problems by carrying out certain elementary, though complicated and initially unfamiliar, computations with ordinary linear algebra.

2.2 The Definitions of Mather's Theory

Now we begin a summary of Mather's theory. The central notion is that of constructing a parameterized differential equation

$$\dot{x} = F(x,\lambda) \text{ with } \begin{cases} x = (x_1,\dots,x_n) \\ F = \left(F_1(x,\lambda),\dots,F_n(x,\lambda)\right) \\ \lambda = (\lambda_1,\dots,\lambda_p) \end{cases} \tag{2.13}$$

from a canonical one

$$\dot{y} = V(y,\mu) \text{ with } \begin{cases} y = (y_1,\dots,y_n) \\ V = \left(V_1(y,\mu),\dots,V_n(y,\mu)\right) \\ \mu = (\mu_1,\dots,\mu_q) \end{cases} \tag{2.14}$$

by means of a contact map; we note that $p \geq q$ so that λ and μ may represent different numbers of parameters in (2.13) and (2.14). A contact map consists of an invertible $n \times n$ matrix $M(x,\lambda)$ depending smoothly on (x,λ), a parameterized coordinate transformation $y = y(x,\lambda)$, satisfying $y(0,0) = 0$, and smooth functions $\mu = \mu(\lambda)$. We may identify the contact map T specified by this data with the data itself,

$$T = \left(M(x,\lambda);\ y(x,\lambda);\ \mu(\lambda)\right) \tag{2.15}$$

and we define its effect on $V(y,\mu)$ by setting

$$(T\ V)\ (x,\lambda) = M(x,\lambda)\cdot V(y(x,\lambda),\ \mu(\lambda)) \tag{2.16}$$

In this case we say that T pulls back V to T V.

We note that our use of the term "contact" does not refer to the preservation of a Hamiltonian structure, but refers instead to preservation of the order of contact, or tangency, between the graphs of V and the linear space $y = 0$. We adopt this usage here because it is standard in the literature of singularity theory.

Suppose that the right sides of (2.13) and (2.14) are related by

$$F = T\ V \tag{2.17}$$

What is the relation between the differential equation $\dot{x} = F(x,\lambda)$ and $\dot{y} = V(y,\mu)$? Upon combining (2.16) and (2.17), we have

$$F(x,\lambda) = M(x,\lambda)\cdot V(y(x,\lambda),\ \mu(\lambda)) \tag{2.18}$$

Because $M(x,\lambda)$ is invertible, we conclude from (2.18) that

$$F(x,\lambda) = 0 \quad \text{if and only if} \quad V(y(x,\lambda),\ \mu(\lambda)) = 0 \tag{2.19}$$

Thus, x_s is a stationary point of $\dot{x} = F(x,\lambda)$ if and only if $y_s = y(x_s,\lambda)$ is a stationary point of $\dot{y} = V(y,\ \mu(\lambda))$. That is, the coordinate transformation $y = y(x,\lambda)$ carries the set of stationary points of $\dot{x} = F(x,\lambda)$ onto precisely that of $\dot{y} = V(y,\ \mu(\lambda))$.

Alternatively, let the inverse coordinate transformation be given by $x = x(y,\lambda)$ so that $x = x(y(x,\lambda),\ \lambda)$; then $x(y,\lambda)$ transforms the set of stationary points of

$\dot{y} = V(y, \mu(\lambda))$ onto precisely that of $\dot{x} = F(x,\lambda)$. Thus, for a fixed value of λ we may identify the stationary phase portrait of $\dot{x} = F(x,\lambda)$ with that of $\dot{y} = V(y, \mu(\lambda))$ via the coordinate transformation $x(y,\lambda)$.

In practice, $\dot{y} = V(y,\mu)$ contains more readily obtainable information about the steady states than does the original system $\dot{x} = F(x,\lambda)$, and the parameterized coordinate transformation $x = x(y,\lambda)$ carries enough of it back to reconstruct all of the stationary point information of $\dot{x} = F(x,\lambda)$. As we find later, T preserves the form of the set of steady states of $\dot{x} = F(x,\lambda)$ but destroys all other information such as linear stability properties or presence of recurrent solutions. Nonetheless, by knowing the singular points of the stationary phase portrait we may often use the principle of exchange of stability to determine the stability of the stationary points (Iooss and Joseph, 1980).

Example 2. A contact map to the cusp: embedding and hysteresis

Let

$$V(y,\mu) = - y^3 + \mu_1 y + \mu_2 \tag{2.20}$$

so that $\dot{y} = V(y,\mu)$ is the re-labeled equation (2.2) of Example 1. We consider the contact map

$$T = \left(M = 1;\ y(x,\lambda) = x;\ \mu_1(\lambda) = 1,\ \mu_2(\lambda) = \lambda\right) \tag{2.21}$$

in which λ now has only one component. Then $F = T\,V$ is given by

$$F(x,\lambda) = - x^3 + x + \lambda \tag{2.22}$$

We have shown the stationary points of $\dot{x} = F(x,\lambda)$ in Fig. 2.3, if we set $\lambda_1 = 1$ and $\lambda = \lambda_2$. As shown in Example 1, this differential equation exhibits hysteresis as we vary the value of λ back and forth between $- 2(1/27)^{1/2} - \varepsilon$ and $2(1/27)^{1/2} + \varepsilon$. Clearly, this hysteresis is pulled back from the hysteresis in the equation $\dot{y} = V(y,\mu)$. This preservation of the stability properties is not guaranteed in higher order differential systems, however, because of the possibility of Hopf bifurcation to a temporally periodic solution. #

Example 3. A contact map to the cusp: embedding and bifurcation

We apply a different contact map,

$$T = \left(M = 1;\ y(x,\lambda) = x;\ \mu_1(\lambda) = \lambda,\ \mu_2(\lambda) = 0\right) \tag{2.23}$$

to (2.20) so that now $F = T\,V$ is given by

$$F(x,\lambda) = - x^3 + \lambda\, x \tag{2.24}$$

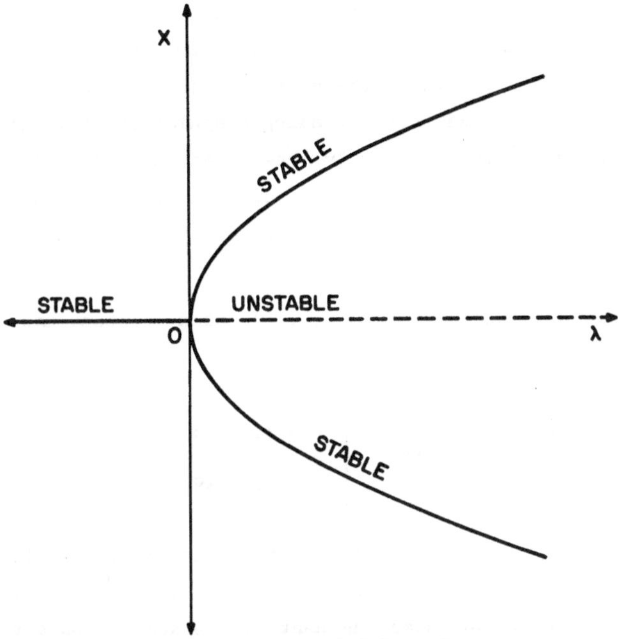

Fig. 2.4 Bifurcation diagram obtained by varying the value of λ in (2.24). Here the stable solutions are denoted by solid lines, unstable ones by dashed lines.

We display the stationary points of (2.24) in Fig. 2.4. Here the system $\dot{x} = F(x,\lambda)$ exhibits a bifurcation at the singularity $\lambda = 0$. Again the branching and stability behavior pulls back from corresponding behavior in the canonical system $\dot{y} = V(y,\mu)$.#

Both of the pulled-back systems in Examples 2 and 3 above are special cases of the more general system (2.20); that is, Figs. 2.3 and 2.4 are cross sections of the steady state solution surface of (2.20) (Fig. 2.2). A system F with more parameters may be pulled back from one V with fewer, but the stationary phase portrait of the pulled-back system F still is derived from that of V.

Example 4. A contact map to the fold: extension

To see the above fact, let

$$V(y,\mu) = y^2 - \mu \tag{2.25}$$

in which now μ has a single component. Define the contact map T by

$$T = \left(M = 1;\ y(x,\lambda) = x;\ \mu(\lambda_1, \lambda_2) = \lambda_1\right) \tag{2.26}$$

with $\lambda = (\lambda_1, \lambda_2)$. Then $F = T\,V$ satisfies

$$F(x, \lambda_1, \lambda_2) = x^2 - \lambda_1 \tag{2.27}$$

and we picture the stationary points of (2.27) in Fig. 2.5. The fact that the singularity $\mu = 0$ is the barrier between those parameters for which there are no stationary points and those for which there are two in the system $\dot{y} = V(y,\mu)$ becomes the fact that $\lambda_1 = 0$ is the corresponding singularity or barrier in the system $\dot{x} = F(x,\lambda)$. The system $\dot{y} = V(y,\mu)$ merely has been extended in an inessential direction (that of λ_2) to obtain $\dot{x} = F(x,\lambda)$; addition of the second parameter λ_2 by the action of T (2.26) introduces no new qualitative information about the branching steady states of (2.25). #

The function $\mu(\lambda)$ in the definition (2.16) of a contact map plays a crucial role. Suppose that the matrix

$$d\mu(0) = \left[\frac{\partial \mu_j}{\partial \lambda_i}(0) \right] \qquad \text{with} \quad \begin{cases} i = 1,\dots,p \\ j = 1,\dots,q \end{cases} \tag{2.28}$$

has maximal rank. (If not, then alter the contact map T via a slight perturbation of $\mu(\lambda)$). In that case, when there are more μ-parameters than λ-parameters (i.e. $q > p$), then $\mu(\lambda)$ is an <u>embedding</u> near $\lambda = 0$ and we have the situation of Examples 2 and 3: the pulled-back system F is a cross section of V. When there are fewer μ-parameters than λ-parameters (i.e. $q < p$), then $\mu(\lambda)$ is a <u>projection</u> and we have the situation of Example 4: the pulled-back system F is an extension of (a

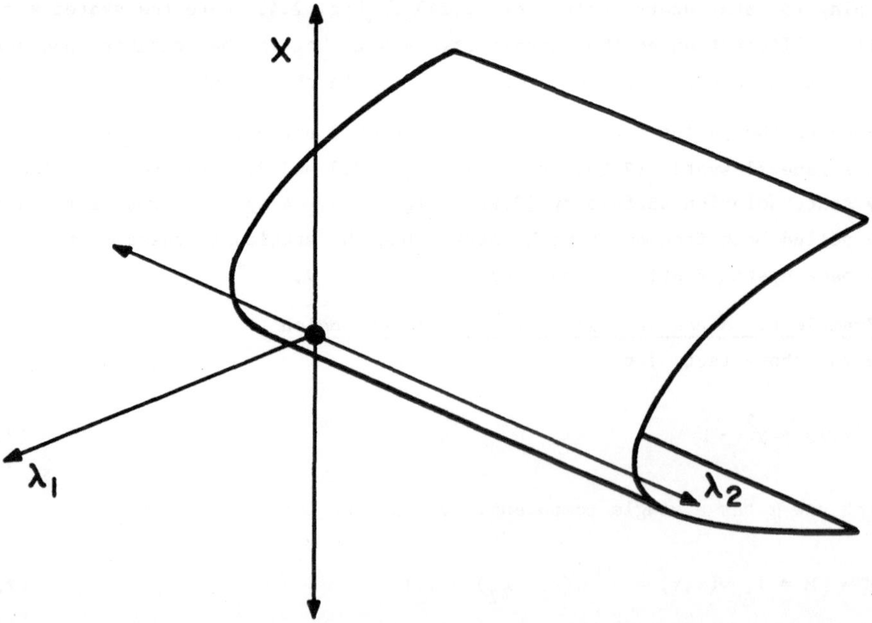

Fig. 2.5 Magnitude x of solutions to (2.27) as functions of λ_1 and λ_2. The resulting surface is a fold stretched in an inessential direction λ_2.

portion of) V in an inessential direction. Finally, when there are exactly as many μ-parameters as λ-parameters (i.e. $q = p$), then $\mu(\lambda)$ is a transformation of coordinates and the pulled-back system F is a portion of V as in the following:

Example 5. A contact map to the cusp: transformation of coordinates

We apply the contact map

$$T = \left(M = 1;\ y(x,\lambda) = x;\ \mu_1(\lambda) = \lambda_1 + 3,\ \mu_2(\lambda) = \lambda_2 + 2\right) \tag{2.29}$$

to (2.20) to obtain

$$F(x,\lambda) = -y^3 + (\lambda_1 + 3)y + \lambda_2 + 2 \tag{2.30}$$

We merely have shifted our region of interest from the neighborhood of the vertex (0,0) of the cusp to the neighborhood of the point (3,2) on it (Fig. 2.6). As we see later, however, a shift of the focus of attention from the vertex of the cusp to some other region leads to less information; we obtain the most information if we shift to the vertex of the cusp because the singularity has the highest order there. #

Returning to the general situation, we note that if there are as many parameters μ as there are parameters λ and if $\mu(\lambda)$ is invertible with $\mu(0) = 0$ and

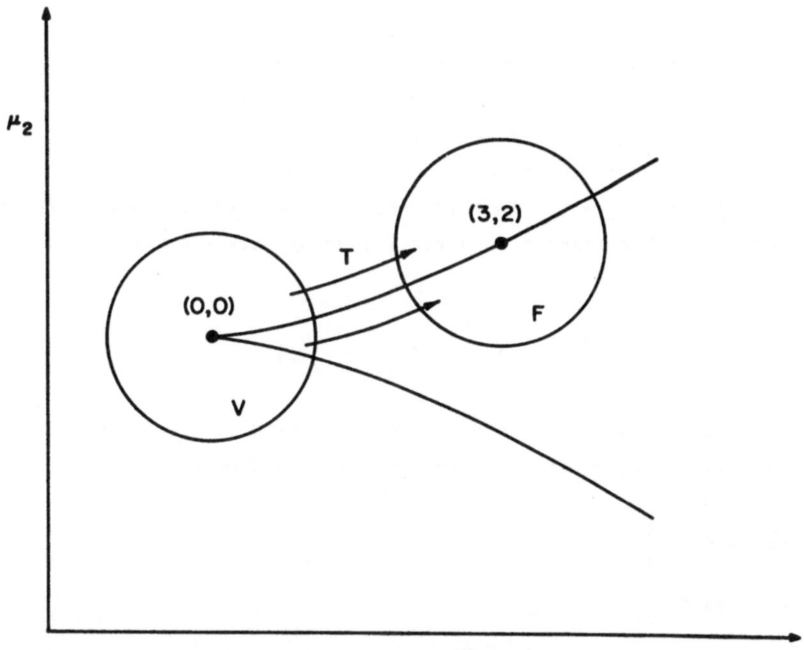

Fig. 2.6 An example of a contact map T from the region centered on the cusp point (0,0) to the region centered on the fold point (3,2).

$y(0,0) = 0$, then T is invertible; therefore for $F = T\,V$ the two systems $\dot{x} = F(x,\lambda)$ and $\dot{y} = V(y,\mu)$ have equivalent stationary phase portraits. In this case, T is called a contact transformation (which, as noted earlier, is not to be confused with the contact or canonical transformations of classical Hamiltonian mechanics), and we say that F and V are contact equivalent. We notice that although a contact transformation T ensures that the stationary phase portraits of F and V are equivalent, the rest of the phase portraits relating to temporal behavior and linear stability do not necessarily correspond. In particular, closed orbits (periodic solutions) and Hopf bifurcations to them may be lost under a contact transformation.

Example 6. Destruction of information: loss of periodic solutions

Let

$$V(y,\mu) = \begin{bmatrix} y_2 + y_1[\mu - (y_1^2 + y_2^2)] \\ -y_1 + y_2[\mu - (y_1^2 + y_2^2)] \end{bmatrix} \qquad (2.31)$$

Then the limit set in the phase portrait of the differential system

$$\left.\begin{aligned} \dot{y}_1 &= y_2 + y_1[\mu - (y_1^2 + y_2^2)] \\ \dot{y}_2 &= -y_1 + y_2[\mu - (y_1^2 + y_2^2)] \end{aligned}\right\} \qquad (2.32)$$

exhibits a transition at $\mu = 0$ from a sink $y = 0$ for $\mu < 0$, to a source $y = 0$ and an attracting periodic solution for $\mu > 0$. Such a transition is called a Hopf bifurcation because a theorem of Hopf (Marsden and McCracken, 1976) reduces the detection of its occurrence to the computation of the eigenvalues of the system (2.32) linearized about $y = 0$. In this case, however, (2.32) is sufficiently simple that we may display this transition directly, using only elementary calculations.

We notice that

$$\frac{d}{dt}(y_1^2 + y_2^2) = 2(y_1^2 + y_2^2)\,[\mu - (y_1^2 + y_2^2)] \qquad (2.33)$$

Thus, $\mu < 0$ implies that $y = 0$ is an attracting fixed point, and $\mu > 0$ implies that $y = 0$ is a repelling fixed point. Moreover, for $\mu > 0$ we find that

$$\left.\begin{aligned} y_1 &= \sqrt{\mu}\ \sin t \\ y_2 &= \sqrt{\mu}\ \cos t \end{aligned}\right\} \qquad (2.34)$$

is an attracting periodic solution that meets $y = 0$ at $\mu = 0$ (Fig. 2.7). These facts follow immediately from inspection of (2.32) and (2.33). To see that $y = 0$ is actually a sink (or attracting whirlpool) for $\mu < 0$ and actually a source for

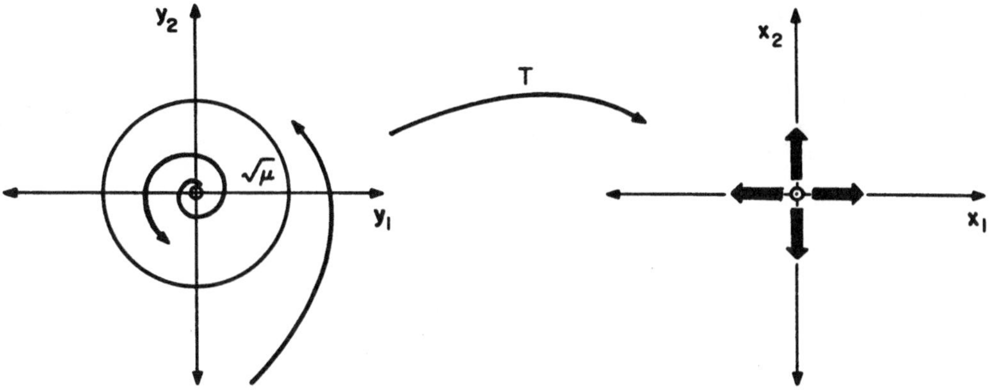

Fig. 2.7 Results of the contact transformation T (2.37) of (2.31) that retains only the existence of the fixed point at the origin but alters its type and destroys the attracting periodic solution (2.34).

$\mu > 0$, we must examine the linearized form of (2.32) about $y = 0$:

$$\begin{bmatrix} \dot{y}_1' \\ \dot{y}_2' \end{bmatrix} = \begin{bmatrix} \mu & 1 \\ -1 & \mu \end{bmatrix} \begin{bmatrix} y_1' \\ y_2' \end{bmatrix} = E\, y' \tag{2.35}$$

The eigenvalues ω of the matrix E are

$$\omega = \mu \pm i \tag{2.36}$$

so that indeed, the solutions of (2.32) spiral in toward $y = 0$ for $\mu < 0$ and out from $y = 0$ for $\mu > 0$ (Hirsch and Smale, 1974). Furthermore, from (2.36) we see that the Hopf Bifurcation Theorem implies that a periodic solution branches from $y = 0$ at $\mu = 0$ because $\mathrm{Re}(\omega) = 0$ and $\partial[\mathrm{Re}(\omega)]/\partial\mu = 1 \neq 0$; of course, we have given this solution already in (2.34).

Now we apply the contact transformation

$$T = \left[M = \frac{1}{(\mu - y_1^2 - y_2^2)^2 + 1} \begin{bmatrix} \mu - y_1^2 - y_2^2 & -1 \\ 1 & \mu - y_1^2 - y_2^2 \end{bmatrix} ; \quad y(x,\lambda) = x;\ \mu(\lambda) = \lambda \right] \tag{2.37}$$

to (2.31) to obtain

$$F(x,\lambda) = \begin{bmatrix} x_1 \\ x_2 \end{bmatrix} \tag{2.38}$$

The limit set of

$$\begin{aligned} \dot{x}_1 &= x_1 \\ \dot{x}_2 &= x_2 \end{aligned} \tag{2.39}$$

exhibits none of the transitions found in the limit set of (2.32), for the limit set of (2.39) contains only a repelling node (Fig. 2.7). Everything is lost by the action of T, except the stationary phase portrait: both (2.32) and (2.39) have the trivial solution $x_1 = x_2 = 0$ or $y_1 = y_2 = 0$. However, we repeat that with use of the principle of exchange of stability, a great deal of information about stability may be recovered from the branching of stationary solutions within the stationary phase portrait. We discuss minor modifications of the notion of contact transformations that are nearly as easy to handle and that preserve more of the information in the limit set regarding stability and Hopf bifurcation in Chapter 6. #

Because we are interested in the system $\dot{x} = F(x,\lambda)$ only for values of λ near the singularity 0, the system $\dot{x} = f(x)$ where $f(x) = F(x,0)$ (cf. (2.12)) plays a central role. We say that $F(x,\lambda)$ is an <u>unfolding</u> of $F(x,0) = f(x)$ about the singularity $\lambda = 0$. An unfolding of $f(x)$ which is maximally complicated with respect to contact maps we will call versal. More precisely, an unfolding $V(y,\mu)$ of $f(y)$ is <u>versal</u> if and only if for <u>any</u> other unfolding $F(x,\lambda)$ of $f(x)$ there is a contact map T such that F = T V. Because the ersatz word "versal" may be irritating, we review the curious reasoning behind its etymology. An unfolding such as the one above may be regarded as "universal" because every other unfolding may be obtained from it. However, a given function may have several different such unfoldings; that is, such unfoldings are not <u>uni</u>que. Consequently, the convention has been to drop the prefix "uni" from "universal" and thus to arrive at the term "versal".

At first sight, it would seem unlikely that versal systems exist: there would seem to be an infinitely great variety of unfoldings $F(x,\lambda)$ of a fixed $f(x)$--too great a variety for them all to be classified by a single canonical form $V(y,\mu)$ containing finitely many parameters $\mu_1,\ldots,\mu_q$. However, consideration of a simple example suggests that versal unfoldings in fact might exist for suitable $f(x)$.

<u>Example 7. Versal unfolding of $f(x) = x$</u>

Again we consider a one-dimensional example; that is, $x = x_1$. We take

$$f(x) = x \tag{2.40}$$

Then it is easy to see that

$$V(y) = y \tag{2.41}$$

is a versal unfolding of $f(y)$. In particular, notice that no parameters μ are needed.

To see that (2.41) is a versal unfolding of (2.40), let $F(x,\lambda)$, with $\lambda = (\lambda_1,\ldots,\lambda_p)$, be an unfolding of $f(x) = x$. That is, $F(x,0) = f(x) = x$. By Taylor's Theorem, we have

$$F(x,\lambda) = x + \sum_{i=1}^{p} \lambda_i \, g_i(x,\lambda) \tag{2.42}$$

where we have used the integral form of the remainder,

$$g_i(x,\lambda) = \int_0^1 \frac{\partial F(x,t,\lambda)}{\partial \lambda_i} \, dt \tag{2.43}$$

Then

$$y(x,\lambda) = x + \sum_{i=1}^{p} \lambda_i \, g_i(x,\lambda) \tag{2.44}$$

is a coordinate transformation for each sufficiently small value of λ, and the contact map

$$T = \left(M = 1;\; y(x,\lambda) = x;\; \text{no function } \mu(\lambda)\right) \tag{2.45}$$

satisfies $F = T\,V$, so that $V(y) = y$ is versal. #

The next most complicated one-dimensional example is given by $f(x) = x^2$. In this case, Mather's theory (or catastrophe theory) implies that $V(y,\mu) = y^2 + \mu$, with μ a single parameter, is a versal unfolding about $\mu = 0$. However, before turning to see how Mather's theory enables us to detect and construct versal unfoldings such as this one, we consider one example of the type discussed in Chapters 3-5 in order to see how versality applies in a concrete example.

<u>Example 8. A versal unfolding of the Lorenz (1963) model: a preview</u>
The spectral model of Lorenz (1963) of the shallow Boussinesq equations (see Chapter 3) at the bifurcation or singular point $r = 1$ is given by

$$\left.\begin{aligned} \dot{x}_1 &= \sigma(x_2 - x_1) \\ \dot{x}_2 &= -\,x_1\,x_3 + x_1 - x_2 \\ \dot{x}_3 &= x_1\,x_2 - b\,x_3 \end{aligned}\right\} \tag{2.46}$$

in which we require that $b > 0$ and $\sigma > 0$. As mentioned earlier, the singular value $r = 1$ can be found by a linear stability analysis of the trivial solution. For the Lorenz model, we have

$$f(x) = \begin{bmatrix} \sigma(x_2 - x_1) \\ - x_1 x_3 + x_1 - x_2 \\ x_1 x_2 - b x_3 \end{bmatrix} \tag{2.47}$$

As we shall see in Section 2.3, Example 10, and apply in Section 3.2, Mather's theory implies that a versal unfolding of $f(x)$ about $r = 1$ is given by

$$V(y,\mu) = \begin{bmatrix} \sigma(y_2 - y_1) \\ - y_1 y_3 + y_1 - y_2 + \mu_2/b \\ y_1 y_2 - b y_3 - \mu_1 \end{bmatrix} \tag{2.48}$$

Here we will use (2.48) to sketch the solution to the following problem. If $\dot{x} = F(x,\lambda)$ is a parameterized perturbation of the Lorenz system (2.46), then how many stationary points does $\dot{x} = F(x,\lambda)$ have near the origin (0,0) when the value of λ is small? This question is surely reasonable. The Lorenz system arises, for example, as a truncated spectral model of the Rayleigh-Bénard problem set in a rectangular tank with rigid boundaries; a two-dimensional flow develops owing to heating from below. But suppose we introduce other parameters λ arising from such small but nonzero physical effects as internal heating, externally-imposed horizontal heating, tilted bottom, or slight compressibility of the fluid. Then, on using the spectral expansion of the Lorenz system, we would obtain a parameterized perturbation $\dot{x} = F(x,\lambda)$ of the Lorenz system, with the property that $F(x,0) = f(x)$. Now, because there are many imperfections such as those cited above that will be present in any actual laboratory experiment, we would not expect that the stationary phase portrait of the Lorenz system will reveal much about the stationary phase portrait of the complete Rayleigh-Bénard problem. This is the case unless (i) we may guarantee in advance that these slight imperfections will not alter drastically the qualitative nature of the stationary phase portrait, or (ii) we may describe in advance exactly how these slight imperfections will affect the nature of the stationary phase portrait. As we will see, it is the second of these cases that holds for the Lorenz system and the others discussed in Chapters 4 and 5 of this monograph. (We note that our interpretation of imperfection parameters differs from that of Golubitsky and Schaeffer, 1979. Their work considers unfoldings of unfoldings, while our work considers only unfoldings themselves.)

Returning to the problem of the number of steady states of the Lorenz system (2.46), we note that because $F(x,\lambda)$ is an unfolding of $f(x)$ (2.47) and $V(y,\mu)$ (2.48) is a versal unfolding of $f(y)$ about $\mu = 0$, then there exist an invertible 3×3 matrix $M(x,\lambda)$ depending smoothly on (x,λ), a parameterized coordinate transformation $y(x,\lambda)$, and smooth functions $\mu_1(\lambda)$ and $\mu_2(\lambda)$ such that

$$F(x,\lambda) = M(x,\lambda)\cdot V(y(x,\lambda), \mu(\lambda)) \tag{2.49}$$

for values of (x,λ) near $(0,0)$. Because $M(x,\lambda)$ is invertible, we have that $F(x,\lambda) = 0$ if and only if $V(y(x,\lambda), \mu(\lambda)) = 0$. Thus, for fixed values of λ, $F(x,\lambda) = 0$ has exactly as many roots near the origin as does $V(y(x,\lambda), \mu(\lambda)) = 0$. But in general $V(y,\mu) = 0$ has one, two or three distinct roots for fixed values of μ, as an easy calculation shows (see Section 3.1). Consequently, the system $\dot{x} = F(x,\lambda)$ has at least one, but no more than three, stationary points near the origin when the value of λ is small.

Having obtained the number of steady states of the Lorenz equations, we may observe further that we may obtain readily much more information from (2.48) and (2.49) without knowing $M(x,\lambda)$, $y(x,\lambda)$ or $\mu(\lambda)$, and certainly without solving (2.46). The stationary phase portrait Stat(V) of (2.48) is very easy to find. We define in general

$$\mathrm{Stat}(V) = \{(y,\mu)\,|\,V(y,\mu) = 0\} \tag{2.50}$$

Then for the Lorenz system (2.48) we have that

$$\mathrm{Stat}(V) = \{(y,\mu)\,|\,y_2 = y_1,\ y_3 = (y_1^{\,2} - \mu_1)/b,\ y_1^{\,3} - \mu_1\, y_1 - \mu_2 = 0\} \tag{2.51}$$

Thus, Stat(V) is described completely by Fig. 2.2 with (y_1,μ_1,μ_2) replacing (x,λ_1,λ_2) respectively (cf. Fig. 3.3). Assuming that our parameterized perturbation $\dot{x} = F(x,\lambda)$ has more than two essentially different imperfection parameters λ, then we may conclude both that two functions $\mu_1(\lambda)$ and $\mu_2(\lambda)$ generated by (2.49) are functionally independent and that $\mu_1(0) = \mu_2(0) = 0$. Then we may use an origin-preserving coordinate transformation in λ-space with the property that $\mu_1(\lambda) = \lambda_1$ and $\mu_2(\lambda) = \lambda_2$. But then we must conclude that among all the imperfection parameters possible, there are just <u>two</u> parameters that completely control the stationary phase portrait of our parameterized perturbation $\dot{x} = F(x,\lambda)$. That is, in the neighborhood of $r = 1$, the system $\dot{x} = F(x,\lambda)$ will have two stable stationary points and one unstable one when $\lambda_1^3/27 - \lambda_2^2/4 > 0$, and one stable stationary point when $\lambda_1^3/27 - \lambda_2^2/4 < 0$ (cf. Example 1). Hysteresis will occur for $\lambda_1 > 0$ as the value of λ_2 is varied between $-\,2(\lambda_1^3/27)^{1/2} - \varepsilon$ and $2(\lambda_1^3/27)^{1/2} + \varepsilon$, as we found in Examples 1 and 2. Bifurcation will occur when $\lambda_2 = 0$ as λ_1 is increased from negative values to positive values, as we saw in Example 3. The rest of the imperfection parameters do not affect the stationary phase portrait, as we noted in Example 4. Of course, we must notice that the two critical parameters μ_1

and μ_2 need not be the natural imperfection parameters λ themselves; in general $\mu(\lambda)$ will be extremely complicated functions. We may use (2.49) to find the power series expansions of μ up to high orders, but this procedure is complicated by the fact that the expression (2.49) is not unique. In any case, we are able to obtain the explicit relationships in the applications we discuss in Chapters 3-5. #

2.3 Mather's Theorems

Although a method for finding power series expansions for $M(x,\lambda)$, $y(x,\lambda)$ and $\mu(\lambda)$ in (2.49) from knowledge of $F(x,\lambda)$ and $V(y,\mu)$ alone awaits further development, we may use Mather's theory to generate a straightforward algorithm for finding a versal unfolding $V(y,\mu)$ of a multi-dimensional $f(x) = F(x,0)$. For our purposes, this theory consists of the definitions discussed previously (e.g. unfolding, contact transformation, and versal unfolding) and two theorems. Typically, definitions are regarded as carrying no information, but in fact useful implicit information is imparted by careful choice of definitions, focusing attention on certain concepts to the exclusion of others. Thus, the definitions of the previous section are certainly important, but the explicit information imparted by Mather's theory is (for us) contained in the two theorems mentioned above.

The amount of information in theorems can be estimated normally by the depth and cleverness of their proofs. Unfortunately, the proofs of Mather's theorems are considerably beyond the scope of this monograph and we cannot furnish in the available space the means to make such an estimate. However, for applications to physical problems, we may see the amount of usable information carried by Mather's theorems by examining what conclusions may be drawn from the theorems rather than examining the work that must be put into producing their proofs. Fortunately, for this purpose these theorems are relatively easy to state and to use.

To state the first theorem, we introduce the differential df of f

$$df(x) = \left[\frac{\partial f_i}{\partial x_j}(x) \right] \quad \text{with} \quad \begin{cases} i = 1,\ldots,n \\ j = 1,\ldots,n \end{cases} \tag{2.52}$$

in which $f(x) = [f_1(x),\ldots,f_n(x)]$ and $x = (x_1,\ldots,x_n)$. We number the rows with index i and the columns with j. We introduce also the following condition relating a given smooth function f(x) and an $n \times q$ matrix N(x) depending smoothly on x:

The Transversality Condition. A smooth $n \times q$ matrix function N(x) satisfies the Transversality Condition with respect to f(x) near the origin $x = 0$ if and only if every smooth n-vector function Y(x) may be written near the origin as

$$Y(x) = df(x)\cdot G(x) + H(x)\cdot f(x) + N(x)\cdot\gamma \tag{2.53}$$

where G(x) is a suitable smooth n-vector function, H(x) is a suitable smooth $n \times n$ matrix function and γ is a suitable constant q-vector.

We note that the above quantities $G(x)$, $H(x)$ and γ all depend on $Y(x)$.

The Transversality Condition may appear to be unverifiable for a given function $f(x)$ and matrix $N(x)$ because it involves verifying (2.53) for each smooth vector function $Y(x)$ near the origin. However, as we will see soon, we may verify it in the cases of interest. Now we may state Mather's Theorem I:

Mather's Theorem I. The unfolding $V(y,\mu)$ with $\mu = (\mu_1,\ldots,\mu_q)$ of $f(y)$ is versal if and only if the matrix

$$N(y) = \left[\frac{\partial V_i}{\partial \mu_j}(y,0) \right]$$

satisfies the Transversality Condition with respect to $f(y)$.

Corollary. Let $L(y,\mu)$ be the first degree Taylor expansion of $V(y,\mu)$ with respect to μ:

$$L(y,\mu) = V(y,0) + N(y)\cdot\mu \tag{2.54}$$

Then $V(y,\mu)$ is versal if and only if $L(y,\mu)$ is versal.

This corollary will be applied in Chapter 5.

Corollary. If a matrix $Q(y)$ satisfies the Transversality Condition with respect to $f(y)$, then

$$V(y,\mu) = f(y) + Q(y)\cdot\mu \tag{2.55}$$

is a versal unfolding of $f(y)$.

In a versal unfolding of the form (2.55), we will call $Q(y)$ the unfolding matrix of (2.55) and its columns the unfolding functions of (2.55).

Example 9. The cusp and Mather's Theorem I

We have encountered already the unfolding

$$V(y,\mu) = -y^3 + \mu_1 y + \mu_2 \tag{2.56}$$

in Examples 2, 3 and 5. For this unfolding, we have $n = 1$, $q = 2$ and

$$\left.\begin{aligned} f(x) &= -x^3 \\ df(x) &= -3x^2 \\ N(x) &= (x,1) \\ \mu &= \begin{bmatrix} \mu_1 \\ \mu_2 \end{bmatrix} \end{aligned}\right\} \tag{2.57}$$

Let Y(x) be an arbitrary smooth one-dimensional vector function near x = 0; in this case, Y(x) is simply a single function of x. By applying Taylor's Theorem with integral remainder to Y(x) near the origin, we see that

$$Y(x) = Y(0) + Y_1(0)\, x + R(x)\, x^2 \tag{2.58}$$

But then we may rewrite (2.58) as

$$Y(x) = (-3x^2)\,[-R(x)/3] + Y_1(0)\, x + Y(0) \tag{2.59}$$

so that Y(x) may be expressed near the origin in the form (2.53) if we choose $G(x) = -R(x)/3$, $H(x) = 0$ and $\gamma = [Y_1(0), Y(0)]^T$. According to Mather's Theorem I, then, the unfolding $V(y,\mu)$ is versal. #

We note that in Example 9, the linear and constant terms are the only ones that cannot be written as $df(x)\cdot G(x) + H(x)\cdot f(x)$. Because both f and df vanish at the singularity, it follows that other functions might be needed to describe fully the steady state behavior nearby the singularity. Mather's Theorem I indicates that in some cases the number of additional required functions is finite. We find this is the case in Section 2.7 in which we classify the type of singularity and specify the canonical functions of N(x) in the versal unfolding about the singularity.

Mather's Theorem I is difficult to apply because it involves the solution of the infinite-dimensional linear problem (2.53). However, Mather (1968) also reduces that problem to a series of finite-dimensional problems. His Theorem II, stated below, constitutes that reduction.

To state Theorem II, we need some additional notation. Let $\xi(n)$ be the set of smooth n-vector functions defined near the origin. Let $P^k(n)$ be those smooth n-vectors in $\xi(n)$ whose components are polynomials of degree $\leq k$. Let $K^k(n)$ be those n-vectors in $P^k(n)$ whose components are homogeneous polynomials of degree exactly equal to k. For each vector Y(x), let $T^kY(x)$ be its Maclaurin expansion of degree k. Thus T^k defines a linear map

$$T^k : \xi(n) \to P^k(n) \tag{2.60}$$

given by

$$T^kY(x) = \sum_{|\alpha|\leq k} \frac{1}{\alpha!} \frac{\partial^\alpha Y}{\partial x^\alpha}(0)\, x^\alpha \tag{2.61}$$

where we have used the usual notation

$$\left.\begin{aligned} &\alpha = (\alpha_1,\ldots,\alpha_n) \\ &\alpha! = \alpha_1!\,\alpha_2!\,\cdots\,\alpha_n! \\ &|\alpha| = \alpha_1 + \cdots + \alpha_n \\ &\frac{\partial^\alpha}{\partial x^\alpha} = \frac{\partial^{\alpha_1}}{\partial x_1^{\alpha_1}}\frac{\partial^{\alpha_2}}{\partial x_2^{\alpha_2}}\cdots\frac{\partial^{\alpha_n}}{\partial x_n^{\alpha_n}} \\ &x^\alpha = x_1^{\alpha_1}\cdots x_n^{\alpha_n} \end{aligned}\right\} \tag{2.62}$$

Let $I^k(f) \subset P^k(n)$ consist of those smooth vectors $Y(x)$ which may be written as

$$Y(x) = T^k[df(x)\cdot G(x) + H(x)\cdot f(x)] \tag{2.63}$$

near the origin, for some n-vector $G(x) \in P^k(n)$ and smooth $n \times n$ matrix $H(x)$ whose entries are all polynomials of degree $k + 1$. Notice that $I^k(f)$ is a finite-dimensional vector space. In computing a basis for it, we recall the rule

$$T^k(A(x)B(x)) = T^k[T^kA(x)\cdot T^kB(x)] \tag{2.64}$$

Now we may state the reduced version of Mather's Theorem I.

Mather's Theorem II. Suppose that for some k we have

$$K^k(n) \subset I^k(f) \tag{2.65}$$

Then f has a versal unfolding. If (2.65) holds, then let $N_1(x),\ldots,N_q(x)$ be members of $P^k(n)$ which together with $I^k(f)$ span all of $P^k(n)$. Then

$$V(x,\mu) = f(x) + \mu_1N_1(x) + \cdots + \mu_qN_q(x) \tag{2.66}$$

is a versal unfolding of $f(x)$. Furthermore, if $f(x)$ has a versal unfolding, then (2.65) holds for some k.

Because the sets $K^k(n)$ and $P^k(n)$ are finite-dimensional vector spaces, application of Mather's Theorem II involves only ordinary linear algebra. Of course, the linear algebra may be somewhat tedious, but it remains essentially elementary. To illustrate the application and notation of Mather's Theorem II, we return to the Lorenz system (2.46).

Example 10. A versal unfolding of the Lorenz model: Mather's Theorem II

We consider again the system $\dot{x} = f(x)$ of the Lorenz model which for convenience we state again as

$$f(x) = \begin{bmatrix} \sigma(x_2 - x_1) \\ - x_1 x_3 + x_1 - x_2 \\ x_1 x_2 - b x_3 \end{bmatrix} \tag{2.67}$$

in which $\sigma > 0$ and $b > 0$. We have asserted already in Example 8 that (e.g., (2.48))

$$V(y,\mu) = \begin{bmatrix} \sigma(y_2 - y_1) \\ - y_1 y_3 + y_1 - y_2 + \mu_2/b \\ y_1 y_2 - b y_3 - \mu_1 \end{bmatrix} \tag{2.68}$$

is a versal unfolding of $f(x)$. Now we verify that this is true by applying Mather's Theorem II in the case $k = 2$. For these systems we have already that $n = 3$.

We begin by observing that $K^2(3)$ is an 18-dimensional vector space spanned by the basis of quadratic forms

$$\begin{bmatrix} x_1^2 \\ 0 \\ 0 \end{bmatrix}, \begin{bmatrix} x_1x_2 \\ 0 \\ 0 \end{bmatrix}, \begin{bmatrix} x_1x_3 \\ 0 \\ 0 \end{bmatrix}, \begin{bmatrix} x_2^2 \\ 0 \\ 0 \end{bmatrix}, \begin{bmatrix} x_2x_3 \\ 0 \\ 0 \end{bmatrix}, \begin{bmatrix} x_3^2 \\ 0 \\ 0 \end{bmatrix}, \begin{bmatrix} 0 \\ x_1^2 \\ 0 \end{bmatrix}, \cdots, \begin{bmatrix} 0 \\ 0 \\ x_3^2 \end{bmatrix} \tag{2.69}$$

$P^2(3)$ is a 30-dimensional vector space with a base consisting of the quadratic forms (2.69) together with the linear and constant forms

$$\begin{bmatrix} x_1 \\ 0 \\ 0 \end{bmatrix}, \begin{bmatrix} x_2 \\ 0 \\ 0 \end{bmatrix}, \begin{bmatrix} x_3 \\ 0 \\ 0 \end{bmatrix}, \cdots, \begin{bmatrix} 0 \\ 0 \\ x_3 \end{bmatrix}, \begin{bmatrix} 1 \\ 0 \\ 0 \end{bmatrix}, \begin{bmatrix} 0 \\ 1 \\ 0 \end{bmatrix}, \begin{bmatrix} 0 \\ 0 \\ 1 \end{bmatrix} \tag{2.70}$$

To compute $I^2(f)$, we begin by using (2.52) to write the differential df of (2.67) as

$$df(x) = \begin{bmatrix} -\sigma & \sigma & 0 \\ 1 - x_3 & -1 & -x_1 \\ x_2 & x_1 & -b \end{bmatrix} \tag{2.71}$$

It is easy to check that for $i = 1, 2, 3$

$$T^2\left[df(x)\begin{bmatrix} \frac{1}{b} x_1x_i \\ \frac{1}{b} x_1x_i \\ -\frac{1}{b} x_3x_i \end{bmatrix}\right] = T^2\left[\begin{bmatrix} -\sigma & \sigma & 0 \\ 1 - x_3 & -1 & -x_1 \\ x_2 & x_1 & -b \end{bmatrix}\begin{bmatrix} \frac{1}{b} x_1x_i \\ \frac{1}{b} x_1x_i \\ -\frac{1}{b} x_3x_i \end{bmatrix}\right]$$

$$= T^2\begin{bmatrix} 0 \\ 0 \\ \frac{1}{b} x_2x_1x_i + \frac{1}{b} x_1{}^2x_i + x_3x_i \end{bmatrix} \tag{2.72}$$

$$= \begin{bmatrix} 0 \\ 0 \\ x_ix_3 \end{bmatrix}$$

because $T^2(x_2x_1x_i) = T^2(x_1{}^2x_i) = 0$. Similarly, we have

$$\begin{bmatrix} 0 \\ 0 \\ x_1x_2 \end{bmatrix} = T^2\left[df(x)\begin{bmatrix} \frac{1}{3} x_1 \\ \frac{1}{3} x_1 \\ -\frac{1}{3} x_3 \end{bmatrix} + \begin{bmatrix} 0 & 0 & 0 \\ 0 & 0 & 0 \\ \frac{x_1}{3\sigma} & 0 & \frac{1}{3} \end{bmatrix} f(x)\right] \tag{2.73}$$

$$\begin{bmatrix} 0 \\ 0 \\ x_1{}^2 \end{bmatrix} = T^2\left[df(x)\begin{bmatrix} \frac{1}{3} x_1 \\ \frac{1}{3} x_1 \\ -\frac{1}{3} x_3 \end{bmatrix} + \begin{bmatrix} 0 & 0 & 0 \\ 0 & 0 & 0 \\ -\frac{2x_1}{3\sigma} & 0 & \frac{1}{3} \end{bmatrix} f(x)\right] \tag{2.74}$$

and

$$\begin{bmatrix} 0 \\ 0 \\ x_2^{\,2} \end{bmatrix} = T^2 \left[df(x) \begin{bmatrix} \frac{1}{3} x_1 \\ \frac{1}{3} x_1 \\ -\frac{1}{3} x_3 \end{bmatrix} + \begin{bmatrix} 0 & 0 & 0 \\ 0 & 0 & 0 \\ \frac{x_1}{3\sigma} + \frac{x_2}{\sigma} & 0 & \frac{1}{3} \end{bmatrix} f(x) \right] \tag{2.75}$$

The remaining basis vectors $[x_1^2,0,0]^T$, $[x_1x_2,0,0]^T,\ldots,[0,x_3^2,0]^T$ of $K^2(3)$ may be expressed in the same way, so that (2.65) holds with $k = 2$. Next we have to find the vectors $N_i(x)$ which, together with $I^2(f)$, are to span $P^2(3)$. Because already $K^2(3) \subset I^2(f)$, we need consider only constant and linear forms for $N_i(x)$. After elementary computations like those above, we find that only the constant vectors $N_2(x) = [0,b,0]^T$ and $N_1(x) = -[0,0,1]^T$ are needed. Thus we conclude that (2.68) is a versal unfolding of (2.67) about the original singularity $r = 1$. #

It may appear from the preceeding example that once we have Mather's Theorem II available, then Theorem I is not necessary. However, as we shall see in the applications in Chapters 3-5 of this monograph, the first versal unfolding we derive for a given system is rarely the most convenient or the easiest to interpret physically. Thus, we will need some way to construct new versal unfoldings from old ones, and Mather's Theorem I serves the purpose effectively.

2.4 Altering Versal Unfoldings

In the preceeding section we saw how to check by means of routine, though lengthy, computations whether the right side of a differential equation $\dot{x} = f(x)$ has a versal unfolding, and if so, how to write one particular unfolding. In the following section, we develop the algorithm that we subsequently apply to our spectral models. This algorithm considerably simplifies the unfolding computations, and it is based on the Lyapunov-Schmidt procedure rather than Mather's Theorems. Unfortunately, use of this method will often lead to the "wrong" versal unfolding, whereupon we are faced with the necessity of altering that unfolding to a more suitable form. It is the methodology of this alteration to which we now turn.

We confine our attention to unfoldings of the form

$$V(y,\mu) = f(y) + N_1(y)\,\mu_1 + \cdots + N_q(y)\,\mu_q \tag{2.76}$$

described in (2.55) of Mather's Theorem I. We wish to replace the given unfolding matrix $N(y)$ with a new one $\tilde{N}(y)$. To do this we choose $G^\circ(y)$, $H^\circ(y)$ and γ° to write a trial column $Y^\circ(y)$ of $\tilde{N}(y)$ in the form (2.53) as

$$Y^\circ(y) = df(y)\cdot G^\circ(y) + H^\circ(y)\cdot f(y) + N_1(y)\,\gamma_1^\circ + \cdots + N_q(y)\,\gamma_q^\circ \tag{2.77}$$

Suppose that $\gamma_i^\circ \neq 0$ for some fixed value of i. Then we may in fact replace the ith member of $N(y)$ with $Y^\circ(y)$, and we define

$$\left.\begin{aligned} \tilde{N}_i(y) &= Y^\circ(y) \\ \tilde{N}_j(y) &= N_j(y) \quad \text{for } j \neq i \end{aligned}\right\} \tag{2.78}$$

Now we have the alternate unfolding

$$\tilde{V}(y,\mu) = f(y) + \tilde{N}_1(y)\,\mu_1 + \cdots + \tilde{N}_q(y)\,\mu_q \tag{2.79}$$

It is easy to verify that $\tilde{V}$ is versal if V is. In fact, we may replace y with x, solve (2.77) for $N_i(x)$ and substitute the result into (2.53) to obtain

$$Y(x) = df(x)\cdot\left[G(x) - \frac{\gamma_i}{\gamma_i^\circ} G^\circ(x)\right] + \left[H(x) - \frac{\gamma_i}{\gamma_i^\circ} H^\circ(x)\right]\cdot f(x) \tag{2.80}$$

$$+ \sum_{j\neq i} \tilde{N}_j(x)\left[\gamma_j - \frac{\gamma_i\gamma_j^\circ}{\gamma_i^\circ}\right] + \tilde{N}_1(x)\,\frac{\gamma_i}{\gamma_i^\circ}$$

Thus, N(y) satisfies the Transversality Condition and so by Mather's Theorem I, we conclude that $\tilde{V}(y,\mu)$ in (2.79) must be versal.

We will refer to the operation just described, of replacing the versal unfolding (2.76) with an equivalent one (2.79) by substituting one unfolding function with another, as an <u>elementary alteration.</u> It is easy to demonstrate that elementary alterations are invertible operations. We have remarked already that if V is versal, then so is $\tilde{V}$. Of course, in practice we will wish to carry out several elementary alterations one after another. The effect of doing so may be obtained more compactly by carrying out a single operation, called an <u>alteration</u>, as follows:

<u>An Alteration.</u> Replace the unfolding matrix N(x) with the unfolding matrix $\tilde{N}(x)$ where

$$\tilde{N}(x) = N(x)\cdot\Lambda + \Omega(x) \tag{2.81}$$

where Λ is a constant invertible $q \times q$ matrix, where

$$\Omega(x) = \left[df(x)\cdot G_1(x) + H_1(x)\cdot f(x),\ldots,df(x)\cdot G_q(x) + H_q(x)\cdot f(x)\right] \tag{2.82}$$

and where $G_1(x),\ldots,G_q(x)$ are smooth n-vector functions near the origin and $H_1(x),\ldots,H_q(x)$ are smooth $n \times n$ matrix functions near the origin. The old versal unfolding V and the new one $\tilde{V}$ have the forms (2.76) and (2.79) respectively.

In order to relate the canonical unfolding parameters μ to physically interpretable ones λ, we must in many cases use alterations because we will need to replace more than one column vector of N(y). For example, in Chapters 4 and 5 we must replace pairs of unfolding functions with other pairs. Operations of the above kinds are illustrated in Examples 12 and 13.

Another detail we must resolve is whether V and $\tilde{V}$ are contact equivalent. A sufficient answer for our purposes is contained in the following proposition.

Proposition 2.1. If $V(y,\mu_1,\ldots,\mu_q)$ and $W(y,\gamma_1,\ldots,\gamma_q)$ are each versal unfoldings of f(y), and if both V and W have the smallest possible number of parameters, then V and W are contact equivalent.

This proposition is a consequence of Mather's Theorem I, but we forgo the proof. What is important for us to recall is that if two unfoldings V and W are contact equivalent, then the stationary phase portraits of V and W are of the same form. Thus, if the unfoldings V and $\tilde{V}$ of (2.76) and (2.79) (with $\tilde{N}(y)$ given by either (2.78) or (2.81)) are versal and if they also have the minimum possible number of parameters, then the stationary phase portraits of the differential systems

$$\dot{y} = V(y,\mu) \tag{2.83}$$

and

$$\dot{y} = \tilde{V}(y,\mu) \tag{2.84}$$

have the same form. Also either an elementary alteration or an alteration in such a case does not change the stationary phase portrait.

Finally, the above role played by the minimum possible number of parameters suggests that this number is an important one to know. We define the codimension d of f(x) to be the minimum possible number of parameters required in a versal unfolding of f(x). We show how to compute d in Section 2.6; we illustrate its importance in the following example.

Example 11. Codimension and the cusp

Let us consider

$$f(x) = -x^3 \tag{2.85}$$

One particular versal unfolding of f(x) is given in Example 9 as

$$V(x,\lambda) = -x^3 + \lambda_1 x + \lambda_2 \tag{2.86}$$

Here the codimension d of $-x^3$ satisfies $d \leq 2$ because (2.86) is a versal unfolding V of (2.85) with two parameters.

Could there be a versal unfolding of (2.85) with only one parameter? To answer this question, we first consider what happens to the stable stationary solutions of

$$\dot{x} = -x^3 + \lambda_1 x + \lambda_2 \tag{2.87}$$

as we vary the value of λ around a circle of radius $\varepsilon > 0$ centered at the origin $\lambda_1 = \lambda_2 = 0$. Suppose we start at the values $\lambda_1 = -\varepsilon$, $\lambda_2 = 0$. By inspecting Fig. 2.2, we see that there is only one solution to (2.87). As we smoothly vary the value of λ around the small circle, the corresponding stable stationary solution also will vary smoothly, tracing a path, say, on the lower pleat. But when this path reaches the fold point on the cusp surface, a further change of the value of λ will cause the solution to jump suddenly from the lower to the upper pleat. Then the solution returns to its initial value on completing the loop. Thus, arbitrarily small loops in the parameter space generate closed hysteresis loops, which are ones that return to their starting position.

Although hysteresis is possible in a one-parameter unfolding, as in Example 2 and Fig. 2.3, it is intuitively clear that in this case the stationary solution will not return to its initial value as the radius of the loop in the parameter space becomes arbitrarily small. But if arbitrarily small loops in the parameter space cannot generate closed hysteresis loops in a one-parameter unfolding, then closed hysteresis loops cannot be pulled back from a one-parameter unfolding. Because, by definition, closed hysteresis loops can be pulled back from a versal unfolding, the required unfolding must have at least two parameters. Thus, it follows that the codimension d of $-x^3$ satisfies $d \geq 2$. Finally, because we have both $d \leq 2$ and $d \geq 2$, we conclude that $d = 2$. #

Now we illustrate elementary alterations and alterations in the following two examples.

Example 12. Versal unfoldings of the Lorenz model: elementary alterations
Again we consider the Lorenz system for which $r = 1$ and

$$f(y) = \begin{bmatrix} \sigma(y_2 - y_1) \\ -y_1y_3 + y_1 - y_2 \\ y_1y_2 - b\,y_3 \end{bmatrix} \tag{2.88}$$

Two versal unfoldings of $f(y)$ are given by

$$V(y,\mu) = \begin{bmatrix} \sigma(y_2 - y_1) \\ -y_1y_3 + y_1 - y_2 + \mu_2/b \\ y_1y_2 - b\,y_3 - \mu_1 \end{bmatrix} \tag{2.89}$$

and

$$W(y,\mu) = \begin{bmatrix} \sigma(y_2 - y_1) \\ - y_1 y_3 + y_1 - y_2 + (\mu_1/b)y_2 + \mu_2/b \\ y_1 y_2 - b\, y_3 \end{bmatrix} \tag{2.90}$$

The physical significance of these two unfoldings is discussed in Chapter 3. We already have verified in Example 10 that the unfolding V is versal. By an argument very similar to that of Example 11 above, using $V(y,\mu)$ and Fig. 2.2, we see that the codimension d of $f(y)$ is again 2.

We wish to show that W in (2.90) is an elementary alteration of V in (2.89). We begin by noticing that

$$N_1(y) = \begin{bmatrix} 0 \\ 0 \\ -1 \end{bmatrix} \tag{2.91}$$

and

$$N_2(y) = \begin{bmatrix} 0 \\ \frac{1}{b} \\ 0 \end{bmatrix} \tag{2.92}$$

are the unfolding functions in the unfolding V. We may write the new unfolding function

$$\begin{bmatrix} 0 \\ \frac{y_2}{b} \\ 0 \end{bmatrix} = df(y)\cdot G(y) + H(y)\cdot f(y) + 1\cdot N_1(y) \tag{2.93}$$

if we choose

$$G(y) = \begin{bmatrix} 0 \\ 0 \\ -\frac{1}{b} \end{bmatrix} \tag{2.94}$$

and

$$H(y) = \begin{bmatrix} 0 & 0 & 0 \\ \frac{1}{\sigma b} & 0 & 0 \\ 0 & 0 & 0 \end{bmatrix} \tag{2.95}$$

The corresponding elementary alteration of V is given by

$$\tilde{V}(y,\mu) = f(y) + \mu_1 \begin{bmatrix} 0 \\ \frac{y_2}{b} \\ 0 \end{bmatrix} + \mu_2 N_2(y) \tag{2.96}$$

which with the aid of (2.88) becomes

$$\tilde{V}(y,\mu) = \begin{bmatrix} \sigma(y_2 - y_1) \\ - y_1 y_3 + y_1 - y_2 + (\mu_1/b)y_2 + \mu_2/b \\ y_1 y_2 - b\, y_3 \end{bmatrix} \tag{2.97}$$

Now we may conclude first that $\tilde{V} = W$ is versal and second, by Proposition 2.1, that V and W are contact equivalent. Thus they have the same stationary phase portraits, and the form of these portraits is displayed in Fig. 2.2.

We also may find an elementary alteration that carries W to V. We begin by writing

$$W(y,\mu) = f(y) + \mu_1 L_1(y) + \mu_2 L_2(y) \tag{2.98}$$

in which

$$L_1(y) = \begin{bmatrix} 0 \\ \frac{y_2}{b} \\ 0 \end{bmatrix} \tag{2.99}$$

and

$$L_2(y) = \begin{bmatrix} 0 \\ \frac{1}{b} \\ 0 \end{bmatrix} \tag{2.100}$$

We may write

$$\begin{bmatrix} 0 \\ 0 \\ -1 \end{bmatrix} = df(y)\cdot G_1(y) + H_1(y)\cdot f(y) + 1\cdot L_1(y) \tag{2.101}$$

when we set

$$G_1(y) = \begin{bmatrix} 0 \\ 0 \\ \frac{1}{b} \end{bmatrix} \tag{2.102}$$

and

$$H_1(y) = \begin{bmatrix} 0 & 0 & 0 \\ -\frac{1}{\sigma b} & 0 & 0 \\ 0 & 0 & 0 \end{bmatrix} \tag{2.103}$$

The corresponding elementary alteration carries W to

$$\tilde{W}(y,\mu) = V(y,\mu) = f(y) + \mu_1 \begin{bmatrix} 0 \\ 0 \\ -1 \end{bmatrix} + \mu_2 L_2(y) \tag{2.104}$$

#

Finally, in the special case of the above example, it is easy to see that $\dot{y} = V(y,\mu)$ and $\dot{y} = W(y,\mu)$ have the same stationary phase portrait because

$$\mathrm{Stat}(V) = \{(y,\mu) | V(y,\mu) = 0\} = \mathrm{Stat}(W) \tag{2.105}$$

In general, Stat(V) and Stat(W) only will be equivalent, and the graph of V will appear as the graph of W only after applying a contact transformation to V. However, in spite of this fortuitous equality in (2.105), and in spite of the explicit elementary transformations found above, it remains a nontrivial task to find a contact transformation carrying V to W. Fortunately, as we see in the next section, we do not need to find such a transformation.

Example 13. Versal unfoldings of the Lorenz model: alterations

Again we compare versal unfoldings of the Lorenz model. The first we recall from Example 12, (2.89)

$$V(y,\mu) = \begin{bmatrix} \sigma(y_2 - y_1) \\ - y_1 y_3 + y_1 - y_2 + \mu_2/b \\ y_1 y_2 - b\, y_3 - \mu_1 \end{bmatrix} \tag{2.106}$$

and the second will be given by

$$U(y,\mu) = \begin{bmatrix} \sigma(y_2 - y_1) + (\sigma/b)\, \mu_2 \\ - y_1 y_3 + y_1 - y_2 + (\mu_1/b)\, y_2 \\ y_1 y_2 - b\, y_3 \end{bmatrix} \tag{2.107}$$

From (2.106) we see that the original unfolding matrix N(y) is

$$N(y) = \begin{bmatrix} 0 & 0 \\ 0 & \frac{1}{b} \\ -1 & 0 \end{bmatrix} \tag{2.108}$$

and from (2.107) that the final unfolding matrix $\tilde{N}(y)$ is

$$\tilde{N}(y) = \begin{bmatrix} 0 & \frac{\sigma}{b} \\ \frac{y_2}{b} & 0 \\ 0 & 0 \end{bmatrix} \tag{2.109}$$

We show that the unfolding $U(y,\mu)$ can be obtained from $V(y,\mu)$ by an alteration by showing that (2.81) is satisfied:

$$\begin{bmatrix} 0 & \frac{\sigma}{b} \\ \frac{y_2}{b} & 0 \\ 0 & 0 \end{bmatrix} = \begin{bmatrix} 0 & 0 \\ 0 & \frac{1}{b} \\ -1 & 0 \end{bmatrix} \begin{bmatrix} 1 & 0 \\ 0 & 1 \end{bmatrix} + \tag{2.110}$$

$$\Bigg[\begin{bmatrix} -\sigma & \sigma & 0 \\ 1-y_3 & -1 & -y_1 \\ y_2 & y_1 & -b \end{bmatrix} \begin{bmatrix} 0 \\ 0 \\ -\frac{1}{b} \end{bmatrix} + \begin{bmatrix} 0 & 0 & 0 \\ \frac{1}{\sigma b} & 0 & 0 \\ 0 & 0 & 0 \end{bmatrix}$$

$$\cdot \begin{bmatrix} \sigma(y_2 - y_1) \\ -y_1y_3 + y_1 - y_2 \\ y_1y_2 - b\,y_3 \end{bmatrix}, \begin{bmatrix} -\sigma & \sigma & 0 \\ 1-y_3 & -1 & -y_1 \\ y_2 & y_1 & -b \end{bmatrix} \begin{bmatrix} -(3b)^{-1} \\ 2(3b)^{-1} \\ y_2(3b^2)^{-1} \end{bmatrix} +$$

$$\begin{bmatrix} 0 & 0 & 0 \\ 0 & 0 & (3b^2)^{-1} \\ 2(3\sigma b)^{-1} & 0 & 0 \end{bmatrix} \begin{bmatrix} \sigma(y_2 - y_1) \\ -y_1y_3 + y_1 - y_2 \\ y_1y_2 - b\,y_3 \end{bmatrix} \Bigg]$$

In this example, Λ is the identity matrix; this will not be the case in general. #

2.5 The Lyapunov-Schmidt Splitting Procedure

We are now ready to complete our algorithmic procedure for identifying both the number of and location of the parameters that control the steady branching behavior of a differential system

$$\dot{x} = f(x) \tag{2.111}$$

for which $f(0) = 0$. In order to find the stationary phase portrait of an arbitrary, smooth parameterized perturbation of (2.111) near the origin $x = 0$, we determine whether $f(x)$ has a versal unfolding and if so what the form of the versal unfolding is. From the stationary phase portrait of the differential equation that has this versal unfolding, we will be able to see immediately all possible types of stationary behavior that might occur near the origin in <u>any</u> parameterized perturbation of (2.111).

Although Mather's Theorem II enables us to carry out the above program, the required computations are still tedious and complicated. In particular, they appear to increase in complexity rapidly as the number n of variables $x_1,\ldots,x_n$ increases.

To simplify the requisite calculations, we use ordinary catastrophe theory to determine the type of lemma we need. In catastrophe theory, the branching behavior of a model is determined by studying the bifurcations that occur in a governing ordinary (potential or Lyapunov) function $\phi(x_1,\ldots,x_n)$, for which $\phi(0) = 0$ and $d\phi(0) = 0$, when it has been subjected to a parameterized perturbation. The first lemma needed for this study is the Gromoll-Meyer Lemma (Gromoll and Meyer, 1969), which supplies new coordinates (u,v) for decomposing $\phi(x)$. Thus we have

$$\phi(x) = \alpha(u) + \beta(v) \tag{2.112}$$

with $\alpha(u)$ a non-degenerate quadratic function and β a function satisfying $\beta(0) = 0$, $d\beta(0) = 0$, and $d^2\beta(0) = 0$. Because the critical points of a quadratic non-degenerate function do not bifurcate, the bifurcative behavior of $f(x)$ is concentrated entirely in $\beta(v)$.

It is exactly the same sort of lemma that we seek here, but we do not have a potential function $\phi(x)$ that governs the solutions to our spectral models. We find that lemma via the classical Lyapunov-Schmidt procedure. We will consider

$$f(x) = \big(f_1(x),\ldots,f_n(x)\big), \quad x = (x_1,\ldots,x_n) \tag{2.113}$$

to be an unfolding with zero parameters. We can do this because we apply the procedure to a differential system evaluated at the singularity $\lambda = 0$. In this case we may still speak of contact transformations of $f(x)$, and contact equivalence of $f(x)$ with other vector functions. We will write contact transformations T in the form (2.15) presented in Section 2.2, but we will omit the last entry because there are now no parameters. Thus we label a contact transformation T with

$$T = \big(M(x);y(x)\big) \tag{2.114}$$

We may write

$$f(x) = \Gamma(x)\cdot x \tag{2.115}$$

where $\Gamma(x)$ is a smooth $n \times n$ matrix. We notice that $df(0) = \Gamma(0)$. We denote the rank of $df(0)$ by r and the corank of f by $n - r$. We recall that singularities are found in practice by performing a linear stability analysis about a stationary point. At the singularity at least one eigenvalue of the linear problem vanishes for bifurcation to another steady state. If one eigenvalue is zero, then $\text{corank}(f) = 1$; in general, however, if m is the number of vanishing eigenvalues, then $\text{corank}(f) \leq m$.

By rearranging coordinates we may assume that the upper left $r \times r$ corner of $df(0)$ is invertible. Now we divide the coordinates into two groups

$$\left.\begin{aligned} u_i &= x_i \;, \quad \text{for } i = 1,\ldots,r \\ v_j &= x_{j+r}, \quad \text{for } j = 1,\ldots,n-r \end{aligned}\right\} \tag{2.116}$$

Then we may write $\Gamma(x)$ in block form

$$\Gamma(x) = \begin{bmatrix} A(u,v) & B(u,v) \\ C(u,v) & D(u,v) \end{bmatrix} \tag{2.117}$$

in which A is an $r \times r$ matrix, invertible near the origin. Combining (2.115)-(2.117), we have

$$f(u,v) = \begin{bmatrix} A(u,v) & B(u,v) \\ C(u,v) & D(u,v) \end{bmatrix} \begin{bmatrix} u \\ v \end{bmatrix} \tag{2.118}$$

We may now state our Lyapunov-Schmidt Splitting Lemma. We include the proof of it because the proof clearly demonstrates the concept of contact transformations, and because the proof itself is useful in computations.

<u>Lyapunov-Schmidt Splitting Lemma.</u> Suppose $f(u,v)$ can be written in the form (2.118). Then $f(u,v)$ is contact equivalent to $\begin{bmatrix} u \\ g(v) \end{bmatrix}$ for some smooth $(n-r)$-component function $g(v)$.

In order to aid the exposition of the proof of this lemma, we write $F \sim G$ to mean "F is contact equivalent to G".

<u>Proof:</u> The contact transformation

$$T = \left[\; M = \begin{bmatrix} A(u,v)^{-1} & 0 \\ -\,C(u,v)A(u,v)^{-1} & 1 \end{bmatrix} \;;\quad \begin{aligned} y_1(u) &= u \\ y_2(v) &= v \end{aligned} \;\right] \tag{2.119}$$

when applied to (2.118) shows that

$$f(u,v) \sim \begin{bmatrix} 1 & A^{-1}B \\ 0 & D - CA^{-1}B \end{bmatrix} \begin{bmatrix} u \\ v \end{bmatrix} \tag{2.120}$$

We write

$$\Delta(u,v) = D(u,v) - C(u,v)A(u,v)^{-1}B(u,v) \tag{2.121}$$

and note that $\Delta(0,0) = 0$ because $\text{corank}(f) = n - r$. Upon defining a coordinate transformation by

$$\bar{u} = u + A(u,v)^{-1}B(u,v)v \tag{2.122}$$

$$\bar{v} = v \tag{2.123}$$

we may rewrite (2.120) as

$$f(u,v) \sim \begin{bmatrix} 1 & A^{-1}B \\ 0 & D-CA^{-1}B \end{bmatrix} \begin{bmatrix} u \\ v \end{bmatrix} = \begin{bmatrix} 1 & 0 \\ 0 & \Delta(u,v) \end{bmatrix} \begin{bmatrix} \bar{u} \\ \bar{v} \end{bmatrix} \tag{2.124}$$

Now we consider u, v to be functions of $\bar{u}$, $\bar{v}$, and these functions are obtained by inverting the coordinate transformation (2.122)-(2.123).

By Taylor's Theorem we may write

$$\Delta(u,v) = S(\bar{v}) + \sum_{i=1}^{n-r} \bar{u}_i \, \Xi_i(\bar{u},\bar{v}) \tag{2.125}$$

in which $S(\bar{v})$ and $\Xi_i(\bar{u},\bar{v})$ are smooth (n-r) × (n-r) matrices. We may write Ξ_i in matrix format as

$$\Xi_i(\bar{u},\bar{v}) = [\Xi_{ik\ell}(\bar{u},\bar{v})] \tag{2.126}$$

with k the row index and ℓ the column index. Now we define the (n-r) × r matrix $K(\bar{u},\bar{v})$ by

$$K(\bar{u},\bar{v}) = \begin{bmatrix} K_{ki}(\bar{u},\bar{v}) \end{bmatrix} = \begin{bmatrix} \sum_{\ell=1}^{n-r} \Xi_{ik\ell}(\bar{u},\bar{v})\bar{v}_\ell \end{bmatrix} \tag{2.127}$$

Then we may write the right side of (2.124) as

$$\begin{bmatrix} 1 & 0 \\ 0 & \Delta(u,v) \end{bmatrix} \begin{bmatrix} \bar{u} \\ \bar{v} \end{bmatrix} = \begin{bmatrix} 1 & 0 \\ K(\bar{u},\bar{v}) & S(\bar{v}) \end{bmatrix} \begin{bmatrix} \bar{u} \\ \bar{v} \end{bmatrix} \tag{2.128}$$

Thus, with the contact transformation

$$T = \left[M = \begin{bmatrix} 1 & 0 \\ 0 & 1 \end{bmatrix} ; \quad \begin{matrix} \bar{u} = u + A(u,v)^{-1}B(u,v)v \\ \bar{v} = v. \end{matrix} \right] \tag{2.129}$$

we find that

$$f(u,v) \sim \begin{bmatrix} 1 & A^{-1}B \\ 0 & D-CA^{-1}B \end{bmatrix} \begin{bmatrix} u \\ v \end{bmatrix} \sim \begin{bmatrix} 1 & 0 \\ K(u,v) & S(v) \end{bmatrix} \begin{bmatrix} u \\ v \end{bmatrix} \tag{2.130}$$

Finally the contact transformation

$$T = \left[M = \begin{bmatrix} 1 & 0 \\ -K(u,v) & 1 \end{bmatrix} \; ; \; \begin{bmatrix} y_1(u) = u \\ y_2(v) = v \end{bmatrix} \right] \tag{2.131}$$

shows that we have

$$f(u,v) \sim \begin{bmatrix} 1 & 0 \\ K & S \end{bmatrix} \begin{bmatrix} u \\ v \end{bmatrix} \sim \begin{bmatrix} 1 & 0 \\ 0 & S \end{bmatrix} \begin{bmatrix} u \\ v \end{bmatrix} = \begin{bmatrix} u \\ S(v)v \end{bmatrix} \tag{2.132}$$

so that the proof is complete when we set $g(v) = S(v)v$. The Lyapunov-Schmidt Lemma is now proved.

We may obtain from the proof a construction of the function $g(v)$. From (2.125) we see that

$$S(\bar{v}) = \Delta(u,v) \qquad \text{when } \bar{u} = 0 \text{ and } \bar{v} = v \tag{2.133}$$

Then from (2.122) we see that $\bar{u} = 0$ provided that

$$u + A(u,v)^{-1}B(u,v)v = 0 \tag{2.134}$$

Thus if (2.134) holds, then $g(v)$ is given by

$$g(v) = S(v)v = \Delta(u,v)v \tag{2.135}$$

Reversing the above argument, we arrive at the following algorithm

<u>Lyapunov-Schmidt Splitting Procedure</u>

i) Let $X(v)$ be the solution of (2.134)

$$X(v) + A(X(v),v)^{-1}B(X(v),v)v = 0 \tag{2.136}$$

We notice that $X(0) = 0$ and that it exists by the Implicit Function Theorem.

ii) Then we have from (2.135) that

$$g(v) = \left[D(X(v),v) - C(X(v),v)A(X(v),v)^{-1}B(X(v),v)\right]v \tag{2.137}$$

or

$$g(v) = D(X(v),v)v + C(X(v),v)X(v) \tag{2.138}$$

It is convenient to have also the following supplement to the Lyapunov-Schmidt Splitting Lemma:

Uniqueness Lemma. If df(0) has rank r and f(x) is contact equivalent to $\begin{bmatrix} u \\ g(v) \end{bmatrix}$ and $\begin{bmatrix} u \\ h(v) \end{bmatrix}$ with r variables in u, then g(v) is contact equivalent to h(v).

We will need the following general lemma.

Invariance Lemma. If $F(x,\mu)$ is a versal unfolding of f(x) and $(M(x),y(x))$ is a contact transformation, then $M(x)F(y(x),\mu)$ is a versal unfolding of $M(x)f(y(x))$.

Now we return to the Lyapunov-Schmidt Splitting Lemma. With it our calculation of versal unfoldings is simplified in the following way.

Reducing Lemma. Suppose that $f(u,v) = \begin{bmatrix} A(u,v) & B(u,v) \\ C(u,v) & D(u,v) \end{bmatrix} \begin{bmatrix} u \\ v \end{bmatrix}$ and g(v) are as in the Lyapunov-Schmidt Splitting Lemma. Suppose further that $g(v) + \mu_1 N_1(v) + \cdots + \mu_q N_q(v)$ is a versal unfolding of g(v). Then

$$F(u,v,\mu) = f(u,v) + \begin{bmatrix} 0 \\ \mu_1 N_1(v) + \cdots + \mu_q N_q(v) \end{bmatrix} \tag{2.139}$$

is a versal unfolding of f(u,v) about $\mu = 0$.

We note that in practice we have changed the parameter variables from λ to μ because the original singularity was specified by $\lambda = 0$. We interpret μ to be the canonical, independent parameters and λ the physically interpretable ones, and in some cases we can relate λ and μ via elementary alterations.

We illustrate the connection between Mather's Theorems and the Lyapunov-Schmidt Splitting Procedure by giving the proof of the Reducing Lemma.

Proof: First we must show that

$$E(u,v,\mu) = \begin{bmatrix} u \\ g(v) + \mu_1 N_1(v) + \cdots + \mu_q N_q(v) \end{bmatrix} \tag{2.140}$$

is a versal unfolding of

$$e(u,v) = \begin{bmatrix} u \\ g(v) \end{bmatrix} \tag{2.141}$$

Because $g(v) + \mu_1 N_1(v) + \cdots + \mu_q N_q(v)$ is a versal unfolding of g(v), Mather's Theorem I implies that any smooth vector V(v) may be written in the form (2.53) as

$$V(v) = dg(v) \cdot G(v) + H(v) \cdot g(v) + \mu_1 N_1(v) + \cdots + \mu_q N_q(v) \tag{2.142}$$

near the origin $v = 0$ for suitable choices of $G(v)$, $H(v)$ and $\mu_1,\ldots,\mu_q$. We notice that

$$de(u,v) = \begin{bmatrix} 1 & 0 \\ 0 & d_v g(v) \end{bmatrix} \tag{2.143}$$

in which we have used the notation

$$d_v g(v) = \left[\frac{\partial g_i}{\partial v_j}(v) \right] , \quad \begin{cases} i = 1,\ldots,n-r \\ j = 1,\ldots,n-r \end{cases} \tag{2.144}$$

Now we let

$$X(u,v) = \begin{bmatrix} U(u,v) \\ V(u,v) \end{bmatrix} \tag{2.145}$$

be any smooth n-dimensional vector. By Taylor's Theorem we may write

$$V(u,v) = V(0,v) + W(u,v)u \tag{2.146}$$

and we may rewrite (2.142) in the form

$$V(0,v) = d_v g(v)\cdot G(v) + H(v)\cdot g(v) + \mu_1 N_1(v) + \cdots + \mu_q N_q(v) \tag{2.147}$$

near the origin. Combining (2.143)-(2.147) we see that

$$X(u,v) = de(u,v) \cdot \begin{bmatrix} U(u,v) \\ G(v) \end{bmatrix} + \begin{bmatrix} 0 & 0 \\ W(u,v) & H(v) \end{bmatrix} \cdot e(u,v) \tag{2.148}$$

$$+ \mu_1 \begin{bmatrix} 0 \\ N_1(v) \end{bmatrix} + \cdots + \mu_q \begin{bmatrix} 0 \\ N_q(v) \end{bmatrix}$$

near the origin. Because an arbitrary function $X(u,v)$ has been written in the form (2.53), from Mather's Theorem I we conclude that $E(u,v,\mu)$ is a versal unfolding of $e(u,v)$.

The Invariance Lemma tells us that if we invert the contact transformation T that carries $f(u,v)$ to $e(u,v)$ and apply T^{-1} to (2.140) then we will obtain a versal unfolding of $f(u,v)$. Applying the inverse of the contact transformation (2.131) to (2.140) and recalling from (2.135) that $g(v) = S(v)v$, we obtain

$$\begin{bmatrix} u \\ g(v) + \mu_1 N_1(v) + \cdots + \mu_q N_q(v) \end{bmatrix} \sim \begin{bmatrix} u \\ K\cdot u + S\cdot v + \mu_1 N_1(v) + \cdots + \mu_q N_q(v) \end{bmatrix} \tag{2.149}$$

According to (2.128) we have

$$\begin{bmatrix} \bar{u} \\ K(\bar{u},\bar{v})\bar{u} + S(\bar{v})\bar{v} + \mu_1 N_1(\bar{v}) + \cdots + \mu_q N_q(\bar{v}) \end{bmatrix} \tag{2.150}$$

$$= \begin{bmatrix} \bar{u} \\ \Delta(\bar{u},\bar{v})\bar{v} + \mu_1 N_1(\bar{v}) + \cdots + \mu_q N_q(\bar{v}) \end{bmatrix}$$

in which we have used the coordinate transformation (2.122)-(2.123). Then, using (2.122), (2.123), and (2.130) in (2.150), we have

$$\begin{bmatrix} \bar{u} \\ K(\bar{u},\bar{v})\bar{u} + S(\bar{v})\bar{v} + \mu_1 N_1(\bar{v}) + \cdots + \mu_q N_q(\bar{v}) \end{bmatrix} \tag{2.151}$$

$$= \begin{bmatrix} u + A(u,v)^{-1}B(u,v)v \\ [D(u,v) - C(u,v)A(u,v)^{-1}B(u,v)]\ v + \mu_1 N_1(v) + \cdots + \mu_q N_q(v) \end{bmatrix}$$

so that

$$\begin{bmatrix} u \\ g(v) + \mu_1 N_1(v) + \cdots + \mu_q N_q(v) \end{bmatrix} \tag{2.152}$$

$$\sim \begin{bmatrix} u + A^{-1}(u,v)B(u,v)v \\ [D(u,v) - C(u,v)A(u,v)^{-1}B(u,v)]\ v + \mu_1 N_1(v) + \cdots + \mu_q N_q(v) \end{bmatrix}$$

Finally, multiplying the right side of (2.152) by $\begin{bmatrix} A(u,v) & 0 \\ C(u,v) & 1 \end{bmatrix}$ we see that

$$\begin{bmatrix} u \\ g(v) + \mu_1 N_1(v) + \cdots + \mu_q N_q(v) \end{bmatrix} \tag{2.153}$$

$$\sim \begin{bmatrix} A(u,v)u + B(u,v)v \\ C(u,v)u + D(u,v)v + \mu_1 N_1(v) + \cdots + \mu_q N_q(v) \end{bmatrix}$$

and the proof is complete.

Example 14. A versal unfolding of the Lorenz model: splitting and reducing lemmas

We turn again to the Lorenz system (2.46). Setting

$$\left.\begin{aligned} u_1 &= x_1 \\ u_2 &= x_3 \\ v &= x_2 \end{aligned}\right\} \qquad (2.154)$$

we may write (2.47) as

$$f(u,v) = \begin{bmatrix} -\sigma & 0 & \sigma \\ 0 & -b & u_1 \\ 1 & -u_1 & -1 \end{bmatrix} \begin{bmatrix} u_1 \\ u_2 \\ v \end{bmatrix} \qquad (2.155)$$

Defining $X(v)$ to be the solution of (2.136), which here is

$$X(v) - \begin{bmatrix} \sigma & 0 \\ 0 & b \end{bmatrix}^{-1} \begin{bmatrix} \sigma \\ X_1(v) \end{bmatrix} v = 0 \qquad (2.156)$$

we see that

$$X(v) = \begin{bmatrix} v \\ \frac{1}{b} v^2 \end{bmatrix} \qquad (2.157)$$

so that $g(v)$ as given by (2.138) becomes

$$g(v) = \frac{1}{b} v^3 \qquad (2.158)$$

According to the Lyapunov-Schmidt Splitting Lemma, $f(u,v)$ is contact equivalent to $\left[{u \atop g(v)}\right]$.

Now, if $x(v)$ is any smooth function, then by Taylor's Theorem, we have

$$x(v) = x(0) + x'(0)v + \frac{x''(0)}{2} v^2 + z(v)v^3 \qquad (2.159)$$

which we may write in the form (2.53) as

$$x(v) = dg(v) \left[\frac{b}{6} x''(0)\right] + \left[b\ z(v)\right] g(v) + x(0) + x'(0)v \qquad (2.160)$$

Thus, from (2.55) we conclude that $v^3/b + \mu_0 + \mu_1 v$ is a versal unfolding of v^3/b, with $\mu_0 = x(0)$ and $\mu_1 = x'(0)$. Then we use the Reducing Lemma to conclude that

$$F(u,v,\mu) = \begin{bmatrix} \sigma(v - u_1) \\ u_1 v - bu_2 \\ - u_1 u_2 + u_1 - v + \mu_0 + \mu_1 v \end{bmatrix} \tag{2.161}$$

is a versal unfolding of (2.155) about $\mu_0 = \mu_1 = 0$. After using (2.154) to redefine the last two variables, we see that

$$V(x,\mu) = \begin{bmatrix} \sigma(x_2 - x_1) \\ x_1 - x_2 - x_1 x_3 + \mu_0 + \mu_1 x_2 \\ - bx_3 + x_1 x_2 \end{bmatrix} \tag{2.162}$$

is a versal unfolding of (2.46). #

Thus, we have found a versal unfolding of (2.46) by applying the Lyapunov-Schmidt Splitting and Reducing Lemmas, which replace a great deal of guessing with what amounts to an algorithm. Nonetheless, the versal unfolding (2.48) is still nicer than (2.162) because the degrees of the coefficients of the control parameters are lower. The problem of passing from one versal unfolding to another via elementary alterations was discussed in Section 2.4; in practice our choice is guided by the physics of the problem rather than the mathematical form of the unfolding. For example, as mentioned previously, we begin by setting some natural parameters λ_i to their singular values λ_{is}, and then find new ones μ_j in the unfolding. Clearly, some of the unfolding parameters μ_j must be associated with $\lambda_i - \lambda_{is}$; this is accomplished via use of elementary alterations.

2.6 Vector Spaces and Contact Computations

The machinery presented so far has for its object the recovery of the stationary phase portrait of any parameterized perturbation $\dot{x} = F(x,\mu)$ of $\dot{x} = f(x)$; this recovery is obtained from a single versal perturbation of $f(x)$. Determination of the existence of a versal unfolding and construction of such an unfolding follow from computations based on Mather's Theorems and are facilitated by the Lyapunov-Schmidt splitting process discussed in Sections 2.3 and 2.5. Also, the form of the resulting versal unfolding may be changed by means of elementary alterations (Section 2.4); moreover, the structure of the stationary phase portrait of the versal unfolding does not change upon alteration, provided that the number of parameters in the versal unfolding is the smallest possible. When a versal unfolding contains the minimum number of unfolding parameters, we will call the unfolding <u>minimal</u>.

Aside from the intuitive discussion presented in Example 11, we have not yet set up any machinery for the computation of this number, the codimension d of $f(x)$. This is a crucial number because it tells us the number of independent parameters needed to describe fully the branching behavior of the parameterized perturbation, and we will use d to aid our classification of the singularity type in Section 2.7. We begin by making two general observations, leaving aside their easy proofs which follow those in Section 2.5.

Proposition 2.2. If f(x) and g(x) are contact equivalent, then codim [f(x)] = codim [g(x)].

We use the convention that codim [f(x)] = ∞ if f(x) has no versal unfoldings.

Proposition 2.3. If $f(x,y) = \begin{bmatrix} x \\ g(y) \end{bmatrix}$, then codim [f(x,y)] = codim [g(y)].

Example 15. Codimension: Propositions 2.2 and 2.3

If f(x) is the right side (2.47) of the Lorenz system, then from Example 14 we see that

$$f(u_1,u_2,v) \sim \begin{bmatrix} u_1 \\ u_2 \\ -v^3 \end{bmatrix} \tag{2.163}$$

so

$$\text{codim }(f) = \text{codim} \begin{bmatrix} u_1 \\ u_2 \\ -v^3 \end{bmatrix} = \text{codim }(-v^3) \tag{2.164}$$

Thus, by using Propositions 2 and 3, we have reduced the computation of d as far as possible to that of finding the codimension of a single function, v^3; we still are faced with the original problem, but in fewer variables. Somewhat surprisingly, the solution of the problem of computing the codimension proceeds in part by interpretation of terminology. We complete the calculation of codim (v^3) in Example 17. #

Let us suppose that we have $f(x) = (f_1(x),\ldots,f_n(x))$ with $x = (x_1,\ldots,x_n)$, and we know that f(x) has a versal unfolding. According to Mather's Theorem I, there exist functions $N_1(x),\ldots,N_q(x)$ such that every smooth vector function Y(x) may be written near the origin in the form (2.53) as

$$Y(x) = df(x)\cdot G(x) + H(x)\cdot f(x) + \mu_1 N_1(x) + \cdots + \mu_q N_q(x) \tag{2.165}$$

in which G is a smooth vector function, H is a smooth n × n matrix function and $\mu_1,\ldots,\mu_q$ are real numbers.

Now we consider the set I(f) of all smooth vector functions J(x) that may be written in the form

$$J(x) = df(x)\cdot G(x) + H(x)\cdot f(x) \tag{2.166}$$

for suitably smooth choices for $G(x)$ and $H(x)$; thus $T^k I(f) = I^k(f)$. Clearly any two members of $I(f)$ may be added to produce another member of $I(f)$, and clearly any member of $I(f)$ may be multiplied by a scalar to produce another member of $I(f)$. Thus, $I(f)$ is a vector space. On the other hand, the set $\xi(n)$ of smooth functions near the origin is also obviously a vector space containing $I(f)$ as a vector subspace. Then the condition (2.165) states exactly that the images of $N_1(x),\ldots,N_q(x)$ in the quotient vector space $\xi(n)/I(f)$ span that vector space. And the value of the number q is the smallest possible if and only if $N_1(x),\ldots,N_q(x)$ determine a basis for $\xi(n)/I(f)$. That is, we see that

$$d = \text{codim}\ [f(x)] = \dim\ [\xi(n)/I(f)] \tag{2.167}$$

or that the codimension of $f(x)$ is exactly the dimension of the space $\xi(n)/I(f)$.

We recall that if Φ is a vector space and Θ is a subspace of Φ, then the <u>quotient vector space</u> Φ/Θ is defined to be that one obtained from Φ by formally setting to zero any member of Θ. Thus, two elements ϕ and $\phi + \theta$ of Φ that differ by an element θ of Θ are regarded as equal in Φ/Θ; moreover, every element η of Φ/Θ is represented by an element ϕ of Φ, and we may write $\eta = [\phi]$. It is easy to check that if Ψ is any subspace of Φ complementary to Θ (that is, $\Psi + \Theta = \Phi$ and $\Psi \cap \Theta = 0$) then there is a one-one linear map from Ψ to Φ/Θ, so Ψ is isomorphic to Φ/Θ. In fact, the correspondence

$$\psi \to [\psi] \tag{2.168}$$

determines such a linear map $\Psi \to \Phi/\Theta$.

We still have not solved the problem of computing codim $[f(x)]$, but we have made a crucial observation concerning what vector space is spanned by the images of the unfolding functions $N_1(x),\ldots,N_q(x)$. Before proceeding let us illustrate these ideas by considering an example of a finite dimensional quotient space of $\xi(n)$.

<u>Example 16. The dimension of $\xi(n)/\xi_2(n)$: quotient spaces</u>

Let $\xi_2(n)$ be the set of smooth vector fields near the origin that vanish to the second order at the origin. That is, a typical member υ of $\xi_2(n)$ has the form

$$\upsilon(x) = \sum_{i,j=1}^{n} X_{ij}(x)\ x_i\ x_j \tag{2.169}$$

Let us see what the vector space $\xi(n)/\xi_2(n)$ is. Let $X(x)$ be an arbitrary smooth vector field near the origin. Using Taylor's Theorem with remainder, we may expand $X(x)$ in the form

$$X(x) = X(0) + \sum_{i=1}^{n} \partial_i X(0)\, x_i + \sum_{i,j=1}^{n} X_{ij}(x)\, x_i\, x_j \tag{2.170}$$

The constant and linear parts of X(x) are determined uniquely by X(x), and their sum is not in $\xi_2(n)$. From (2.169) we see that the remainder

$$R(x) = \sum_{i,j=1}^{n} X_{ij}(x)\, x_i\, x_j \tag{2.171}$$

is in $\xi_2(n)$ and thus R is exactly what we may neglect when finding elements of the quotient space $\xi(n)/\xi_2(n)$. That is, X(x) is represented uniquely in $\xi(n)/\xi_2(n)$ by

$$\eta(x) = X(0) + \sum_{i=1}^{n} \partial_i X(0)\, x_i \tag{2.172}$$

A basis for $\xi(n)/\xi_2(n)$ easily is seen to be represented uniquely by

$$\begin{bmatrix} 1 \\ 0 \\ \vdots \\ 0 \end{bmatrix};\ldots;\begin{bmatrix} 0 \\ 0 \\ \vdots \\ 1 \end{bmatrix};\begin{bmatrix} x_i \\ 0 \\ \vdots \\ 0 \end{bmatrix}, i = 1,\ldots,n;\begin{bmatrix} 0 \\ 0 \\ \vdots \\ x_i \end{bmatrix}, i = 1,\ldots,n \tag{2.173}$$

Thus, we conclude that

$$\dim\left[\xi(n)/\xi_2(n)\right] = n + n^2 = n(n+1) \tag{2.174}$$

#

More generally, let $\xi_k(n)$ be the smooth vector functions near the origin that vanish to the order k. Thus, all coefficients of products with fewer terms than k are zero (cf. (2.169) for the quadratic case). Let $P^k(n)$ be the polynomial vector functions of degree $\leq$ k. Then with use of Taylor's Theorem, we may split the vector space $\xi(n)$ into two complementary components $P^k(n)$ and $\xi_{k+1}(n)$,

$$\xi(n) = P^k(n) + \xi_{k+1}(n) \tag{2.175}$$

Then it follows that the quotient vector space $\xi(n)/\xi_{k+1}(n)$ is isomorphic to $P^k(n)$ so that

$$\dim\left[\xi(n)/\xi_{k+1}(n)\right] = \dim\left[P^k(n)\right] = n\begin{bmatrix} n+k \\ n \end{bmatrix} \tag{2.176}$$

With this information we can complete the computation of the codimension of the Lorenz system.

Example 17. Codimension of x^3: versal unfoldings

From Example 15, we take n = 1 and

$$f(x) = - x^3 \tag{2.177}$$

What is I(f) in this case? I(f) consists of all the smooth functions J(x) that may be written in the form (2.166) as

$$J(x) = - 3x^2g(x) + h(x)x^3 \tag{2.178}$$

Because g(x) and h(x) are arbitrary, J(x) may be any function whose form is

$$J(x) = x^2m(x) \tag{2.179}$$

From (2.169) we see that J(x) is any function that vanishes to the second order at the origin, and we have in this case

$$I(f) = \xi_2(1) \tag{2.180}$$

Thus, with the aid of (2.174) we find that

$$\text{codim } (- x^3) = \dim \left[\xi(1)/I(f)\right] = \dim \left[\xi(1)/\xi_2(1)\right] = 1 + 1^2 = 2 \tag{2.181}$$

Moreover, we see from (2.173) that a basis for $\xi(1)/I(f)$ is represented by the constant function 1 and the linear function x. Thus we take $N_1(x) = 1$ and $N_2(x) = x$ in a minimal versal unfolding of $- x^3$. #

In general we are not so fortunate as to have $I(f) = \xi_k(n)$ for some k. However, if we discover by means of Mather's Theorem II that f(x) has a versal unfolding, then we automatically have that

$$K^k(n) \subset I(f) \tag{2.182}$$

for some value of k. Let us re-examine the condition that J(x) be in I(f) given by

$$J(x) = df(x)\cdot G(x) + H(x)\cdot f(x) \tag{2.183}$$

If $\alpha(x)$ is any smooth scalar function, then we have

$$\alpha(x)J(x) = df(x)\cdot\alpha(x)G(x) + H(x)\cdot\alpha(x)f(x) \tag{2.184}$$

so that we may conclude that any member of I(f) may be multiplied by a smooth scalar function to produce another member of I(f).

On the other hand, there is a close connection between $K^k(n)$ and the space of smooth vector functions that vanish to the order k at the origin. In fact, an easy application of Taylor's Theorem tells us that if Y(x) is a smooth vector function that vanishes to the order k at the origin, then we may write

$$Y(x) = \sum_{\text{finite}} \alpha_i(x)\, Y_i(x) \tag{2.185}$$

in which $\alpha_i(x)$ are scalar functions and $Y_i(x)$ is homogeneous of order k; but when $Y_i(x)$ is homogeneous of order k, then $Y_i(x)$ is in $K^k(n)$. Now, (2.182) implies that each $Y_i(x)$ is a member of I(f). Thus, from (2.184) we conclude that each $\alpha_i(x)Y_i(x)$ is in I(f) and from (2.185) that Y(x) is a member of I(f). To summarize, we have found that every smooth vector function that vanishes to the order k at the origin is in I(f), and we may write

$$\xi_k(n) \subset I(f) \tag{2.186}$$

Now we recall from elementary linear algebra that

$$\frac{\xi(n)}{I(f)} \text{ is isomorphic to } \frac{\xi(n)/\xi_k(n)}{I(f)/\xi_k(n)} \tag{2.187}$$

because $\xi(n) \supset I(f) \supset \xi_k(n)$. So we have

$$\begin{aligned} \text{codim}\,[f(x)] &= \dim\,[\xi(n)/I(f)] = \dim\left[\frac{\xi(n)/\xi_k(n)}{I(f)/\xi_k(n)}\right] \\ &= \dim\,[\xi(n)/\xi_k(n)] - \dim\,[I(f)/\xi_k(n)] \end{aligned} \tag{2.188}$$

We already know that $\dim\,[\xi(n)/\xi_k(n)]$ is finite, and from (2.176) equal to $\dim\,[P^{k-1}(n)]$. The dimension of $I(f)/\xi_k(n)$ may be calculated routinely though painfully by finding the dimension of the vector space $I^{k-1}(f)$ of polynomial vector functions Q(x) that may be written in the form

$$Q(x) = T^{k-1}\,[T^{k-1}(df(x))\cdot G(x) + H(x)\cdot T^{k-1}f(x)] \tag{2.189}$$

in which $T^{k-1}g(x)$ denotes the Taylor polynomial expansion of degree k-1 of g(x) at the origin, and G(x) and H(x) are restricted to be polynomial vector functions of degree $\leq$ k-1. Because $\dim\,[I^{k-1}(f)] = \dim\,[I(f)/\xi_k(n)]$, it is clear that codim (f) may be found in a finite but large number of steps.

We illustrate these computations in the following two examples.

<u>Example 18. Unfoldings of $\pm\, x^k$, $k \geq 2$: minimal versal forms in codimension 1</u>
In this example we give the unfoldings of all the functions that we will encounter in Chapters 3-5 of this monograph.

We take n = 1 and set

$$f(x) = \pm\, x^k \quad , \quad k \geq 2 \tag{2.190}$$

Then we have that

$$df(x) = \pm\, kx^{k-1} \tag{2.191}$$

Thus, I(f) is the set of all functions that may be written as $x^{k-1}y(x) + \ell(x)x^k$, and that is the set of all multiples of x^{k-1}. Thus, we may write $I(f) = \xi_{k-1}(1)$. Then, we recall from Taylor's Theorem (2.175) that

$$\xi(1) = P^{k-2}(1) + \xi_{k-1}(1) \tag{2.192}$$

so that a basis for $P^{k-2}(1)$ determines one for $\xi(1)/I(f) = \xi(1)/\xi_{k-1}(1)$. Such a basis is given by $\{1, x, x^2, \ldots, x^{k-2}\}$ so that finally Mather's Theorem I tells us that

$$V(x,\mu) = \pm\, x^k + \mu_{k-2}x^{k-2} + \cdots + \mu_1 x + \mu_0 \tag{2.193}$$

is a minimal versal unfolding of $f(x) = \pm\, x^k$. #

Example 19. The hyperbolic umbilic: minimal versal unfoldings

Now we take n = 2, and we consider

$$f(x_1,x_2) = \begin{bmatrix} x_1 x_2 \\ x_1^2 - x_2^2 \end{bmatrix} \tag{2.194}$$

Then we have

$$df(x) = \begin{bmatrix} x_2 & x_1 \\ 2x_1 & -2x_2 \end{bmatrix} \tag{2.195}$$

We will apply Mather's Theorem II with n = k = 2 to see that f(x) has a versal unfolding. Thus, we must check that

$$K^2(2) \subset I(f) \tag{2.196}$$

or more straightforwardly, that every vector function

$$\begin{bmatrix} x_1^2 \\ 0 \end{bmatrix}, \begin{bmatrix} x_1x_2 \\ 0 \end{bmatrix}, \begin{bmatrix} x_2^2 \\ 0 \end{bmatrix}, \begin{bmatrix} 0 \\ x_1^2 \end{bmatrix}, \begin{bmatrix} 0 \\ x_1x_2 \end{bmatrix}, \begin{bmatrix} 0 \\ x_2^2 \end{bmatrix} \tag{2.197}$$

may be written in the form (2.166) as

$$J(x) = df(x)\cdot G(x) + H(x)\cdot f(x) \tag{2.198}$$

Accordingly, we find that

$$\begin{bmatrix} x_1^2 \\ 0 \end{bmatrix} = \begin{bmatrix} x_2 & x_1 \\ 2x_1 & -2x_2 \end{bmatrix} \begin{bmatrix} 0 \\ x_1 \end{bmatrix} + \begin{bmatrix} 0 & 0 \\ 2 & 0 \end{bmatrix} \begin{bmatrix} x_1x_2 \\ x_1^2 - x_2^2 \end{bmatrix} \tag{2.199}$$

$$\begin{bmatrix} x_1x_2 \\ 0 \end{bmatrix} = \begin{bmatrix} 1 & 0 \\ 0 & 0 \end{bmatrix} \begin{bmatrix} x_1x_2 \\ x_1^2 - x_2^2 \end{bmatrix} \tag{2.200}$$

$$\begin{bmatrix} x_2^2 \\ 0 \end{bmatrix} = \begin{bmatrix} x_2 & x_1 \\ 2x_1 & -2x_2 \end{bmatrix} \begin{bmatrix} 0 \\ x_1 \end{bmatrix} + \begin{bmatrix} 0 & -1 \\ 2 & 0 \end{bmatrix} \begin{bmatrix} x_1x_2 \\ x_1^2 - x_2^2 \end{bmatrix} \tag{2.201}$$

$$\begin{bmatrix} 0 \\ x_1^2 \end{bmatrix} = \begin{bmatrix} x_2 & x_1 \\ 2x_1 & -2x_2 \end{bmatrix} \begin{bmatrix} \frac{1}{2}x_1 \\ 0 \end{bmatrix} + \begin{bmatrix} -\frac{1}{2} & 0 \\ 0 & 0 \end{bmatrix} \begin{bmatrix} x_1x_2 \\ x_1^2 - x_2^2 \end{bmatrix} \tag{2.202}$$

$$\begin{bmatrix} 0 \\ x_1x_2 \end{bmatrix} = \begin{bmatrix} 0 & 0 \\ 1 & 0 \end{bmatrix} \begin{bmatrix} x_1x_2 \\ x_1^2 - x_2^2 \end{bmatrix} \tag{2.203}$$

$$\begin{bmatrix} 0 \\ x_2^2 \end{bmatrix} = \begin{bmatrix} x_2 & x_1 \\ 2x_1 & -2x_2 \end{bmatrix} \begin{bmatrix} \frac{1}{2}x_1 \\ 0 \end{bmatrix} + \begin{bmatrix} -\frac{1}{2} & 0 \\ 0 & -1 \end{bmatrix} \begin{bmatrix} x_1x_2 \\ x_1^2 - x_2^2 \end{bmatrix} \tag{2.204}$$

We conclude that $\xi_2(2) \subset I(f)$ and consequently that $\xi_2(2) \subset I^2(f)$ so that $f(x)$ has a versal unfolding.

Now we must find a basis for $\xi(2)/I(f)$. Because $\xi_2(2) \subset I(f)$ we note that

$$\frac{\xi(2)}{I(f)} = \frac{[\xi(2)/\xi_2(2)]}{[I(f)/\xi_2(2)]} \tag{2.205}$$

and as a result we may consider the two finite dimensional vector spaces $\xi(2)/\xi_2(2)$ and $I(f)/\xi_2(2)$ separately. First, a basis for $\xi(2)/\xi_2(2)$ is represented by the constant and linear vector functions

$$e_1 = \begin{bmatrix} 1 \\ 0 \end{bmatrix} \tag{2.206}$$

$$e_2 = \begin{bmatrix} 0 \\ 1 \end{bmatrix} \tag{2.207}$$

$$e_3 = \begin{bmatrix} x_1 \\ 0 \end{bmatrix} \tag{2.208}$$

$$e_4 = \begin{bmatrix} x_2 \\ 0 \end{bmatrix} \tag{2.209}$$

$$e_5 = \begin{bmatrix} 0 \\ x_1 \end{bmatrix} \tag{2.210}$$

$$e_6 = \begin{bmatrix} 0 \\ x_2 \end{bmatrix} \tag{2.211}$$

Thus, we find that $\xi(2)/\xi_2(2)$ is six-dimensional. Next we consider $I(f)/\xi_2(2)$. This vector space is composed of functions of the form (2.189), which is the Taylor's expansion containing only the first terms. Thus, $I(f)/\xi_2(2)$ consists of functions whose terms of degree greater than or equal to two have been erased. But then we see that the part $H \cdot f(x)$ can be erased automatically because $f(x)$ (2.194) is homogeneous of degree 2. On the other hand, from writing

$$df(x) \cdot G(x) = df(x) \cdot G(0) + df(x) \cdot [G(x) - G(0)] \tag{2.212}$$

we see that the expression $df(x) \cdot [G(x) - G(0)]$ can be erased because it contains no terms of degree ≤ 1. Finally, then, we have that $I(f)/\xi_2(2)$ is spanned by the two columns of $df(x)$,

$$\begin{bmatrix} x_2 \\ 2x_1 \end{bmatrix} = e_4 + 2e_5 \tag{2.213}$$

and

$$\begin{bmatrix} x_1 \\ -2x_2 \end{bmatrix} = e_3 - 2e_6 \tag{2.214}$$

These are linearly independent, so we see that $\dim \left[I(f)/\xi_2(2) \right] = 2$, and then that

$$\operatorname{codim}(f) = \dim \left[\xi(2)/\xi_2(2) \right] - \dim \left[I(f)/\xi_2(2) \right] = 6 - 2 = 4 \tag{2.215}$$

How do we use the above information to obtain a basis for $\xi(2)/I(f)$, given that (2.205) holds? Comparing the two lists (2.206)-(2.211) and (2.213)-(2.214) and recalling the construction of quotient spaces, we see that $[\xi(2)/\xi_2(2)]/[I(f)/\xi_2(2)]$ is spanned by $e_1,\ldots,e_6$ with $e_4 + 2e_5$ and $e_3 - 2e_6$ set equal to zero. That is, our quotient space is spanned by $e_1,\ldots,e_6$ with $e_4 = -2e_5$ and $e_3 = 2e_6$. Thus, four immediate choices of bases for $\xi(2)/I(f)$ are

$$\{e_1, e_2, e_3, e_4\} \tag{2.216}$$

$$\{e_1, e_2, e_6, e_4\} \tag{2.217}$$

$$\{e_1, e_2, e_3, e_5\} \tag{2.218}$$

$$\{e_1, e_2, e_6, e_5\} \tag{2.219}$$

The first choice (2.216), for example, determines the minimal versal unfolding

$$V(x,\mu) = \begin{bmatrix} x_1x_2 \\ x_1^2 - x_2^2 \end{bmatrix} + \mu_1e_1 + \mu_2e_2 + \mu_3e_3 + \mu_4e_4 \tag{2.220}$$

$$= \begin{bmatrix} x_1x_2 + \mu_1 + \mu_3x_1 + \mu_4x_2 \\ x_1^2 - x_2^2 + \mu_2 \end{bmatrix}$$

The second choice (2.217) produces a second minimal versal unfolding

$$W(x,\mu) = \begin{bmatrix} x_1x_2 + \mu_1 + \mu_4x_2 \\ x_1^2 - x_2^2 + \mu_2 + \mu_3x_2 \end{bmatrix} \tag{2.221}$$

There are two other minimal versal unfoldings given by (2.218)-(2.219). #

Example 20. The elliptic umbilic: minimal versal unfoldings

Now we consider

$$f(x_1,x_2) = \begin{bmatrix} x_1^2 \\ x_2^2 \end{bmatrix} \tag{2.222}$$

This example proceeds along the lines of Example 19, except that the present one is much easier, and we merely list the results. We begin by using the basis $e_1,\ldots,e_6$ (2.206)-(2.211) for $\xi(2)/\xi_2(2)$ that was given for the preceeding example. Again $I(f) \supset \xi_2(2)$ and a basis for $I(f)/\xi_2(2)$ is given by

$$2e_3 = \begin{bmatrix} 2x_1 \\ 0 \end{bmatrix} \tag{2.223}$$

and

$$2e_6 = \begin{bmatrix} 0 \\ 2x_2 \end{bmatrix} \tag{2.224}$$

As before, a basis for $\xi(2)/I(f)$ is given by $\{e_1, e_2, e_3, e_4\}$ so that a minimal versal unfolding of f(x) is given by

$$V(x,\mu) = \begin{bmatrix} x_1^2 + \mu_1 + \mu_3 x_1 + \mu_4 x_2 \\ x_2^2 + \mu_2 \end{bmatrix} \tag{2.225}$$

Alternate unfoldings may be found easily by choosing other bases. #

2.7 Classification of Singularities

We have described a family of routine computations, prescribed by Mather's Theorem II, to decide whether a given vector function has a versal unfolding; to apply these computations to systems of ordinary differential equations, we have described a simplifying splitting scheme given by the Lyapunov-Schmidt method. In the case a versal unfolding exists, we have described also relatively easy computations that enable us to find such a versal unfolding, and still further ones to alter the unfolding to a more suitable form.

However, an objection remains: perhaps only very exceptional vector functions f(x) have versal unfoldings. Actually a theorem of Thom (1964) states that the reverse is true: only very exceptional vector functions fail to have versal unfoldings. To state the theorem more precisely, we consider the space $P^k(n)$ of polynomial vector functions $f(x) = (f_1(x),\ldots,f_n(x))$ with $x = (x_1,\ldots,x_n)$ and f(x) a scalar polynomial of degree $\leq k$. Then we have the following theorem.

Theorem (Thom). There is an open dense subset U of $P^k(n)$ such that $T^k f(x) \in U$ implies that f(x) has a versal unfolding.

This theorem can be proved using the methods of Mather (1968). The complement of U is actually a smooth hypersurface, which must then have measure zero. Consequently, almost every polynomial function of degree k is the kth Taylor polynomial of vector functions having versal unfoldings. In fact, any vector function having such a kth degree Taylor polynomial has a versal unfolding. Thus the objection above is countered in a very strong sense.

A related theorem is the following (Mather, 1968).

<u>Theorem (Mather).</u> If $I(f) \supset \xi_k(n)$ and $h(x)$ is in $\xi_{k+2}(n)$, then $f(x)$ and $f(x) + h(x)$ are contact equivalent.

<u>Corollary.</u> If $f(x)$ has a versal unfolding, then $f(x)$ is contact equivalent to a polynomial vector function.

The corollary can be easily proved: Because $f(x)$ has a versal unfolding, we have $I(f) \supset \xi_k(n)$ for some k. Using Taylor's Theorem with remainder, we write

$$f(x) = T^{k+1}f(x) + R(x) \tag{2.226}$$

in which $T^{k+1}f(x)$ is the (k+1)th Taylor polynomial of $f(x)$. Then $R(x) \in \xi_{k+2}(n)$. Thus $f(x)$ is contact equivalent to $f(x) - R(x) = T^{k+1}f(x)$.

It follows from the corollary that we may confine our attention to polynomial vector functions, as indeed we do in the sequel.

<u>Example 21. A versal unfolding of a nonpolynomial function: contact equivalence to a polynomial</u>

We take n = 1 and

$$f(x) = e^x - 1 - x - x^2/2 \tag{2.227}$$

In Examples 9 and 18 we found that

$$g(x) = x^3/6 \tag{2.228}$$

has a versal unfolding, given by

$$g(x,\mu) = x^3/6 + \mu_2 x + \mu_1 \tag{2.229}$$

If f and g are contact equivalent, then we immediately have an unfolding for x near x = 0. From (2.227) we see that $I(f) \supset \xi_2(1)$. Moreover, because

$$h(x) = g(x) - f(x) \tag{2.230}$$

is in $\xi_4(1)$, the above theorem of Mather implies that $f(x)$ and $g(x)$ are contact equivalent. Thus a versal unfolding of $f(x)$ is

$$f(x,\mu) = f(x) + \mu_2 x + \mu_1 \tag{2.231}$$

In this particular case we may see this contact equivalence directly without recourse to Mather's theorem. We may write

$$f(x) = x^3\, \gamma(x)/6 \tag{2.232}$$

Then we notice that $\gamma(0) = 1$ so that the function

$$\beta(x) = [\gamma(x)]^{1/3} \tag{2.233}$$

is smooth near $x = 0$ and $\beta(0) = 1$. Then

$$y(x) = x\,\beta(x) \tag{2.234}$$

is a coordinate transformation. Applying the contact transformation

$$T = \left(M = 1;\; y(x) = x\,\beta(x)\right) \tag{2.235}$$

to $g(x)$, we obtain

$$M \cdot g(y(x)) = 1\frac{y(x)^3}{6} = x^3\beta^3(x)/6 = f(x) \tag{2.236}$$

showing that $f(x)$ and $g(x)$ are contact equivalent.

Of course, if $n > 1$, then such an argument will not apply and we must appeal to Mather's theorem. #

In Section 4.1 we will encounter a situation similar to the one cited in Example 21, in which we use the contact equivalence of a fifth degree polynomial

$$f(x) = x^5 + a_1x^4 + a_2x^3 \quad , \quad a_2 \neq 0 \tag{2.237}$$

with

$$g(x) = x^3 \tag{2.238}$$

to obtain a minimal versal unfolding of $f(x)$,

$$f(x,\mu) = x^5 + a_1x^4 + a_2x^3 + \mu_2x + \mu_1 \tag{2.239}$$

Thus in general we classify the corank 1 singularities by finding the lowest nonvanishing terms of the governing polynomials.

Finally, a contact catastrophe version of Thom's celebrated theorem on elementary catastrophes is the following.

<u>Theorem</u> <u>(Thom)</u>. Suppose that codim $[f(x)] \leq 4$. Then $f(x)$ is contact equivalent to one of the following

$$\begin{bmatrix} x_1^k \\ x_2 \\ \vdots \\ x_n \end{bmatrix}, \quad \text{with } k = 2, 3, 4, \text{ or } 5 \tag{2.240}$$

$$\begin{bmatrix} x_1 x_2 \\ x_1^2 - x_2^2 \\ x_3 \\ \vdots \\ x_n \end{bmatrix} \tag{2.241}$$

or

$$\begin{bmatrix} x_1^2 \\ x_2^2 \\ x_3 \\ \vdots \\ x_n \end{bmatrix} \tag{2.242}$$

Minimal versal unfoldings are given in Examples 18–20 and are summarized in Tables 2.1 and 2.2. The corank 1 singularities (2.240) are the fold, cusp, swallowtail and butterfly repectively and we note that the codimension of these singularities is the same in catastrophe and contact catastrophe theories. This is not the case for the two corank 2 singularities (2.241) and (2.242), however. These last two singularities correspond to Thom's hyperbolic and elliptic umbilic catastrophes. Their codimension in contact catastrophe theory is 4, whereas the codimension of the corresponding catastrophes is 3. In these two cases, then, one more independent parameter is needed when no Lyapunov function is known (or exists) than when a Lyapunov function is known.

TABLE 2.1

CORANK 1 UNFOLDINGS

FORM	TYPE	CODIMENSION d OF SINGULARITY	UNFOLDING
x^2	FOLD	1	$x^2 - \mu_0$
x^3	CUSP	2	$x^3 - \mu_1 x - \mu_0$
x^4	SWALLOWTAIL	3	$x^4 - \mu_2 x^2 - \mu_1 x - \mu_0$
x^5	BUTTERFLY	4	$x^5 - \mu_3 x^3 - \mu_2 x^2 - \mu_1 x - \mu_0$

TABLE 2.2

CORANK 2 UNFOLDINGS

FORM	TYPE	CODIMENSION d OF SINGULARITY	UNFOLDING
$\begin{bmatrix} x_1 x_2 \\ x_1^2 - x_2^2 \end{bmatrix}$	HYPERBOLIC UMBILIC	4	$\begin{bmatrix} x_1 x_2 + \mu_1 + \mu_4 x_2 \\ x_1^2 - x_2^2 + \mu_2 + \mu_3 x_2 \end{bmatrix}$
$\begin{bmatrix} x_1^2 \\ x_2^2 \end{bmatrix}$	ELLIPTIC UMBILIC	4	$\begin{bmatrix} x_1^2 + \mu_1 + \mu_3 x_1 + \mu_4 x_2 \\ x_2^2 + \mu_2 \end{bmatrix}$

We are now ready to apply the contact catastrophe theory to spectral models of three different physical systems. For later reference, we summarize the discussion of this chapter in the following section so that the essential steps of our procedure can be clearly seen.

2.8 Summary

The theory described above may be applied to particular problems in many different ways. In order to indicate the kind of work that will have to be performed in such applications, we outline here one possible procedure. It handles slightly more general cases than those discussed earlier in this chapter, and is the one we apply in Chapters 3-5.

We begin with a system of n autonomous parameterized differential equations

$$\frac{dx_i}{dt} = f_i(x_1,\dots,x_n;\ \lambda_1,\dots,\lambda_p) \qquad i = 1,\dots,n \tag{2.243}$$

which arises from a physical problem. We may write (2.243) more briefly as

$$\frac{dx}{dt} = f(x, \lambda) \tag{2.244}$$

We do not assume that every essentially relevant physical parameter is some function of the (multi-) parameter λ appearing explicitly in (2.244); essential but hidden parameters may be affecting transitions within the physical system. We do suppose that we are given a stationary solution $x = x^0$ for some value λ^0 of the parameters that appear explicitly in (2.244); in practice, λ^0 will be the singular point having the greatest available codimension d (see Tables 2.1 and 2.2). We are interested in both the number and the location of nearby stationary solutions that arise when the values of the physical parameters are varied slightly; in particular, λ is near λ^0.

In order to describe the form of these nearby solutions, we will have to discover, as well, essentially all of the hidden parameters. We will do so in two steps. The first is clear-cut and mathematical; it generates enough additional parameters to represent all the qualitatively different structures for the nearby stationary solutions. The second step is not entirely clear-cut: it gives physical interpretations to the additional parameters. However, this step is facilitated by a clear-cut mathematical procedure which enables the replacement of one mathematically adequate set of parameters with another equally adequate set; in this way, we may shift to a physically less awkward set of parameters. When the procedure is complete, all further parameters will act on the physical stationary phase portrait only as functions of the known parameters.

The first step in the procedure is to translate coordinates in the dynamical variables and the parameters so that $\lambda^0 = 0$ and $x^0 = 0$. Now we may rewrite the system (2.244) in the form

$$\frac{dx}{dt} = \Gamma_1(x, \lambda)\cdot x \tag{2.245}$$

where $\Gamma_1(x, \lambda)$ is an $n \times n$ matrix depending smoothly on (x, λ). At this point we must relinquish the temporal information in (2.245) in order to study its stationary behavior; we perform contact transformations of $\Gamma_1(x, \lambda)\cdot x$ (Section 2.2) to bring it into a normal form.

The first transformation consists of permuting independently the rows and columns of $\Gamma_1(0, 0)$ so that the top left $r \times r$ corner has rank r, where $r = \text{rank}\,[\Gamma_1(0, 0)]$; that is, the top left corner carries all the rank. More formally, we find permutation matrices P and Q such that the upper left $r \times r$ corner of $P\,\Gamma_1(0, 0)\,Q$ has rank r. Then we may write

$$\Gamma(x, \lambda)\cdot x = P\,\Gamma_1(Qx, \lambda)\cdot Qx \tag{2.246}$$

so that $\Gamma(x, \lambda)\cdot x$ is contact equivalent to $\Gamma_1(x, \lambda)\cdot x$ and the upper left $r \times r$

corner of $\Gamma(0, 0)$ has rank r. We denote the first r components of x by v and the last n - r by w so that we may write $\Gamma\cdot x$ in block form

$$\Gamma(x, \lambda)\cdot x = \begin{bmatrix} A(v, w, \lambda) & B(v, w, \lambda) \\ C(v, w, \lambda) & D(v, w, \lambda) \end{bmatrix} \cdot \begin{bmatrix} v \\ w \end{bmatrix} \tag{2.247}$$

with A an invertible $r \times r$ matrix and D an $(n - r) \times (n - r)$ matrix. In practice, we verify that we have made the correct choice in (2.247) by checking that

$$\Delta(0, 0, 0) = D(0, 0, 0) - C(0, 0, 0)\, A^{-1}(0, 0, 0)\, B(0, 0, 0) = 0 \tag{2.248}$$

The second transformation arises from the solution $X(w, \lambda)$ of the following equation

$$X(w, \lambda) + A\big(X(w, \lambda), w, \lambda\big)^{-1} \cdot B\big(X(w, \lambda), w, \lambda\big)\cdot w = 0 \tag{2.249}$$

in which $X(w, \lambda)$ is an r-vector function. We set

$$p(w, \lambda) = D\big(X(w, \lambda), w, \lambda\big)\cdot w + C\big(X(w, \lambda), w, \lambda\big)\cdot X(w, \lambda) \tag{2.250}$$

so that there exists a contact transformation carrying $\Gamma(x, \lambda)\cdot x$ into $\begin{bmatrix} v \\ p(w, \lambda) \end{bmatrix}$ as in Section 2.5. However we will not need this transformation, only a versal unfolding

$$V(w, 0, \mu) = p(w, 0) + \mu_1 N_1(w) + \cdots + \mu_q N_q(w) \tag{2.251}$$

of $p(w, 0)$. Once this versal unfolding has been found, a versal unfolding of $\Gamma(x, 0)\cdot x$ is given by

$$U(x, 0, \mu) = \Gamma(x, 0)\cdot x + \begin{bmatrix} 0 \\ \mu_1 N_1(w) + \mu_2 N_2(w) + \cdots + \mu_q N_q(w) \end{bmatrix} \tag{2.252}$$

where $x = \begin{bmatrix} v \\ w \end{bmatrix}$ as before. Finally, then, a versal unfolding of $\Gamma_1(x, 0)\cdot x$ is given by

$$W(x, 0, \mu) = P^{-1}\, U(Q^{-1}x, 0, \mu) \tag{2.253}$$

and it is this unfolding which contains all the information about the number and location of stationary solutions of (2.243) near $x = x^0$ when $\lambda = \lambda^0$. Some of the parameters μ are the parameters which were "hidden" in (2.243) but which are necessary to describe fully all the perturbations of the stationary phase portrait of (2.243).

But a question remains: how do we find the versal unfolding (2.251)? This question has two answers. The first answer applies customarily in practice and the second applies more generally in principle; the first employs Thom's catastrophes, the second Mather's Theorems. Often $p(w, \lambda)$ and w are 1-vectors or 2-vectors; that is, the problem is one of corank 1 or 2. In these cases, a simple contact transformation will express $p(w, 0)$ as one of Thom's catastrophes, and then we may read a versal unfolding from Table 2.1 or 2.2. From inspection of the form of $p(w, \lambda)$ (or, in the cases discussed in Chapters 3-5, of the form of the numerator of $p(w, \lambda)$), we may find the value λ^0 of λ for which the codimension d is the greatest. This value of λ can be translated to $\lambda = 0$ in (2.251) and the necessary unfolding functions $N_1, \ldots, N_q$ can be found from Tables 2.1 or 2.2. If we cannot contact transform $p(w, 0)$ to one of Thom's singularities in the corank 1 or 2 case, then a versal unfolding does not exist and our procedure comes to a halt; but, in such a situation $p(w, 0)$ is highly exceptional and almost any perturbation will carry it into one of Thom's catastrophes.

In the general case, the following problem remains: we must first check that a versal unfolding exists, and we do so by applying Mather's Theorem II. We may apply this theorem either to $f(x, 0)$ or to $p(w, 0)$, but application to the latter is easier. The theorem prescribes an infinite sequence of tests, each of which terminates after finitely many steps; if any one of these tests is successful, then the function has a versal unfolding and we may then use the computations in that one test to find a versal unfolding. To carry out the kth test, we find the set $I^k(p)$ of all kth-degree polynomial $(n - r)$-vectors which may be written

$$Y(w) = T^k[dp(w, 0)\cdot G(w) + H(w)\cdot p(w, 0)] \tag{2.254}$$

where $G(w)$ is a kth degree polynomial vector and $H(w)$ is a $(k + 1)$st degree polynomial matrix. The set $I^k(p)$ is a finite-dimensional vector space; accordingly, it may be produced in finitely many steps. Let $K^k(n - r)$ be the set of polynomial $(n - r)$-vectors whose entries are homogeneous of degree k. Then $K^k(n - r)$ is also a finite-dimensional vector space, and it is a matter of finitely many steps to check whether each member of a basis of $K^k(n - r)$ is in $I^k(p)$. In the event that every such member is in $I^k(p)$, that is, in the event that

$$K^k(n - r) \subset I^k(p) \tag{2.255}$$

Mather's Theorem II states that $p(w, 0)$ has a versal unfolding, and the kth test has met with success. Let $P^k(n - r)$ be the kth degree polynomial $(n - r)$-vectors; to write a versal unfolding of $p(w, 0)$, we choose a minimal set of kth-degree polynomial $(n - r)$-vectors $N_1(w), \ldots, N_q(w)$ with the property that the set represented by

$$T^k \frac{\partial p}{\partial \lambda_1}(w, 0), \ldots, T^k \frac{\partial p}{\partial \lambda_p}(w, 0), N_1(w), \ldots, N_q(w)$$

in $P^k(n - r)/I^k(p)$ spans the whole of that quotient vector space. Then

$$V(w, 0, \mu) = p(w, 0) + \mu_1 N_1(w) + \cdots + \mu_q N_q(w) \tag{2.256}$$

is a versal unfolding of $p(w, 0)$ and (2.253) determines a versal unfolding of $\Gamma_1(x, 0)\cdot x$.

We arrive finally at the second of the two steps mentioned above. If the parameters μ in (2.253) do not have readily available physical interpretations, then how may we replace (2.253) with another versal unfolding, with perhaps more amenable parameters? Specifically, which of the parameters μ_j are associated with λ near λ^o? Which of the μ_j are the new hidden parameters? The clear-cut and mathematical procedure for doing so involves simply choosing a new basis for the appropriate vector space, but in practice this vector space is rather remote from the versal unfolding (2.253). Instead, the versal unfolding itself may be transformed by means of an alteration. We suppose that we have a linear unfolding

$$W(x, \mu) = f(x) + N_1(x)\mu_1 + \cdots + N_q(x)\mu_q \tag{2.257}$$

or

$$W(x, \mu) = f(x) + N(x)\cdot\mu \tag{2.258}$$

where $N(x)$ is an $n \times q$ matrix. Choose Λ to be any invertible $q \times q$ constant matrix, $G_1(x), \ldots, G_q(x)$ to be any n-vector functions and $H_1(x), \ldots, H_q(x)$ to be any $n \times n$ matrix functions. Define $\tilde{N}(x)$ by setting

$$\tilde{N}(x) = \left[df(x)\cdot G_1(x) + H_1(x)\cdot f(x), \ldots, df(x)\cdot G_q(x) + H_q(x)\cdot f(x)\right] \tag{2.259}$$

and then define $\tilde{W}(x, \mu)$ by setting

$$\tilde{W}(x, \mu) = f(x) + \tilde{N}(x)\cdot\mu \tag{2.260}$$

The linear unfolding $\tilde{W}(x, \mu)$ will be versal, and in fact any minimal linear versal unfolding of $f(x)$ may be obtained from $W(x, \mu)$ in this way. Of course the physical interpretation of μ will change correspondingly under such an alteration. The principal difficulty with this approach arises from the great freedom it allows; there are many choices to be made when we may prefer to deal with only a single offending parameter. But the apparently great freedom is reduced drastically in practice because at least some elements of $\tilde{N}(x)$ will be those associated with λ near λ^o.

This observation suggests that we could have left the parameters λ in (2.251)-(2.252) instead of setting them equal to the singular value zero; in fact, this approach works well in some cases. What is involved is a complication of the Lyapunov-Schmidt Splitting Procedure, so that a trade-off results: Either we may invest more time splitting the initial equations (2.244) into (2.247) to reduce the cost of altering the final versal unfolding via (2.258), or we may prefer to minimize the time spent getting started and concentrate our efforts on the alteration procedure at the end. An advantage of the latter approach is that in the search for suitable physical interpretations of the parameters, we may indeed require the freedom to re-interpret all the parameters.

A somewhat more restricted operation is an elementary alteration. In this procedure, we choose an n-vector function $G^o(x)$, an $n \times n$ matrix function $H^o(x)$ and parameter values $\mu_1^o, \ldots, \mu_q^o$ not all zero. Let i be an index for which $\mu_i^o \neq 0$. Define $\tilde{N}(x)$ from $N(x)$ by replacing the ith column of $N(x)$ with the column

$$Y^o(x) = df(x) \cdot G^o(x) + H^o(x) \cdot f(x) + N_1(x)\, \mu_1^{\,o} + \cdots + N_q(x)\, \mu_q^{\,o} \tag{2.261}$$

and leaving all the other columns the same. Then $\tilde{W}(x, \mu)$ defined by (2.260) is a versal unfolding, and again, every minimal linear versal unfolding of $f(x)$ may be obtained from $W(x, \mu)$ by performing a sequence of elementary alterations.

At this point we arrive at the irreducible difficulty of interpreting physically the hidden parameters. The Lyapunov-Schmidt procedure simplifies our system; Mather's Theorems discover the hidden parameters; and alterations re-cast the role played by these parameters into any other admissible role. But nowhere in this machinery is there a mechanism for interpreting physically the parameters. This interpretation must be supplied by us as shown in Chapters 3-5 by extending the original equations of motion to accommodate known physical parameters in such a way that these parameters then appear as the parameters in one of the versal unfoldings generated by Mather's theory.

To put it another way, we inevitably have to make an educated guess; Mather's Machine tells us when the guess is correct but it cannot generate the guess itself.

CHAPTER 3

RAYLEIGH-BÉNARD CONVECTION

As we discussed in Chapter 1, Rayleigh-Bénard convection is an example of a physical system exhibiting distinct transitions from laminar to turbulent flow as the value of a parameter, the Rayleigh number R, is increased. The first branching solution is steady and two-dimensional (Krishnamurti, 1970a); it is natural to seek the number of parameters necessary for describing completely this first transition that occurs at $R = R_s$. Determination of these parameters is a necessary first step in the eventual development of a model capable of reproducing the entire hierarchy of transitions.

The simplest spectral model that captures the form of this first observed convective flow is the three-coefficient model of Lorenz (1963). The adequacy of this truncation (at least near $R = R_s$) was demonstrated by Saltzman (1962) who performed several numerical integrations of a larger, seven component system and found that in some cases four of the spectral components decayed toward zero while the other three approached nontrivial solutions. The fact that the steady solutions to the Lorenz model have the correct form can be seen by noting that these solutions are proportional to $(R - R_s)^{1/2}$; this is the form deduced by Chandrasekhar (1961, Appendix I) from consideration of only the energetics of the governing partial differential system.

A schematic bifurcation diagram for the steady states of the Lorenz model is given in Fig. 3.1b, in which w^* represents the amplitude of the steady states, $d_1 = (R - R_s)/R_s$ and $d_1 = 0$ is the value of the singular point. Both upper and lower branches are locally stable and hence are equally likely to be observed; physically, one branch represents clockwise, the other counterclockwise circulation.

However, transitions from conductive to convective states are observed to occur more smoothly than those depicted in Fig. 3.1b (Tavantzis et al., 1978). A much better representation is given by the monotonic curves in either Fig. 3.1a or c; the primary difference between the two curves is in the circulation sense that they represent. Actually, Fig. 3.1a-c are three parallel cross-sections through the standard cusp surface (Fig. 3.1d), one passing through the cusp point, and the other two passing through points on either side.

From these observations, we might suspect that the singularity of the Lorenz model is of cusp type. In this case, we would need two parameters for describing the branching behavior in the neighborhood of $R = R_s$. In this chapter, we apply to the Lorenz model the procedure discussed in Chapter 2 for classifying a singularity, unfolding about it, and interpreting physically the necessary parameters. The exposition follows closely that of Shirer and Wells (1982); we have corrected here a minor algebraic mistake in their modified Lorenz model that does not affect the qualitative nature of their results.

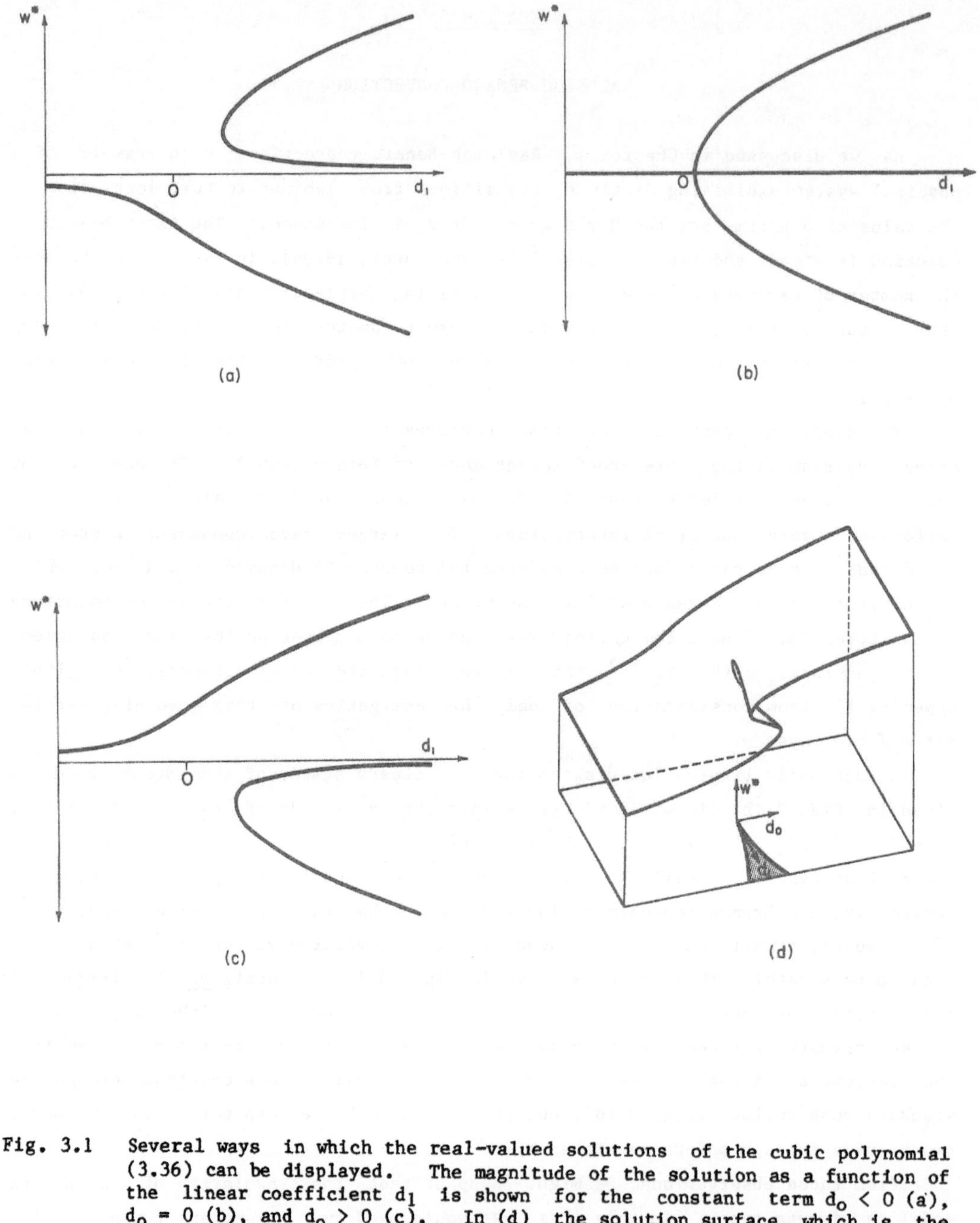

Fig. 3.1 Several ways in which the real-valued solutions of the cubic polynomial (3.36) can be displayed. The magnitude of the solution as a function of the linear coefficient d_1 is shown for the constant term $d_0 < 0$ (a), $d_0 = 0$ (b), and $d_0 > 0$ (c). In (d) the solution surface, which is the standard cusp surface, is shown as a function of both coefficients d_0 and d_1.

3.1 Classification of the Singularity

Convection develops from a motionless, basic state whose vertical temperature distribution is linear. Thus, it is appropriate to use perturbation variables for representing the convective state. Accordingly, we write the temperature field as

$$T = T_o + \Delta_z T\,(z/H) + \theta \tag{3.1}$$

in which $\Delta_z T$ is the difference between the temperatures of the top and bottom surfaces, H is the distance between these surfaces, and θ is a perturbation that vanishes on them. We note that the basic temperature field $T_o + \Delta_z T\,(z/H)$ satisfies the conduction relation $\nabla^2 T = 0$.

The two-dimensional shallow Boussinesq equations can be written in the form

$$\frac{\partial}{\partial t^*}\tilde{\nabla}^2\psi^* = -K(\psi^*, \tilde{\nabla}^2\psi^*) + \sigma(1+a^2)^{-1}\tilde{\nabla}^4\psi^* + \sigma(1+a^2)\frac{\partial\theta^*}{\partial x^*} \tag{3.2}$$

$$\frac{\partial\theta^*}{\partial t^*} = -K(\psi^*, \theta^*) + r\frac{\partial\psi^*}{\partial x^*} + (1+a^2)^{-1}\tilde{\nabla}^2\theta^* \tag{3.3}$$

in which the asterisk denotes a nondimensional variable and the velocity components satisfy $u^* = -\partial\psi^*/\partial z^*$ and $w^* = \partial\psi^*/\partial x^*$. Here we have adopted the nondimensionalization used by Lorenz (1963); the necessary forms are

$$t = t^* H^2 \pi^{-2}(1+a^2)^{-1}\kappa^{-1} \tag{3.4}$$

$$x = x^* H \pi^{-1} a^{-1} \tag{3.5}$$

$$z = z^* H \pi^{-1} \tag{3.6}$$

$$\psi = \psi^* (1+a^2)\kappa a^{-1} \tag{3.7}$$

$$\theta = -\theta^* \pi^3 (1+a^2)^3 \Delta_z T\, a^{-2} R^{-1} \tag{3.8}$$

and the nondimensional parameters are the Rayleigh number,

$$R = -g\,\Delta_z T\, H^3 T_o^{-1} \nu^{-1}\kappa^{-1} \tag{3.9}$$

the critical Rayleigh number,

$$R_s = (1+a^2)^3 \pi^4 a^{-2} \tag{3.10}$$

the Prandtl number,

$$\sigma = \nu/\kappa \tag{3.11}$$

the aspect ratio,

$$a = H/L \tag{3.12}$$

and the normalized Rayleigh number,

$$r = R/R_s \tag{3.13}$$

Also, L is the domain width, ν is the viscosity, κ is the thermometric conductivity, K is the Jacobian operator in the variables x* and z*,

$$K(f^*, g^*) = \frac{\partial f^*}{\partial x^*}\frac{\partial g^*}{\partial z^*} - \frac{\partial f^*}{\partial z^*}\frac{\partial g^*}{\partial x^*} \tag{3.14}$$

and $\tilde{\nabla}^2$ is a nondimensional Laplacian defined by

$$\tilde{\nabla}^2 \equiv a^2 \frac{\partial^2}{\partial x^{*2}} + \frac{\partial^2}{\partial z^{*2}} \tag{3.15}$$

For boundary conditions we require that the normal velocity derivatives vanish on each boundary so that $\psi^* = 0$ at $x = 0, n\pi$ $(n = 1, 2, \ldots)$ and $z = 0, \pi$. We also assume that there are no thermal perturbations at the top and bottom and no lateral heat flux at the sides, so that $\theta^* = 0$ at $z^* = 0, \pi$ and $\partial\theta^*/\partial x^* = 0$ at $x^* = 0, n\pi$. These allow thermal perturbations on the side walls where upward motion may occur. Alternate boundary conditions for θ^* are $\theta^* = 0$ at $x^* = 0, n\pi$ and $\partial\theta^*/\partial z^* = 0$ at $z^* = 0, \pi$, and these were used to develop the model of Yost and Shirer (1982). These boundary conditions maintain a constant temperature difference in the horizontal and a constant heat flux in the vertical, although the compatibility of these thermal boundary conditions with those on the motion field is hard to justify.

Upon substitution of

$$\psi^* = \sqrt{2}\, x_1 \sin x^* \sin z^* \tag{3.16}$$

$$\theta^* = \sqrt{2}\, x_2 \cos x^* \sin z^* - x_3 \sin 2z^* \tag{3.17}$$

into (3.2)-(3.3), multiplication by the appropriate basis function and integration of the result over the domain $0 \leq x^* \leq n\pi$, $0 \leq z^* \leq \pi$, we obtain the Lorenz system

$$\dot{x}_1 = -\sigma x_1 + \sigma x_2 = F_1 \tag{3.18}$$

$$\dot{x}_2 = -x_1 x_3 + r x_1 - x_2 = F_2 \tag{3.19}$$

$$\dot{x}_3 = x_1 x_2 - b x_3 = F_3 \tag{3.20}$$

in which we have defined

$$b = 4(1 + a^2)^{-1} \tag{3.21}$$

Bifurcation occurs at the value of the normalized Rayleigh number given by $d_1 = r_s - 1 = 0$ (Fig. 3.1b).

In order to apply the contact catastrophe method of Chapter 2, we first write the right side of (3.18)-(3.20) in the form (2.247)

$$\underset{\sim}{F}(\underset{\sim}{v},\underset{\sim}{w},\underset{\sim}{\alpha}_s) = \begin{bmatrix} A(\underset{\sim}{v},\underset{\sim}{w}) & B(\underset{\sim}{v},\underset{\sim}{w}) \\ C(\underset{\sim}{v},\underset{\sim}{w}) & D(\underset{\sim}{v},\underset{\sim}{w}) \end{bmatrix} \begin{bmatrix} \underset{\sim}{v} \\ \underset{\sim}{w} \end{bmatrix} \tag{3.22}$$

in which the steady solution is $(\underset{\sim}{v},\underset{\sim}{w}) = (0,0,0)$ and

$$\underset{\sim}{\alpha}_s = [r_s,\sigma,b] = [1,\sigma,b] \tag{3.23}$$

$$\underset{\sim}{v} = \begin{bmatrix} v_1 \\ v_2 \end{bmatrix} = \begin{bmatrix} x_1 \\ x_3 \end{bmatrix} \tag{3.24}$$

$$\underset{\sim}{w} = w = [x_2] \tag{3.25}$$

$$A(\underset{\sim}{v},\underset{\sim}{w}) = \begin{bmatrix} -\alpha_2 & 0 \\ 0 & -\alpha_3 \end{bmatrix} \tag{3.26}$$

$$B(\underset{\sim}{v},\underset{\sim}{w}) = \begin{bmatrix} \alpha_2 \\ v_1 \end{bmatrix} \tag{3.27}$$

$$C(\underset{\sim}{v},\underset{\sim}{w}) = [1, -v_1] \tag{3.28}$$

$$D(\underset{\sim}{v},\underset{\sim}{w}) = [-1] \tag{3.29}$$

We note that $\alpha_2 = \sigma > 0$ and $\alpha_3 = b > 0$ so that A is nonsingular as required.

To verify that we have made the correct choices for the matrices A-D, we substitute (3.26)-(3.29) evaluated at $[v_1, v_2, w] = [0,0,0]$ into (2.248) to obtain

$$\Delta(0,\underset{\sim}{\alpha}_s) = -1 - [1,0] \begin{bmatrix} -\alpha_2^{-1} & 0 \\ 0 & -\alpha_3^{-1} \end{bmatrix} \begin{bmatrix} \alpha_2 \\ 0 \end{bmatrix} = -1 + 1 = 0 \tag{3.30}$$

Because A is nonsingular and $\Delta(0,\underset{\sim}{\alpha}_s) = 0$, the corank of $\underset{\sim}{F}$ is 1. This is important for determining the class of the singularity and therefore how many parameters must be added to unfold it. In this monograph, we discuss only corank 1 examples, and their unfoldings are shown in Table 2.1 in Chapter 2. Examples of the more difficult corank 2 unfoldings, some of which are discussed in Chapters 2 and 6, will be the subject of a future report.

Next we must calculate the solutions $\underset{\sim}{v}^*(\underset{\sim}{w}^*)$ of (2.249). They are given by

$$\begin{bmatrix} v_1^* \\ v_2^* \end{bmatrix} + \begin{bmatrix} -\alpha_2^{-1} & 0 \\ 0 & -\alpha_3^{-1} \end{bmatrix} \begin{bmatrix} \alpha_2 \\ v_1^* \end{bmatrix} \cdot w^* = 0 \tag{3.31}$$

or

$$v_1^* = w^* \tag{3.32}$$

and

$$v_2^* = w^{*2}/\alpha_3 \tag{3.33}$$

The function $p(w^*)$ is found by substituting (3.32)-(3.33) into (3.28)-(3.29) and the result into (2.250). Thus we obtain

$$p(w^*) = [-1]\cdot w^* + [1, -w^*] \begin{bmatrix} w^* \\ w^{*2}/\alpha_3 \end{bmatrix} = -w^{*3}/\alpha_3 \tag{3.34}$$

and we conclude that the steady states of the Lorenz model (3.18)-(3.20) are the same as those of the transformed system

$$\underset{\sim}{F}^*(\underset{\sim}{v}^*, \underset{\sim}{w}^*, \underset{\sim}{\alpha}_s) = \begin{bmatrix} v_1^* \\ v_2^* \\ -w^{*3}/\alpha_3 \end{bmatrix} \tag{3.35}$$

In this case we note that $p(w^*)$ already is a polynomial and so we need not consider the numerator separately. As we will see in Chapters 4 and 5, this is an exceptional case.

From Table 2.1 we see that two parameters are needed to describe completely solutions of a steady state polynomial that consists of only a cubic term. These two parameters are associated with the constant and linear terms of the cubic polynomial; we unfold about the singularity $\underset{\sim}{\alpha} = \underset{\sim}{\alpha}_s$ of the Lorenz model when we write

$$p(w^*) = -(w^{*3}/\alpha_3 - d_1 w^* - d_0) \tag{3.36}$$

Solutions of (3.36) for three different values of d_0 are given in Fig. 3.1; the case $d_0 = 0$ (Fig. 3.1b) corresponds to that of the Lorenz model with r evidently playing the role of d_1. But the parameter d_0 is new and demands physical interpretation; as mentioned previously, it is apparently important for an accurate explanation of the smooth development of a convective flow. Application of Mather's Theorem I toward this end is the subject of the next section.

3.2 Physical Interpretation of the Unfolding

Merely determining the number of independent parameters needed to describe branching behavior in the neighborhood of a singularity is not sufficient for application of the results to actual systems. The locations of the new parameters given by the contact catastrophe theory are often not obtainable via arbitrary insertion of a new term into the governing partial differential equations and calculation of a new Fourier coefficient for the spectral equations. But with use of elementary alterations discussed in Section 2.4, we may find those locations in the spectral systems that are amenable to physical interpretation.

In the original variables, the unfolding (3.36) of the Lorenz model (3.18)-(3.20) is (for $\alpha_1 = r = 1$)

$$F_1 = -\alpha_2 x_1 + \alpha_2 x_2 \tag{3.37}$$

$$F_2 = -x_1 x_3 + x_1 + (d_1 - 1) x_2 + d_o \tag{3.38}$$

$$F_3 = x_1 x_2 - \alpha_3 x_3 \tag{3.39}$$

The new parameter d_0 would represent a diabatic heating term in the thermodynamic equation (see Section 5.4); but in its present location, d_1 cannot represent the normalized Rayleigh number r. Thus, we prefer another form of $\underset{\sim}{F}$ that is still a versal unfolding.

In order to find alternate terms as factors for d_0 and d_1, we write (3.37)-(3.39) as

$$\begin{bmatrix} F_1 \\ F_2 \\ F_3 \end{bmatrix} = \begin{bmatrix} -\alpha_2 x_1 + \alpha_2 x_2 \\ -x_1 x_3 + x_1 - x_2 \\ x_1 x_2 - \alpha_3 x_3 \end{bmatrix} + d_1 \begin{bmatrix} 0 \\ x_2 \\ 0 \end{bmatrix} + d_o \begin{bmatrix} 0 \\ 1 \\ 0 \end{bmatrix} \tag{3.40}$$

With combination of (3.40) and (2.261) we may write an arbitrary function $\underset{\sim}{Y}(\underset{\sim}{x})$ as

$$\begin{bmatrix} Y_1 \\ Y_2 \\ Y_3 \end{bmatrix} = \begin{bmatrix} -\alpha_2 & \alpha_2 & 0 \\ -x_3+1 & -1 & -x_1 \\ x_2 & x_1 & -\alpha_3 \end{bmatrix} \begin{bmatrix} g_1(\underset{\sim}{x}) \\ g_2(\underset{\sim}{x}) \\ g_3(\underset{\sim}{x}) \end{bmatrix}$$

$$+ \begin{bmatrix} h_{11}(\underset{\sim}{x}) & h_{12}(\underset{\sim}{x}) & h_{13}(\underset{\sim}{x}) \\ h_{21}(\underset{\sim}{x}) & h_{22}(\underset{\sim}{x}) & h_{23}(\underset{\sim}{x}) \\ h_{31}(\underset{\sim}{x}) & h_{32}(\underset{\sim}{x}) & h_{33}(\underset{\sim}{x}) \end{bmatrix} \begin{bmatrix} -\alpha_2 x_1 + \alpha_2 x_2 \\ -x_1 x_3 + x_1 - x_2 \\ x_1 x_2 - \alpha_3 x_3 \end{bmatrix}$$

$$+ \gamma_1 \begin{bmatrix} 0 \\ x_2 \\ 0 \end{bmatrix} + \gamma_o \begin{bmatrix} 0 \\ 1 \\ 0 \end{bmatrix} \tag{3.41}$$

in which the choices of the elements of $\underset{\sim}{G}$, $\underset{\sim}{H}$, and $\underset{\sim}{\gamma}$ depend on Y.

We know that r = 1 must correspond to $d_1 = 0$. Thus we must be able to move d_1 from a product with x_2 in (3.38) to a product with x_1 in (3.38). In order to accomplish this we must find suitable choices for $g_i(\underset{\sim}{x})$ and $h_{ij}(\underset{\sim}{x})$ in (3.41). If we set $Y_1 = 0$, $Y_2 = x_1$ and $Y_3 = 0$ in (3.41), then we see immediately that the only nonzero value for g_i or h_{ij} is $h_{21}(\underset{\sim}{x}) = -\alpha_2^{-1}$. Upon interchanging $[0, x_1, 0]^T$ and $[0, x_2, 0]^T$, we find that

$$\begin{bmatrix} 0 \\ x_2 \\ 0 \end{bmatrix} = \begin{bmatrix} 0 & 0 & 0 \\ \alpha_2^{-1} & 0 & 0 \\ 0 & 0 & 0 \end{bmatrix} \begin{bmatrix} -\alpha_2 x_1 + \alpha_2 x_2 \\ -x_1 x_3 + x_1 - x_2 \\ x_1 x_2 - \alpha_3 x_3 \end{bmatrix} + \begin{bmatrix} 0 \\ x_1 \\ 0 \end{bmatrix} \tag{3.42}$$

This will be a more convenient way to write Mather's theorem, for now it is apparent that we may replace the column vector on the left side of (3.42) with the new one on the right side. Thus, from the conclusions of Mather's theory we may replace (3.37)-(3.39) with the equivalent form

$$F_1 = -\alpha_2\, x_1 + \alpha_2\, x_2 \tag{3.43}$$

$$F_2 = -x_1\, x_3 + (d_1 + 1)\, x_1 - x_2 + d_o \tag{3.44}$$

$$F_3 = x_1\, x_2 - \alpha_3\, x_3 \tag{3.45}$$

The steady states of spectral models with either (3.37)-(3.39) or (3.43)-(3.45) as right sides are governed by a cusp surface (Fig. 3.1d). Now d_1 appears where it must, and as expected, we have the relation $r = d_1 + 1$.

If we had no a priori reason to replace the column $[0, x_2, 0]^T$ with the column $[0, x_1, 0]^T$ in (3.42), then we could choose several other column vectors which would lead to the linear term in the unfolded steady state cubic (3.36). For example, we may write (now that we know that $[0, x_1, 0]^T$ and $[0, x_2, 0]^T$ are equivalent via (3.42))

$$\begin{bmatrix} 0 \\ x_1 \\ 0 \end{bmatrix} = \begin{bmatrix} -\alpha_2 & \alpha_2 & 0 \\ 1 - x_3 & -1 & -x_1 \\ x_2 & x_1 & -\alpha_3 \end{bmatrix} \begin{bmatrix} 0 \\ 0 \\ -1 \end{bmatrix} - \alpha_3 \begin{bmatrix} 0 \\ 0 \\ 1 \end{bmatrix} \tag{3.46}$$

and so we have the alternate unfolding of (3.37)-(3.39) or (3.43)-(3.45) (cf. Example 10 in Chapter 2)

$$F_1 = -\alpha_2 x_1 + \alpha_2 x_2 \tag{3.47}$$

$$F_2 = -x_1 x_3 + x_1 - x_2 + d_o \tag{3.48}$$

$$F_3 = x_1 x_2 - \alpha_3 x_3 - \alpha_3 d_1 \tag{3.49}$$

From (3.47)-(3.49) we may conclude that horizontally-independent internal (diabatic) heating of a neutrally stratified fluid leads to the same qualitative steady state behavior as does external heating applied to either the bottom or top surfaces.

From the above analysis, we conclude that external heating d_1 and internal heating d_o that varies spatially in at least the horizontal (cf. Section 5.4) are independent effects because their contributions to the branching states are manifested by different terms in the unfolding (3.43)-(3.45). But from (3.47)-(3.49) we find that vertically varying internal heating and horizontally varying internal heating are independent effects as well. We conclude this because no relationship of the form (3.41) between $[0, 1, 0]^T$ and $[0, 0, 1]^T$ can be found.

If we do not wish to interpret d_o as a diabatic heating term, then we must find an alternate unfolding function to serve as a factor for d_o. We find that there is only one other factor: the role of the inhomogeneous term d_o is to eliminate the trivial solution from the system (3.43)-(3.45), and this causes the branching diagram to change from that shown in Fig. 3.1b to that given in either Figs. 3.1a or 3.1c. Thus we consider only inhomogeneous terms in F_1 or F_3; but because we have shown above that $[0, 0, 1]^T$ is equivalent to $[0, x_1, 0]^T$ we need only check the possibility that $[1, 0, 0]^T$ might replace $[0, 1, 0]^T$ in (3.43)-(3.45). This is indeed the case, because we find that

$$\begin{bmatrix} 0 \\ 1 \\ 0 \end{bmatrix} = \begin{bmatrix} -\alpha_2 & \alpha_2 & 0 \\ 1 - x_3 & -1 & -x_1 \\ x_2 & x_1 & -\alpha_3 \end{bmatrix} \begin{bmatrix} 1/3 \\ -2/3 \\ -x_2(3\alpha_3)^{-1} \end{bmatrix} \tag{3.50}$$

$$+ \begin{bmatrix} 0 & 0 & 0 \\ 0 & 0 & -(3\alpha_3)^{-1} \\ -2(3\alpha_2)^{-1} & 0 & 0 \end{bmatrix} \begin{bmatrix} -\alpha_2 x_1 + \alpha_2 x_2 \\ -x_1 x_3 + x_1 - x_2 \\ x_1 x_2 - \alpha_3 x_3 \end{bmatrix} + \alpha_2 \begin{bmatrix} 1 \\ 0 \\ 0 \end{bmatrix}$$

Thus we may replace (3.43)-(3.45) by the equivalent system

$$F_1 = -\alpha_2 x_1 + \alpha_2 x_2 + d_0 \alpha_2 \tag{3.51}$$

$$F_2 = -x_1 x_3 + (d_1 + 1) x_1 - x_2 \tag{3.52}$$

$$F_3 = x_1 x_2 - \alpha_3 x_3 \tag{3.53}$$

The steady states represented by (3.51)-(3.53) are governed by a cusp surface, too, but the parameter d_0 can be interpreted differently.

When $d_0 \neq 0$, we see from (3.51)-(3.53) that x_1 cannot vanish in any of the steady states. From (3.16), we see that the Fourier coefficient x_1 represents the intensity of the fluid circulation, so motion must occur, even in the conductive case, if $d_0 \neq 0$. Because the fluid is unaccelerated in the steady case, we know from Jeffrey's Theorem (Dutton, 1976b) that motion must occur whenever there are horizontal temperature gradients on a level surface. Thus, we conclude that an externally imposed horizontal temperature difference $\Delta_x T$ might serve as a candidate for the physical interpretation of d_0.

Accordingly, we rewrite the temperature field (3.1) in the form

$$T = T_0 + \Delta_z T(z/H) + \Delta_x T(ax/H) + \theta \tag{3.54}$$

Because $\Delta_x T$ is constant, (3.54) still satisfies the conduction relation $\nabla^2 T = 0$. Now the shallow Boussinesq system is written as

$$\frac{\partial}{\partial t^*} \tilde{\nabla}^2 \psi^* = -K(\psi^*, \tilde{\nabla}^2 \psi^*) + \sigma(1 + a^2)^{-1} \tilde{\nabla}^4 \psi^* + \sigma(1 + a^2) \frac{\partial \theta^*}{\partial x^*} + \sigma(1 + a^2) h \tag{3.55}$$

$$\frac{\partial \theta^*}{\partial t^*} = -K(\psi^*, \theta^*) + r \frac{\partial \psi^*}{\partial x^*} + h \frac{\partial \psi^*}{\partial z^*} + (1 + a^2)^{-1} \tilde{\nabla}^2 \theta^* \tag{3.56}$$

in which the nondimensional horizontal heating parameter, the Hadley number h (Yost and Shirer, 1982), is defined here by

$$h = -\frac{\Delta_x T}{\Delta_z T} r \tag{3.57}$$

The inhomogeneous term in (3.55) survives in the spectral system for solutions of the form (3.16) only if the horizontal domain width corresponds to an odd multiple of π. This, however, is a natural physical requirement when there is horizontal heating (Fig. 3.2). If the right boundary is slightly warmer than the left one, then we expect rising motion at the right and sinking motion at the left wall. Once this direct circulation has been established, then by continuity an odd number of cells must occur as shown (Fig. 3.2). The actual value of n is determined by the value of a; when the magnitude of h is small, then $n \sim \sqrt{2}\, L(2H)^{-1}$.

The simplest case, in which the domain is $0 \leq x^* \leq \pi$ and $0 \leq z^* \leq \pi$, was studied by Yost and Shirer (1982). For this domain, we may substitute (3.16)-(3.17) into (3.55)-(3.56) to obtain

$$\dot{x}_1 = -\sigma x_1 + \sigma x_2 - 8\sqrt{2}\,\sigma h/\pi^2 \tag{3.58}$$

$$\dot{x}_2 = -x_1 x_3 + r x_1 - x_2 \tag{3.59}$$

$$\dot{x}_3 = x_1 x_2 - b x_3 - \frac{16\sqrt{2}\, h}{3\pi^2} x_1 \tag{3.60}$$

In (3.60) we have corrected an error in Shirer and Wells (1982). By comparing (3.51)-(3.53) and (3.58)-(3.60), we can see that h and d_0 are related apparently by

$$d_o = -8\sqrt{2}\, h/\pi^2 \tag{3.61}$$

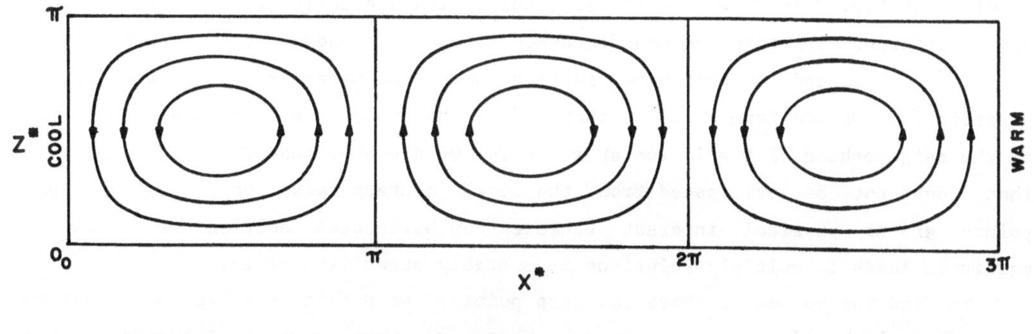

Fig. 3.2 An example showing schematically a three-celled direct circulation that would be expected if the right wall is heated, but the left wall is cooled. Note that an odd number of cells is required if fluid rises at the warm boundary and descends at the cool one.

To verify this, we use Mather's Theorem I (2.261) to rewrite (3.50) as

$$\begin{bmatrix} 0 \\ 1 \\ 0 \end{bmatrix} = \begin{bmatrix} -\alpha_2 & \alpha_2 & 0 \\ 1 - x_3 & -1 & -x_1 \\ x_2 & x_1 & -\alpha_3 \end{bmatrix} \begin{bmatrix} 1/9 \\ -8/9 \\ -x_2(9\alpha_3)^{-1} \end{bmatrix} \tag{3.62}$$

$$+ \begin{bmatrix} 0 & 0 & 0 \\ 0 & 0 & -(9\alpha_3)^{-1} \\ -2(9\alpha_2)^{-1} & 0 & 0 \end{bmatrix} \begin{bmatrix} -\alpha_2 x_1 + \alpha_2 x_2 \\ -x_1 x_3 + x_1 - x_2 \\ x_1 x_2 - \alpha_3 x_3 \end{bmatrix}$$

$$+ \alpha_2 \begin{bmatrix} 1 \\ 0 \\ 0 \end{bmatrix} + \frac{2}{3} \begin{bmatrix} 0 \\ 0 \\ x_1 \end{bmatrix}$$

Now we may write (3.51)-(3.53) in the following form

$$F_1 = -\alpha_2 x_1 + \alpha_2 x_2 + d_o \alpha_2 \tag{3.63}$$

$$F_2 = -x_1 x_3 + (d_1 + 1) x_1 - x_2 \tag{3.64}$$

$$F_3 = x_1 x_2 - \alpha_3 x_3 + (2/3) d_o x_1 \tag{3.65}$$

This is precisely the form of the right side of (3.58)-(3.60), and by comparing (3.58)-(3.60) with (3.63)-(3.65) we see that indeed h and d_o are related by (3.61).

To demonstrate that the steady states of (3.58)-(3.60) actually have the form of a cusp surface, we display them in Fig. 3.3. This figure is a slightly modified version of Fig. 3 in Yost and Shirer (1982); although their three-component model has a different form than the one discussed here, the steady states of both systems can be represented on the same surface once the parameter axes are labeled accordingly. We see from Fig. 3.3 that indeed the steady states form a cusp surface in the neighborhood of r = 1; but at r = - 26 we discover that two new cusps exist that could not be anticipated from the local contact catastrophe theory. These points are of physical interest because now horizontal heating of sufficient magnitude leads to multiple solutions in a stably stratified fluid.

To find the values of these new cusp points, we notice from Fig. 3.3 that the critical values of x_1 are nonzero. Thus, we must rewrite the steady state polynomial for (3.58)-(3.60) in a new form whose triple zeros correspond to triple nontrivial values of x_1.

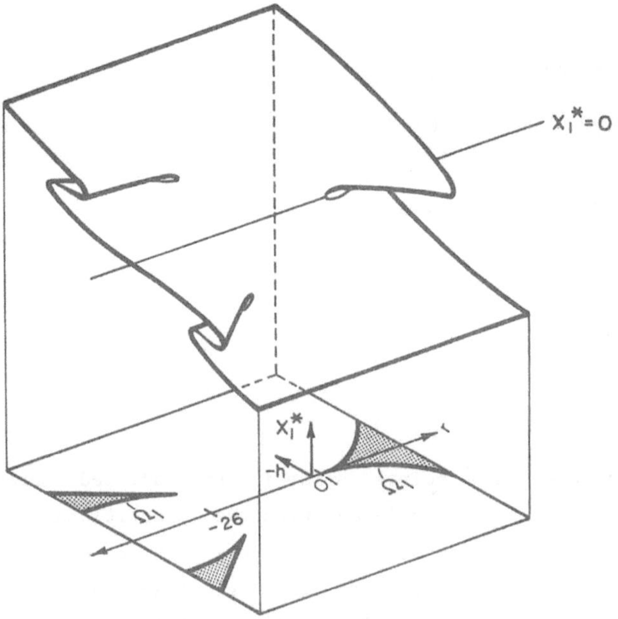

Fig. 3.3 A schematic diagram showing the relationship between the folded solution surface of the Lorenz model (3.58)-(3.60) and the singularity set Ω_1 in the r - h plane. For values of r and h inside the cusps of Ω_1 (shaded regions), three distinct real roots of the steady state polynomial exist, but for values of r and h outside Ω_1, only one real root exists. Note the extra two cusps when $r \leq -26$. (After Yost and Shirer, 1982).

The stationary solutions of (3.58)-(3.60) are governed by

$$x_1^{\,3} + 8\sqrt{2}\,(3\pi^2)^{-1}\,h\,x_1^{\,2} - b\,(r-1)\,x_1 + 8\sqrt{2}\,b\,h\,\pi^{-2} \qquad (3.66)$$

We translate x_1 to a new variable δ via

$$x_1 = X + \delta \qquad (3.67)$$

Now trivial solutions $\delta = 0$ will correspond to nontrivial solutions $x_1 = X$. Substitution of (3.67) into (3.66) produces the generalized steady polynomial

$$\begin{aligned} &\delta^3 + [3\,X + 8\sqrt{2}\,(3\pi^2)^{-1}\,h]\,\delta^2 \qquad (3.68)\\ &+ [3\,X^2 + 16\sqrt{2}\,(3\pi^2)^{-1}\,X\,h - b\,(r-1)]\,\delta \\ &+ [X^3 + 8\sqrt{2}\,(3\pi^2)^{-1}\,h\,X^2 - b\,(r-1)\,X + 8\sqrt{2}\,b\,\pi^{-2}\,h] = 0 \end{aligned}$$

Cusp points are those given by the vanishing separately of the expressions in each bracketed coefficient in (3.68). This occurs provided

$$\left.\begin{aligned} r &= -26 \\ h &= \mp 27\,\pi^2 \sqrt{b}\,(8\sqrt{2})^{-1} \\ X &= \pm 3\sqrt{b} \end{aligned}\right\} \qquad (3.69)$$

which corresponds to the values

$$\left.\begin{aligned} x_1 &= \pm 3\sqrt{b} \\ x_2 &= \mp 24\sqrt{b} \\ x_3 &= -18 \end{aligned}\right\} \qquad (3.70)$$

Although three cusp points exist in the modified Lorenz model, different physical meaning must be associated with the transitions between two flows that occur in their neighborhoods. To see the distinction, we recall that direct circulations are ones in which warm fluid rises and cold fluid sinks (Fig. 3.2); here that corresponds to $x_1 h < 0$. Indirect circulations are ones in which warm fluid sinks and cold fluid rises; here that corresponds to $x_1 h > 0$.

In the neighborhood of $r = 1$, stable convective solutions may satisfy either $x_1 h < 0$ or $x_1 h > 0$ (Fig. 3.3). Thus, as the value of $h \propto \Delta_x T$ changes, we find that sudden transitions occur between indirect and direct circulations (Fig. 3.4a) of equal intensities; moreover, as the value of h is increased and decreased alternately, these transitions form a closed hysteresis loop (Fig. 2.3). Because the fluid in this case circulates in response to thermal forcing in the vertical, the circulation can have either a clockwise or a counterclockwise sense whenever the horizontal heating rates are small. But once the value of h becomes sufficiently large, only a direct circulation is possible.

When the fluid is stably stratified, however, the fluid circulates primarily in response to horizontal thermal forcing. In the neighborhood of the cusp point $r = -26$, we see from (3.69) that the flow is direct because $X h < 0$. Thus, the transitions in this regime are between direct circulations of different intensities (Fig. 3.4b).

Thus, we conclude that two regimes exhibiting multiple flows are modeled by the modified Lorenz system (3.58)-(3.60); these regimes are identified by the sign of $\Delta_z T$ giving the vertical stratification of the basic thermal field. We will explore the implications of this further in Chapter 5, in which we add the effects of rotation to the Boussinesq system in order to study rotating axisymmetric flows. For now, we note that use of Mather's theory to determine locally the number and meaning of the necessary parameters in the neighborhood of a singularity leads sometimes to some global insight into the flow characteristics far from the singularity. That new results were obtained might be labeled fortuitous, but they demonstrate that requiring the solutions to a model to satisfy certain topological

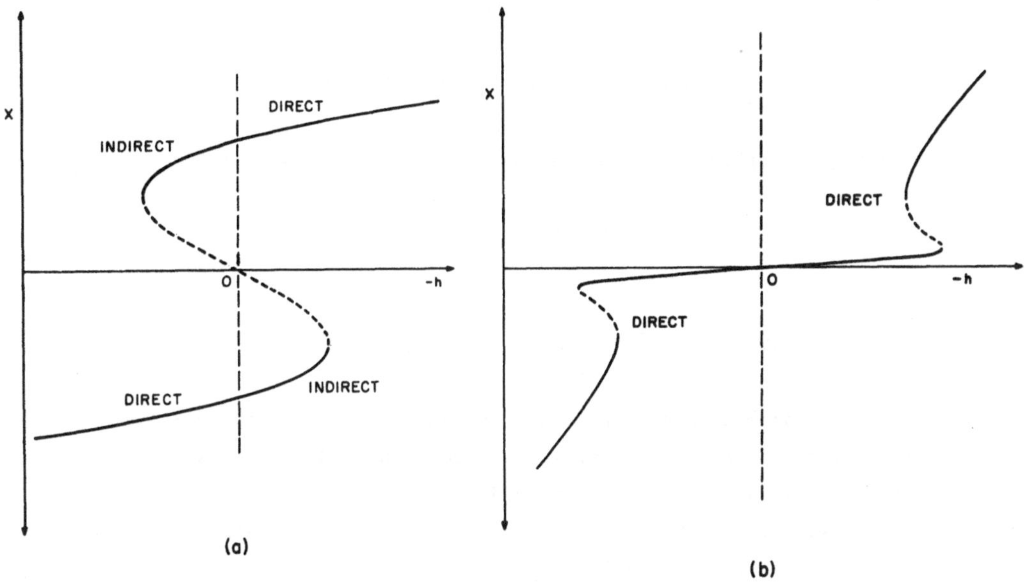

Fig. 3.4 Two cross-sections through the steady state surface of the Lorenz model (3.58)-(3.60); stable solutions are denoted by solid lines, unstable solutions by dashed lines. When $r > 0$ (a) sudden transitions between direct and indirect circulations occur; but when $r < 0$ (b) sudden transitions occur only between direct circulations of varying intensity.

properties is related to an appropriate, more general view of a physical problem. In the case discussed in this chapter, we have shown that shallow convection is viewed properly as being forced in both the horizontal and the vertical.

CHAPTER 4

QUASI-GEOSTROPHIC FLOW IN A CHANNEL

The quasi-geostrophic equation has long been viewed as a suitable simple model of large-scale flow in the atmosphere. The horizontal, β-plane, forced, dissipative version of this equation can be written in terms of a stream function ψ as (Dutton, 1976a)

$$\frac{\partial}{\partial t}(\nabla_H^2\psi) + J(\psi,\nabla_H^2\psi) + \beta\frac{\partial\psi}{\partial x} = H + \nu\nabla_H^4\psi \tag{4.1}$$

in which J is the Jacobian operator in the variables x and y (cf. (3.14)), β is the latitudinal variation of the Coriolis parameter, $H(x,y)$ is the Newtonian heating rate at the level on which (4.1) applies and ν is the eddy viscosity.

The simplest nonlinear spectral model of (4.1) was studied by Vickroy and Dutton (1979) who chose the domain $0 \le x \le 2\pi$, $-\pi/2 \le y \le \pi/2$. They wrote the stream function $\psi(x,y,t)$ as

$$\begin{aligned}\psi(x,y,t) = a_1(t)\pi^{-1}\sin y + a_2(t)\sqrt{2\pi}^{-1}\cos y \quad \sin \ell x \\ + a_3(t)\sqrt{2\pi}^{-1}\cos 3y \quad \cos \ell x\end{aligned} \tag{4.2}$$

so that their spectral model takes the form

$$\dot{a}_1 = \Lambda_1 a_2 a_3 - \nu\lambda_1 a_1 - H_1/\lambda_1 \tag{4.3}$$

$$\dot{a}_2 = \Lambda_2 a_1 a_3 - \nu\lambda_2 a_2 - H_2/\lambda_2 \tag{4.4}$$

$$\dot{a}_3 = \Lambda_3 a_1 a_2 - \nu\lambda_3 a_3 - H_3/\lambda_3 \tag{4.5}$$

in which H_i are the Fourier coefficients of $H(x,y)$ that drives the flow. Here $\lambda_1 = 1$, $\lambda_2 = 1+\ell^2$, $\lambda_3 = 9+\ell^2$, $D = -8\ell/15\pi^2$ and

$$\Lambda_1 = (\lambda_2 - \lambda_3)D/\lambda_1 \tag{4.6}$$

$$\Lambda_2 = (\lambda_3 - \lambda_1)D/\lambda_2 \tag{4.7}$$

$$\Lambda_3 = (\lambda_1 - \lambda_2)D/\lambda_3 \tag{4.8}$$

Two fundamental properties of the unforced, nondissipative version of the spectral system (4.3)-(4.5) are that $\lambda_1\Lambda_1 + \lambda_2\Lambda_2 + \lambda_3\Lambda_3 = 0$, which ensures that energy is conserved, and $\lambda_1^2\Lambda_1 + \lambda_2^2\Lambda_2 + \lambda_3^2\Lambda_3 = 0$, which ensures that enstrophy is conserved.

We note that a five-coefficient truncation is the smallest one in which the latitudinal variation β of the Coriolis parameter is retained in a physically acceptable nonlinear model (Vickroy and Dutton, 1979; Mitchell and Dutton, 1981). For different values of the thermal forcing, Vickroy and Dutton (1979) found that one, three or five steady states were possible. Although they were unable to find magnitudes of the parameters at which the five solutions met simultaneously, they nevertheless concluded that a butterfly surface modeled the steady states of (4.3)-(4.5). We may use the contact catastrophe procedure discussed in Chapter 2 to check their assertion. We note that Mitchell and Dutton (1981) show that inclusion of β leads to periodic solutions rather than butterfly points.

4.1 Heating at the Middle Wavenumber Only

Mitchell and Dutton (1981) demonstrated that when both energy and enstrophy are conserved in the unforced, inviscid case, only inhomogeneous forcing of the intermediate component of the interacting triad will lead to bifurcations. Indeed, if $H_1 = H_3 = 0$, then (4.3)-(4.5) admits of the basic steady solution $a_i = A_i$ given by:

$$\left.\begin{aligned} A_1 &= 0 \\ A_2 &= -H_2/\nu\lambda_2^2 \\ A_3 &= 0 \end{aligned}\right\} \tag{4.9}$$

At a critical value $\hat{H}_2$ of H_2 two solutions bifurcate from the basic one; we may write

$$\hat{H}_2 = \nu^2 \lambda_2^2 (\lambda_1\lambda_3/\Lambda_1\Lambda_3)^{1/2} \tag{4.10}$$

Thus $H_2 = \hat{H}_2$ is the singular point about which we will unfold the steady states of the model. Because $\lambda_1 < \lambda_2 < \lambda_3$, the heating coefficient H_2 is the component of the Newtonian heating rate $H(x,y)$ at the middle wavenumber.

Before we can apply our contact catastrophe theory to classify the singularity $H_2 = \hat{H}_2$, we must first write (4.3)-(4.5) in a homogeneous form. This is accomplished by setting

$$\left.\begin{aligned} \alpha_1 &= a_1 \\ \alpha_2 &= a_2 - A_2 \\ \alpha_3 &= a_3 \end{aligned}\right\} \tag{4.11}$$

If we set $H_2 = \hat{H}_2$, and therefore $A_2 = \hat{A}_2$ given by

$$\hat{A}_2 = -\nu[\lambda_1\lambda_3(\Lambda_1\Lambda_3)^{-1}]^{1/2} \tag{4.12}$$

then at the singular point we may write (4.3)-(4.5) as

$$\dot{\alpha}_1 = \Lambda_1\alpha_2\alpha_3 + \Lambda_1\hat{A}_2\alpha_3 - \nu\lambda_1\alpha_1 \tag{4.13}$$

$$\dot{\alpha}_2 = \Lambda_2\alpha_1\alpha_3 - \nu\lambda_2\alpha_2 \tag{4.14}$$

$$\dot{\alpha}_3 = \Lambda_3\alpha_1\alpha_2 + \Lambda_3\hat{A}_2\alpha_1 - \nu\lambda_3\alpha_3 \tag{4.15}$$

We may write (4.13)-(4.15) in the form (2.247) as

$$\underset{\sim}{v} = \begin{bmatrix} \alpha_2 \\ \alpha_3 \end{bmatrix} \tag{4.16}$$

$$\underset{\sim}{w} = [\alpha_1] \tag{4.17}$$

$$A(\underset{\sim}{v},\underset{\sim}{w}) = \begin{bmatrix} -\nu\lambda_2 & 0 \\ 0 & -\nu\lambda_3 \end{bmatrix} \tag{4.18}$$

$$B(\underset{\sim}{v},\underset{\sim}{w}) = \begin{bmatrix} \Lambda_2 v_2 \\ \Lambda_3\hat{A}_2 + \Lambda_3 v_1 \end{bmatrix} \tag{4.19}$$

$$C(\underset{\sim}{v},\underset{\sim}{w}) = [0, \Lambda_1\hat{A}_2 + \Lambda_1 v_1] \tag{4.20}$$

$$D(\underset{\sim}{v},\underset{\sim}{w}) = [-\nu\lambda_1] \tag{4.21}$$

By combining (2.248) and (4.18)-(4.21) we find that $\Delta(0,0)=0$ so that the singularity is of corank 1.

We insert (4.18) and (4.19) into (2.249) to obtain the solutions

$$v_1^*(w^*) = \Lambda_2\Lambda_3\hat{A}_2 w^{*2} (\nu^2\lambda_2\lambda_3 - \Lambda_2\Lambda_3 w^{*2})^{-1} \tag{4.22}$$

$$v_2^*(w^*) = \nu\lambda_2\Lambda_3\hat{A}_2 w^* (\nu^2\lambda_2\lambda_3 - \Lambda_2\Lambda_3 w^{*2})^{-1} \tag{4.23}$$

After combining (2.250) and (4.20)-(4.23) we may write the singular function $p(w^*)$ as

$$p(w^*) = (-\nu\lambda_1\Lambda_2^2\Lambda_3^2 w^{*5} + 2\nu^3\lambda_1\lambda_2\lambda_3\Lambda_2\Lambda_3 w^{*3}) (\nu^2\lambda_2\lambda_3 - \Lambda_2\Lambda_3 w^{*2})^{-2} \tag{4.24}$$

Because $\Lambda_2\Lambda_3 < 0$, the denominator of (4.24) cannot vanish for any values of w*, and all the singular behavior in p(w*) is contained in its numerator

$$q(w^*) = -\nu\lambda_1\Lambda_2^2\Lambda_3^2 w^{*5} + 2\nu^3\lambda_1\lambda_2\lambda_3\Lambda_2\Lambda_3 w^{*3} \tag{4.25}$$

Because the viscosity $\nu > 0$, the coefficient of the cubic term of p(w*) can never be zero. Thus the singularity $H_2=\hat{H}_2$ is a cusp point and the versal unfolding (2.251) involves two parameters (Table 2.1). Accordingly, the unfolded version of (4.25) about $H_2=\hat{H}_2$ is

$$q(w^*) = -\nu\lambda_1\Lambda_2^2\Lambda_3^2 w^{*5} + 2\nu^3\lambda_1\lambda_2\lambda_3\Lambda_2\Lambda_3 w^{*3} + \mu_2 w^* + \mu_1 \tag{4.26}$$

In the original system (4.13)-(4.15) the unfolding (4.26) leads via (2.252) to

$$\dot{\alpha}_1 = \Lambda_1\alpha_2\alpha_3 + \Lambda_1\hat{A}_2\alpha_3 - \nu\lambda_1\alpha_1 + (\mu_2\alpha_1 + \mu_1)\,(\nu^2\lambda_2\lambda_3 - \Lambda_2\Lambda_3\alpha_1^2)^{-2} = F_1 \tag{4.27}$$

$$\dot{\alpha}_2 = \Lambda_2\alpha_1\alpha_3 - \nu\lambda_2\alpha_2 = F_2 \tag{4.28}$$

$$\dot{\alpha}_3 = \Lambda_3\alpha_1\alpha_2 + \Lambda_3\hat{A}_2\alpha_1 - \nu\lambda_3\alpha_3 = F_3 \tag{4.29}$$

The steady states of (4.27)-(4.29) are given by the real roots of (4.26) with $w^* = \alpha_1$.

Physical interpretation of the parameters μ_1 and μ_2 in the unfolded system (4.27)-(4.29) proceeds in two stages. First, we must remove the denominators in the factors of μ_1 and μ_2 in (4.27); these denominators cannot be obtained from substitution of the expansion (4.2) into any reasonably modified version of (4.1). This is accomplished by finding a contact equivalent form of the function $\underset{\sim}{F}^*(\underset{\sim}{v}^*,w^*,0) = \left[\begin{smallmatrix} v^* \\ p(w^*)\end{smallmatrix}\right]$. Because the denominator $(\nu^2\lambda_2\lambda_3 - \Lambda_2\Lambda_3\alpha_1^2)^2 \neq 0$ for any value of α_1, we may define the 3 × 3 invertible matrix M in (2.16) to be

$$M(x,\lambda) = M(\underset{\sim}{v}^*,w^*,0) = \begin{bmatrix} 1 & 0 & 0 \\ 0 & 1 & 0 \\ 0 & 0 & (\nu^2\lambda_2\lambda_3 - \Lambda_2\Lambda_3 w^{*2})^2 \end{bmatrix} \tag{4.30}$$

Thus the right side $\underset{\sim}{U}(\underset{\sim}{v}^*,w^*,0) = M(\underset{\sim}{v}^*,w^*,0)\cdot\underset{\sim}{F}^*(\underset{\sim}{v}^*,w^*,0)$ is contact equivalent to $\underset{\sim}{F}^*(\underset{\sim}{v}^*,w^*,0)$, and we may write (4.27)-(4.29) in the equivalent form (2.252)

$$\dot{\alpha}_1 = \Lambda_1\alpha_2\alpha_3 + \Lambda_1\hat{A}_2\alpha_3 - \nu\lambda_1\alpha_1 + \mu_2\alpha_1 + \mu_1 \tag{4.31}$$

$$\dot{\alpha}_2 = \Lambda_2\alpha_1\alpha_3 - \nu\lambda_2\alpha_2 \tag{4.32}$$

$$\dot{\alpha}_3 = \Lambda_3\alpha_1\alpha_2 + \Lambda_3\hat{A}_2\alpha_1 - \nu\lambda_3\alpha_3 \tag{4.33}$$

We note that (4.31)-(4.33) must still be a versal unfolding of (4.13)-(4.15).

This application of Mather's Theorem I can always be used to eliminate the nonsingular denominator that originally appears in the steady state rational function $p(w^*)$. However, the steady states of the new system (4.31)-(4.33) are governed by the quintic polynomial

$$\begin{aligned}(\mu_2\Lambda_2^2\Lambda_3^2 - \nu\lambda_1\Lambda_2^2\Lambda_3^2)\,\alpha_1^5 + \mu_1\Lambda_2^2\Lambda_3^2\alpha_1^4 \\ + (2\nu^3\lambda_1\lambda_2\lambda_3\Lambda_2\Lambda_3 - 2\nu^2\lambda_2\lambda_3\Lambda_2\Lambda_3\mu_2)\,\alpha_1^3 \\ - 2\nu^2\lambda_2\lambda_3\Lambda_2\Lambda_3\mu_1\alpha_1^2 + \nu^4\lambda_2^2\lambda_3^2\mu_2\alpha_1 + \nu^4\lambda_2^2\lambda_3^2\mu_1 = 0\end{aligned} \tag{4.34}$$

This is a versal unfolding of (4.25), and we still obtain (4.25) at the singular point $\mu_1 = \mu_2 = 0$. Introduction of μ_1 and μ_2 into the higher order terms of (4.34) leads to no new branching behavior because the coefficient of the cubic term does not change sign in the neighborhood of the singular point $\mu_1 = \mu_2 = 0$. Only the quartic, quadratic, linear and constant terms of (4.26) and (4.34) vanish at the singularity so that both unfoldings are of cusp type, although at first glance they appear to be of butterfly type because the polynomial (4.26) is a quintic.

Relating the parameters of the problem to μ_1 and μ_2 proceeds as follows. By comparing (4.3)-(4.5), (4.13)-(4.15) and (4.31)-(4.33), we see that μ_1 and H_1 are proportional. Because setting $H_2 = \hat{H}_2$ led us to the cusp point given by $\mu_2 = 0$, an obvious choice is to associate μ_2 with H_2 (or A_2). Because homogenization (4.11) causes $\hat{A}_2$ to appear in both (4.13) and (4.15), the required unfolding function is more complicated than that needed for the Lorenz model discussed in Chapter 3. Nevertheless, using (2.261) and (4.13)-(4.15) we have that

$$\begin{bmatrix}\alpha_1\\0\\0\end{bmatrix} = \begin{bmatrix}-\nu\lambda_1 & \Lambda_1\alpha_3 & \Lambda_1(\alpha_2+\hat{A}_2)\\ \Lambda_2\alpha_3 & -\nu\lambda_2 & \Lambda_2\alpha_1\\ \Lambda_3(\alpha_2+\hat{A}_2) & \Lambda_3\alpha_1 & -\nu\lambda_3\end{bmatrix}\begin{bmatrix}0\\ \alpha_2(2\nu\lambda_1)^{-1}\\ \alpha_3(2\nu\lambda_1)^{-1}\end{bmatrix}$$
$$+ \begin{bmatrix}-(\nu\lambda_1)^{-1} & 0 & 0\\ 0 & -(2\nu\lambda_1)^{-1} & 0\\ 0 & 0 & -(2\nu\lambda_1)^{-1}\end{bmatrix}\begin{bmatrix}-\nu\lambda_1\alpha_1 + \Lambda_1(\alpha_2+\hat{A}_2)\alpha_3\\ -\nu\lambda_2\alpha_2 + \Lambda_2\alpha_1\alpha_3\\ -\nu\lambda_3\alpha_3 + \Lambda_3(\alpha_2+\hat{A}_2)\alpha_1\end{bmatrix}$$
$$+ \frac{\hat{A}_2}{2\nu\lambda_1}\begin{bmatrix}\Lambda_1\alpha_3\\0\\ \Lambda_3\alpha_1\end{bmatrix} \tag{4.35}$$

so that (4.31)-(4.33) is equivalent to

$$\dot{\alpha}_1 = \Lambda_1\alpha_2\alpha_3 + \Lambda_1\hat{A}_2\,[1 + \mu_2(2\nu\lambda_1)^{-1}]\,\alpha_3 - \nu\lambda_1\alpha_1 + \mu_1 \tag{4.36}$$

$$\dot{\alpha}_2 = \Lambda_2\alpha_1\alpha_3 - \nu\lambda_2\alpha_2 \tag{4.37}$$

$$\dot{\alpha}_3 = \Lambda_3\alpha_1\alpha_2 + \Lambda_3\hat{A}_2\,[1 + \mu_2(2\nu\lambda_1)^{-1}]\,\alpha_1 - \nu\lambda_3\alpha_3 \tag{4.38}$$

which in the original variables a_i is

$$\dot{a}_1 = \Lambda_1 a_2 a_3 - \nu\lambda_1 a_1 + \mu_1 \tag{4.39}$$

$$\dot{a}_2 = \Lambda_2 a_1 a_3 - \nu\lambda_2 a_2 + \nu\lambda_2\hat{A}_2 + \mu_2\lambda_2(2\lambda_1)^{-1}\hat{A}_2 \tag{4.40}$$

$$\dot{a}_3 = \Lambda_3 a_1 a_2 - \nu\lambda_3 a_3 \tag{4.41}$$

In (4.36), (4.38), and (4.40), we have expressed the steady state A_2 in the form

$$A_2 = \hat{A}_2\,[1 + \mu_2(2\nu\lambda_1)^{-1}] \tag{4.42}$$

Upon comparing (4.3) and (4.39) we discover that

$$\mu_1 = -H_1/\lambda_1 \tag{4.43}$$

After combining (4.9) and (4.42), we obtain

$$\mu_2 = (H_2/\hat{H}_2 - 1)\,2\nu\lambda_1 \tag{4.44}$$

Thus the independent parameters μ_1 and μ_2 in the unfolding about the cusp point correspond to heating rates having the largest and middle wavenumbers.

The canonical form for the cusp surface (Fig. 3.1d) of steady states is obtained by plotting the magnitude of α_1 as a function of the coefficients μ_1 and μ_2 in the unfolding (4.26) or (4.34). Moreover, we know from our discussion of the cubic polynomial (3.36) that the folded pleats of the cusp surface, and therefore multiple solutions, are possible when $\mu_2 > 0$. But because μ_1 and μ_2 are each directly proportional to H_1 and $H_2 - \hat{H}_2$, respectively, then the form of the steady state surface also will be the standard cusp one when the steady states are displayed as functions of H_1 and H_2; furthermore, the condition $\mu_2 > 0$ for multiple solutions becomes $H_2 > \hat{H}_2$. These deductions are correct, as shown in Fig. 4.1 taken from Vickroy and Dutton (1979), in which the cusp singularity set exists for sufficiently large values of H_2.

We note that if the canonical parameters μ_i and physical parameters of the unfolding are not separately proportional, then graphing the solution surface as a function of the physical parameters will not produce one of the standard forms of catastrophe theory. These forms are obtained by displaying the solution surface as a function of μ_i, and our contact catastrophe procedure identifies transformations of the original physical parameters to the canonical ones that describe the independent branching behavior in the model. We will see examples of this in the following.

Relating μ_1 to H_1 is not the only possible physical interpretation. As in the Lorenz model unfolding discussed in Chapter 3, we also may interpret the constant term of the unfolding as one of two forcing terms. Here we find that $\mu_1 \propto H_3$ in (4.5) is also possible so that a second cusp surface arises from heating of both the smallest and middle wavenumbers. We demonstrate this by using Mather's Theorem (2.261) to write

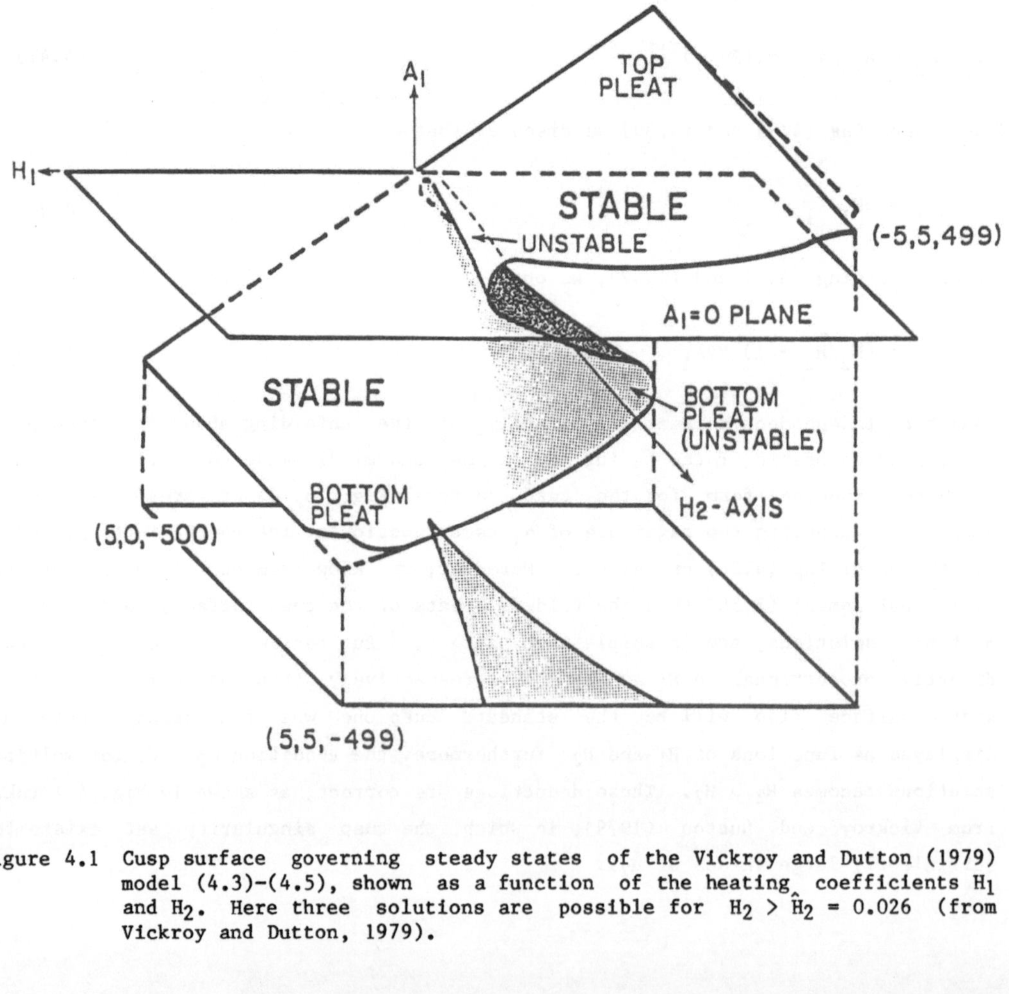

Figure 4.1 Cusp surface governing steady states of the Vickroy and Dutton (1979) model (4.3)-(4.5), shown as a function of the heating coefficients H_1 and H_2. Here three solutions are possible for $H_2 > \hat{H}_2 = 0.026$ (from Vickroy and Dutton, 1979).

$$\begin{bmatrix} 1 \\ 0 \\ 0 \end{bmatrix} = \begin{bmatrix} -\nu\lambda_1 & \Lambda_1\alpha_3 & \Lambda_1(\alpha_2+\hat{A}_2) \\ \Lambda_2\alpha_3 & -\nu\lambda_2 & \Lambda_2\alpha_1 \\ \Lambda_3(\alpha_2+\hat{A}_2) & \Lambda_3\alpha_1 & -\nu\lambda_3 \end{bmatrix} \begin{bmatrix} g_{10} + g_{12}\,\alpha_2 + g_{15}\,\alpha_1^{\,2} \\ g_{21}\,\alpha_1 + g_{23}\,\alpha_3 \\ g_{30} + g_{32}\,\alpha_2 \end{bmatrix}$$

$$+ \begin{bmatrix} h_{111}\,\alpha_1 & h_{120} + h_{122}\,\alpha_2 & h_{133}\,\alpha_3 \\ h_{210} & h_{221}\,\alpha_1 & h_{230} \\ h_{311}\,\alpha_1 & h_{320} + h_{322}\,\alpha_2 & h_{331}\,\alpha_1 \end{bmatrix} \begin{bmatrix} -\nu\lambda_1\alpha_1 + \Lambda_1(\alpha_2+\hat{A}_2)\,\alpha_3 \\ -\nu\lambda_2\alpha_2 + \Lambda_2\alpha_1\alpha_3 \\ -\nu\lambda_3\alpha_3 + \Lambda_3(\alpha_2+\hat{A}_2)\,\alpha_1 \end{bmatrix}$$

$$+ \frac{\nu\lambda_3}{\Lambda_1\hat{A}_2} \begin{bmatrix} 0 \\ 0 \\ 1 \end{bmatrix} \tag{4.45}$$

in which we have used (4.12) and set

$$g_{10} = \frac{-1}{2\nu\lambda_1} \tag{4.46}$$

$$g_{12} = \frac{1}{4\nu\lambda_1\hat{A}_2} \tag{4.47}$$

$$g_{15} = \frac{5\Lambda_2\Lambda_3}{8\nu^3\lambda_1\lambda_2\lambda_3} \tag{4.48}$$

$$g_{21} = \frac{-\Lambda_2}{4\nu\lambda_2\Lambda_1\hat{A}_2} \tag{4.49}$$

$$g_{23} = \frac{\Lambda_2}{4\nu^2\lambda_1\lambda_2} \tag{4.50}$$

$$g_{30} = \frac{1}{2\Lambda_1\hat{A}_2} \tag{4.51}$$

$$g_{32} = \frac{3\Lambda_3}{8\nu^2\lambda_1\lambda_3} \tag{4.52}$$

$$h_{111} = \frac{-5\Lambda_2\Lambda_3}{8\nu^3\lambda_1\lambda_2\lambda_3} \tag{4.53}$$

$$h_{120} = \frac{5}{8\nu\lambda_2\hat{A}_2} \tag{4.54}$$

$$h_{122} = \frac{3\Lambda_1\Lambda_3}{8\nu^3\lambda_1\lambda_2\lambda_3} \tag{4.55}$$

$$h_{133} = \frac{\Lambda_1\Lambda_2}{4\nu^3\lambda_1\lambda_2\lambda_3} \tag{4.56}$$

$$h_{210} = \frac{-\Lambda_2}{4\nu\lambda_1\Lambda_1\hat{A}_2} \tag{4.57}$$

$$h_{221} = \frac{-5\Lambda_2\Lambda_3}{8\nu^3\lambda_1\lambda_2\lambda_3} \tag{4.58}$$

$$h_{230} = \frac{-\Lambda_2}{\nu^2\lambda_1\lambda_3} \tag{4.59}$$

$$h_{311} = \frac{-\Lambda_2\Lambda_3}{4\nu^2\lambda_1\lambda_2\Lambda_1\hat{A}_2} \tag{4.60}$$

$$h_{320} = \frac{-5\Lambda_3}{8\nu^2\lambda_1\lambda_2} \tag{4.61}$$

$$h_{322} = \frac{\Lambda_3}{4\nu^2\lambda_1\lambda_2\hat{A}_2} \tag{4.62}$$

$$h_{331} = \frac{-5\Lambda_2\Lambda_3}{8\nu^3\lambda_1\lambda_2\lambda_3} \tag{4.63}$$

Thus, with the aid of (2.260) we discover that another unfolding of (4.39)-(4.41) is

$$\dot{a}_1 = \Lambda_1 a_2 a_3 - \nu\lambda_1 a_1 \tag{4.64}$$

$$\dot{a}_2 = \Lambda_2 a_1 a_3 - \nu\lambda_2 a_2 + \nu\lambda_2\hat{A}_2 + \mu_2\lambda_2(2\lambda_1)^{-1}\hat{A}_2 \tag{4.65}$$

$$\dot{a}_3 = \Lambda_3 a_1 a_2 - \nu\lambda_3 a_3 + \frac{\nu\lambda_3}{\Lambda_1\hat{A}_2}\mu_1 \tag{4.66}$$

and in this case we find by comparing (4.5) and (4.66) that

$$\mu_1 = -H_3\Lambda_1\hat{A}_2/\nu\lambda_3^{\,2} \tag{4.67}$$

The above calculation is not an easy one, but with it we demonstrate that the steady states of the complete model form two cusp surfaces, one depending on the heating components H_1 and H_2 and the other depending on the heating components H_2 and H_3. However, the singularity at $A = [0,\hat{A}_2,0]^T$ is in fact a cusp with one too many parameters. That is, it is a cusp stretched out in the same manner as was the fold in Example 4 and Fig. 2.5 of Chapter 2. The two apparently distinct cusps are actually two cross sections through the singularity set. Nevertheless, Vickroy and Dutton (1979) found values of H_1, H_2, and H_3 for which five distinct stationary solutions occurred. Thus, either a swallowtail point at which four solutions meet or a butterfly point at which five solutions meet might be possible. The above analysis shows that these singularities, if they exist, cannot be near the cusp point $H_2 = \hat{H}_2$. We find all singularities in the Vickroy and Dutton (1979) model (4.3)-(4.5) in the next section.

4.2 Singularities in the Vickroy and Dutton Model

We found cusp points in the Vickroy and Dutton (1979) model (4.3)-(4.5) by finding values of the forcing parameters H_i for which the quadratic, linear and constant terms of the steady state polynomial all vanished. These points existed for the special case $H_1 = H_3 = 0$, but higher order singularities and more general branching behavior might be possible when two or three of the heating components are nonzero. To investigate this possibility, we write the coefficients of each of the terms of the steady state quintic polynomial as functions of H_1, H_2 and H_3. Then we seek a transformation of variables for which values of H_1, H_2 and H_3 exist that either cause all but the quartic and quintic terms to vanish (swallowtail point) or cause all but the quintic term to vanish (butterfly point).

To follow the above program, we first look for swallowtail points. The stationary solutions of (4.3)-(4.5) are governed by

$$e_1 a_1^5 + e_2 H_1' a_1^4 + e_3 a_1^3 + (e_4 H_1' + e_5 H_2' H_3') a_1^2 + (e_6 + e_7 H_2'^2 + e_8 H_3'^2) a_1 + e_9 H_1' + e_{10} H_2' H_3' = 0 \tag{4.68}$$

in which $H_1' = H_1/\lambda_1$, $H_2' = H_2/\lambda_2$, $H_3' = H_3/\lambda_3$ and

$$e_1 = -\Lambda_2^2 \Lambda_3^2 \nu \lambda_1 \tag{4.69}$$

$$e_2 = -\Lambda_2^2 \Lambda_3^2 \tag{4.70}$$

$$e_3 = 2\nu^3 \lambda_1 \lambda_2 \lambda_3 \Lambda_2 \Lambda_3 \tag{4.71}$$

$$e_4 = 2\nu^2 \lambda_2 \lambda_3 \Lambda_2 \Lambda_3 \tag{4.72}$$

$$e_5 = \Lambda_1 \Lambda_2 \Lambda_3 \tag{4.73}$$

$$e_6 = -\nu^5\lambda_1\lambda_2{}^2\lambda_3{}^2 \tag{4.74}$$

$$e_7 = \nu\lambda_3\Lambda_1\Lambda_3 \tag{4.75}$$

$$e_8 = \nu\lambda_2\Lambda_1\Lambda_2 \tag{4.76}$$

$$e_9 = -\nu^4\lambda_2{}^2\lambda_3{}^2 \tag{4.77}$$

$$e_{10} = \nu^2\lambda_2\lambda_3\Lambda_1 \tag{4.78}$$

The swallowtail point might correspond to some nonzero value y of a_1, so we must first change variables via

$$a_1 = x + y \tag{4.79}$$

After substitution of (4.79) into (4.68) we find that the steady state polynomial becomes

$$\begin{aligned} &e_1x^5 + (5e_1y + e_2H_1')x^4 + (10e_1y^2 + 4e_2H_1'y + e_3)x^3 \\ &\quad + \left[10e_1y^3 + 6e_2H_1'y^2 + 3e_3y + (e_4H_1' + e_5H_2'H_3')\right]x^2 \\ &\quad + \left[5e_1y^4 + 4e_2H_1'y^3 + 3e_3y^2 + 2(e_4H_1' + e_5H_2'H_3')y \right. \\ &\qquad\qquad \left. + e_6 + e_7H_2'^2 + e_8H_3'^2\right]x + \left[e_1y^5 + e_2H_1'y^4 + e_3y^3 \right. \\ &\quad \left. + (e_4H_1' + e_5H_2'H_3')y^2 + (e_6 + e_7H_2'^2 + e_8H_3'^2)y + e_9H_1' + e_{10}H_2'H_3'\right] = 0 \end{aligned} \tag{4.80}$$

For swallowtail points, we require that the cubic, quadratic, linear and constant terms of (4.80) each vanish. The cubic term vanishes if

$$H_1' = -(e_3 + 10e_1y^2)/(4ye_2) \tag{4.81}$$

the quadratic term vanishes if

$$H_2' = (5e_1y^4 + e_3y^2 + e_6)/(e_5H_3'y) \tag{4.82}$$

and the linear and constant terms vanish if

$$\begin{aligned} &(e_5{}^2e_8y^2)H_3'^4 + e_5{}^2(5e_1y^6 - e_3y^4 + e_6y^2)H_3'^2 \\ &\quad + (25e_1{}^2e_7y^8 + 10e_1e_3e_7y^6 + \tfrac{7}{2}e_3{}^2e_7y^4 + 2e_3e_6e_7y^2 + e_6{}^2e_7) = 0 \end{aligned} \tag{4.83}$$

and

$$(e_5{}^2 e_8 y^2)H_3'^4 + [e_5{}^2(\frac{7}{2}e_1 y^6 - \frac{13}{4}e_3 y^4 - \frac{7}{2}e_6 y^2) + \frac{3}{2}e_6 e_7 e_8]H_3'^2 \tag{4.84}$$
$$+ (25e_1{}^2 e_7 y^8 + 10e_1 e_3 e_7 y^6 + \frac{7}{2}e_3{}^2 e_7 y^4 + 2e_3 e_6 e_7 y^2 + e_6{}^2 e_7) = 0$$

We have simplified (4.83)-(4.84) with use of the following relations:
$e_1e_4 = e_2e_3$, $e_3e_4 = 4e_2e_6$, $e_3{}^2 = 4e_1e_6$, $e_3e_5 = -2e_1e_{10}$, $e_1e_9 = e_2e_6$, $e_3e_{10} = -2e_5e_6$, and $e_5e_{10} = e_7e_8$.

We seek common roots of the two quadratic equations (4.83) and (4.84), which can be written in the form

$$f_1(H_3'^2)^2 + f_2(H_3'^2) + f_3 = 0 \tag{4.85}$$

$$f_1(H_3'^2)^2 + f_4(H_3'^2) + f_3 = 0 \tag{4.86}$$

Common roots of (4.85) and (4.86) exist if and only if the following determinant, called the eliminant, vanishes (Richards, 1959)

$$\begin{vmatrix} f_3 & f_2 & f_1 & 0 \\ 0 & f_3 & f_2 & f_1 \\ f_3 & f_4 & f_1 & 0 \\ 0 & f_3 & f_4 & f_1 \end{vmatrix} = f_1 f_3 (f_2 - f_4)^2 = 0 \tag{4.87}$$

With use of the definitions of f_i provided by (4.83)-(4.84), we conclude that the eliminant (4.87) vanishes when

$$(e_5{}^2 e_8)(25e_1{}^2 e_7 y^8 + 10e_1 e_3 e_7 y^6 + \frac{7}{2}e_3{}^2 e_7 y^4 + 2e_3 e_6 e_7 y^2 + e_6{}^2 e_7) \tag{4.88}$$
$$\times (\frac{3}{2}e_1 e_5{}^2 y^6 + \frac{9}{4}e_3 e_5{}^2 y^4 + \frac{9}{2}e_6 e_5{}^2 y^2 - \frac{3}{2}e_6 e_7 e_8)^2 = 0$$

After substitution of (4.69)-(4.78) into (4.88), we find that (4.88) becomes

$$(y^2 - \nu^2\lambda_2\lambda_3\Lambda_2{}^{-1}\Lambda_3{}^{-1})^6 (5\Lambda_2{}^2\Lambda_3{}^2 y^4 - 2\nu^2\lambda_2\lambda_3\Lambda_2\Lambda_3 y^2 + \nu^4\lambda_2{}^2\lambda_3{}^2)^2 = 0 \tag{4.89}$$

Finally, the roots of (4.89) are

$$y^2 = \frac{\nu^2\lambda_2\lambda_3}{\Lambda_2\Lambda_3} = -\frac{\nu^2\lambda_2{}^2\lambda_3{}^2}{D^2(\lambda_2-\lambda_1)(\lambda_3-\lambda_1)} < 0 \tag{4.90}$$

and

$$y^2 = (\nu^2\lambda_2\lambda_3 \pm i\ 2\nu^2\lambda_2\lambda_3)\ (5\Lambda_2\Lambda_3)^{-1} \tag{4.91}$$

in which we have used (4.7), (4.8) and $\lambda_1 < \lambda_2 < \lambda_3$ to obtain the inequality in (4.90). Thus, the roots (4.90) and (4.91) of (4.89) are both complex, and no transformation (4.79) of a_1 to y exists that leads to swallowtail points in the Vickroy and Dutton (1979) model (4.3)-(4.5). We conclude from (4.80) that the highest order singularity in their model is the cusp point $y = H_1' = H_3' = 0$ whose unfolding we discussed in the previous section.

4.3 Butterfly Points in the Rossby Regime

In its present form, the Vickroy and Dutton model is not as general as possible because its quintic steady state polynomial does not admit of five simultaneous real-valued solutions. Thus, we must introduce one or two parameters into the system (4.3)-(4.5), but we cannot use our local contact catastrophe procedure to find them. Rather, we need to use a bit of physical intuition to include the proper additional forcing.

There are several possible physical effects that may provide an explanation for the missing parameter. One possibility is to include the latitudinal variation β of the Coriolis parameter. However, a five-coefficient truncation is the smallest physically acceptable one in which β is still retained (Vickroy and Dutton, 1979; Wiin-Nielsen, 1979). In addition, Mitchell and Dutton (1981) show that the β-effect leads to periodic solutions in a six-coefficient truncation rather than butterfly points. Thus, the β-effect does not represent the missing parameter in the present three-component system.

The effects of barotropic instability of a basic zonal current U(y) and of orographic forcing also have been investigated with two-dimensional quasi-geostrophic models. Recently, Charney and DeVore (1979) investigated the latter effect with a truncated spectral model. If $\beta = 0$, then we will show in the following that addition of either a nonlinear basic current U(y) or a sinusoidally varying lower boundary into the quasi-geostrophic model (4.1) will produce a three-component system of the same form. Then we will demonstrate that the necessary four parameters for the complete description of transitions among the steady states are the three Newtonian heating coefficients plus the new one. We note that the singular values of the new parameters will be nonzero. We may anticipate this result because the values of these new parameters are zero in the original Vickroy and Dutton model, and we have found in Section 4.1 that only two parameters were needed because the singularity was of cusp type. The new model, therefore is the correct one to use for study of more complicated quasi-geostrophic systems (e.g. Clark, 1983).

Quasi-geostrophic flow forced by bottom topography was studied by Charney and DeVore (1979). They used a modified version of (4.1), which can be written in nondimensional form as

$$\frac{\partial}{\partial t}(\nabla_H^2\psi - \frac{\psi}{\lambda^2}) + J(\psi,\nabla_H^2\psi + h) + \beta\,\frac{\partial\psi}{\partial x} = -\,k[\nabla_H^2\,(\psi - \psi^*)] \qquad (4.92)$$

in which

$$\lambda^2 = gHf_o^{-2}L^{-2} \qquad (4.93)$$

$$k = D_E H^{-1}/2 \qquad (4.94)$$

$$h = h_o H^{-1}\cos\frac{nx}{L}\,\sin\frac{y}{L} \qquad (4.95)$$

Here g is the acceleration of gravity, H is the mean height of the domain, h_o is the amplitude of the lower boundary elevation h, f_o is the value of the Coriolis parameter at the β-plane latitude, πL is the distance between the side walls, D_E is the Ekman depth $(2\,\nu_E/f_o)^{1/2}$, ν_E is the bulk eddy viscosity and ψ^* is a forcing term representing a momentum source, created for example by the radiation field. We note that the case $\lambda^{-2} = 0$ corresponds to replacing the upper free boundary with a rigid horizontal boundary, and this is the case studied by Charney and DeVore (1979).

For the case $\beta = 0$ and $\lambda^{-2} = 0$, an appropriate three-component truncation for ψ is one which contains the same mode as that forced by orography via (4.95). A suitable nonlinear model is obtained therefore from the choice

$$\psi = \psi_K 2\cos\frac{nx}{L}\,\sin\frac{y}{L} + \psi_C\sqrt{2}\,\cos\frac{2y}{L} + \psi_N 2\sin\frac{nx}{L}\,\sin\frac{2y}{L} \qquad (4.96)$$

in which a similar expansion is used for ψ^*. Upon substituting (4.96) into (4.92) and integrating the result over the domain $0 \le x \le 2\pi L$, $0 \le y \le \pi L$, we obtain the truncated spectral model

$$\dot{\psi}_K = -\,\delta_{n1}\psi_C\psi_N - k\psi_K + k\psi_K^* \qquad (4.97)$$

$$\dot{\psi}_C = \varepsilon_n\psi_K\psi_N - k\psi_C + h_{02}\psi_N + k\psi_C^* \qquad (4.98)$$

$$\dot{\psi}_N = \delta_{n2}\psi_C\psi_K - k\psi_N - h_{n2}\psi_C + k\psi_N^* \qquad (4.99)$$

in which

$$\delta_{n1} = \frac{64\sqrt{2}\,n^3}{(n^2+1)15\pi} \qquad (4.100)$$

$$\delta_{n2} = \frac{64\sqrt{2}\,n(n^2-3)}{15\pi(n^2+4)} \qquad (4.101)$$

$$\varepsilon_n = \frac{16\sqrt{2}\,n}{5\pi} \qquad (4.102)$$

$$h_{02} = \frac{8\sqrt{2}\, n\, h_o}{15\pi\, H} \tag{4.103}$$

$$h_{n2} = \frac{32\sqrt{2}\, n\, h_o}{15\pi\, H(n^2 + 4)} \tag{4.104}$$

We note that a typographical error in (16) of Charney and DeVore (1979) has been corrected in (4.99). The coefficients h_{02} and h_{n2} are each proportional to the amplitude h_o of the sinusoidally varying bottom h (4.95).

By comparing (4.3)-(4.5) and (4.97)-(4.99) we see that the form of the Charney and DeVore model reduces to the form of the Vickroy and Dutton model when the amplitude h_o of the orography vanishes. Thus, the Charney and DeVore model provides a candidate with a stationary phase portrait generalizing that of the Vickroy and Dutton system.

A second modification of the Vickroy and Dutton model is obtained by considering the interaction of barotropic instability and thermal forcing. The truncated spectral model that we obtain is of the same form as the Charney and DeVore system (4.97)-(4.99).

The heating function H(x,y) that forces dissipative quasi-geostrophic motion will in general contain a nonperiodic, latitudinally varying component $H_o(y)$ (differential heating) and a cyclic component H'(x,y) (internal heating). Vickroy and Dutton (1979) considered only the latter effect H'(x,y)--the Fourier components H_i of the Newtonian heating represent some of the effects introduced by the spatially periodic portion of the total heating H(x,y).

However, the latitudinal component $H_o(y)$ produces a basic current $U(y) = - \partial\Psi(y)/\partial y$, and the (linear) barotropic instability of U(y) is thought to be one mechanism by which Rossby waves are generated and intensified in the atmosphere. In a nonlinear model, this instability is manifested by bifurcation and the existence of multiple states. Thus, a logical generalization of the Vickroy and Dutton model might be obtained by introducing a time-independent zonal current U(y) into the problem.

Accordingly, we write

$$\psi(x,y,t) = \Psi(y) + \psi'(x,y,t) \tag{4.105}$$

$$H(x,y) = H_o(y) + H'(x,y) \tag{4.106}$$

and substitute these expressions into the quasi-geostrophic equation (4.1) to obtain

$$\frac{\partial}{\partial t} \nabla_H^2\psi' + J(\psi', \nabla_H^2\psi') - \frac{\partial\psi'}{\partial x}\frac{\partial^2 U}{\partial y^2} + U\frac{\partial}{\partial x}\nabla_H^2\psi' + \beta\frac{\partial\psi'}{\partial x} - \nu\nabla_H^4\psi' = H'(x,y) \tag{4.107}$$

in which we require that

$$H_o(y) = \nu\partial^3U/\partial y^3 \tag{4.108}$$

Thus, the basic state fields $\Psi(y), H_o(y)$ satisfy (4.1).

Two basic states of physical interest obey (4.108). The simplest nonlinear solution (the Fourier coefficients of a linear form for U(y) vanish) is a quadratic form that vanishes on the side boundaries $y = \pm\ \pi/2$

$$\left.\begin{aligned} H_o(y) &= 0 \\ U(y) &= |U|_1\ (\pi^2/4 - y^2) \end{aligned}\right\} \tag{4.109}$$

In this case the zonal current is not driven by thermal forcing.

But a zonal current is often considered to develop from differential latitudinal heating, so we might choose

$$\left.\begin{aligned} H_o(y) &= -\ (\Delta H)y \\ U(y) &= |U|_2\ (\pi^4/16 - y^4) \end{aligned}\right\} \tag{4.110}$$

in which U(y) again vanishes at the side walls $y = \pm\ \pi/2$ and we require that the amplitude $|U|_2$ of U satisfies

$$|U|_2 = \Delta H/(24\nu) \tag{4.111}$$

Both of the above choices for U(y) lead to butterfly points, although the critical magnitudes of $|U|_1$ and $|U|_2$ are different in the two cases. We note that for unforced quasi-geostrophic flow perturbations, these zonal currents are stable because U_{yy} does not change sign in the domain for either choice (Dutton, 1976b). However, because (4.109) or (4.110) will lead to branching behavior, the presence of a zonal flow in which $U_{yy} \neq 0$ everywhere can still lead to instability once thermal forcing is included.

Upon substitution of (4.2) into (4.107), we obtain the revised model

$$\dot{a}_1 = \Lambda_1 a_2 a_3 - \nu\lambda_1 a_1 - H_1/\lambda_1 \tag{4.112}$$

$$\dot{a}_2 = \Lambda_2 a_1 a_3 - \nu\lambda_2 a_2 + \Gamma_1 a_3 - H_2/\lambda_2 \tag{4.113}$$

$$\dot{a}_3 = \Lambda_3 a_1 a_2 - \nu\lambda_3 a_3 - \Gamma_2 a_2 - H_3/\lambda_3 \tag{4.114}$$

in which the Fourier coefficients of (4.109) are

$$\Gamma_1 = 3\ell\lambda_3|U|_1/8\lambda_2 \tag{4.115}$$

$$\Gamma_2 = 3\ell\lambda_2|U|_1/8\lambda_3 \tag{4.116}$$

and the Fourier coefficients of (4.110) are

$$\Gamma_1 = [(\frac{3}{16}\pi^2 - \frac{45}{32})\lambda_3 + \frac{9}{2}]\frac{|U|_2\ell}{\lambda_2} \tag{4.117}$$

$$\Gamma_2 = [(\frac{3}{16}\pi^2 - \frac{45}{32})\lambda_2 + \frac{9}{2}]\frac{|U|_2\ell}{\lambda_3} \tag{4.118}$$

In the Vickroy and Dutton (1979) model (4.3)-(4.5) the bifurcation point is on the H_2 axis. This can be anticipated because forcing of only the largest or smallest scale can not lead to instability when the unforced inviscid model conserves both energy and enstrophy (Mitchell and Dutton, 1981); hence, bifurcations can occur only when there is intermediate forcing H_2 of sufficient magnitude.

However, the enstrophy constraint does not apply to (4.112)-(4.114). In this case, five steady states meet simultaneously at the singular point on the H_1 axis. When a zonal current U of sufficient magnitude is introduced, then, forcing H_1 at the largest scale leads to more complicated instabilities than does forcing H_2 at the middle scale.

We may calculate the critical values of H_1 and $|U|_1$ or $|U|_2$ by noting that if $H_2 = H_3 = 0$, then (4.112)-(4.114) admits of the solution $a_1 = A_1$ given by

$$\left.\begin{aligned} A_1 &= - H_1/\nu\lambda_1^2 \\ A_2 &= 0 \\ A_3 &= 0 \end{aligned}\right\} \tag{4.119}$$

Upon defining

$$\left.\begin{aligned} \alpha_1 &= a_1 - A_1 \\ \alpha_2 &= a_2 \\ \alpha_3 &= a_3 \end{aligned}\right\} \tag{4.120}$$

we may write (4.112)-(4.114) in the homogeneous form

$$\dot{\alpha}_1 = -\nu\lambda_1\alpha_1 + \Lambda_1\alpha_2\alpha_3 \tag{4.121}$$

$$\dot{\alpha}_2 = -\nu\lambda_2\alpha_2 + \alpha_3(\Lambda_2 A_1 + \Gamma_1) + \Lambda_2\alpha_1\alpha_3 \tag{4.122}$$

$$\dot{\alpha}_3 = -\nu\lambda_3\alpha_3 + \alpha_2(\Lambda_3 A_1 - \Gamma_2) + \Lambda_3\alpha_1\alpha_2 \tag{4.123}$$

The characteristic equation governing the stability of the trivial solution of (4.121)-(4.123) may be written as

$$\omega^2 + \nu(\lambda_2 + \lambda_3)\,\omega + \nu^2\lambda_2\lambda_3 - (\Lambda_2 A_1 + \Gamma_1)(\Lambda_3 A_1 - \Gamma_2) = 0 \tag{4.124}$$

Neutral stability, or $\omega = 0$, occurs when the constant term of (4.124) vanishes. When this occurs, we obtain two singular points of (4.112)-(4.114). The order of these singular points is greatest when the two individual singular points coalesce. Because the constant term is quadratic in A_1, this occurs when the discriminant of this quadratic vanishes. Thus the two solutions come together with

$$\hat{A}_1 = (\hat{\Gamma}_2\Lambda_2 - \hat{\Gamma}_1\Lambda_3)/(2\Lambda_2\Lambda_3) \tag{4.125}$$

where $\hat{\Gamma}_1$ and $\hat{\Gamma}_2$ are determined by the condition that the discriminant vanish. This condition is exactly

$$\hat{\Gamma}_1\Lambda_3 + \hat{\Gamma}_2\Lambda_2 = \pm\, 2\nu(-\Lambda_2\Lambda_3\lambda_2\lambda_3)^{1/2} = 2d \tag{4.126}$$

But we already know that A_1 satisfies (4.119). Using (4.125) we obtain

$$\hat{\Gamma}_1\Lambda_3 - \hat{\Gamma}_2\Lambda_2 = 2\Lambda_2\Lambda_3\hat{H}_1/(\nu\lambda_1{}^2) \tag{4.127}$$

Now, given $\hat{H}_1$ and $\hat{H}_2 = \hat{H}_3 = 0$, the two equations (4.126) and (4.127) uniquely determine $\hat{\Gamma}_1$ and $\hat{\Gamma}_2$ so that (4.125) in turn determines a unique stationary point $[\hat{A}_1,0,0]^T$ exhibiting double neutral stability. It is this stationary point that we will show is a butterfly point. For brevity we have defined $d = \pm\,\nu(-\Lambda_2\Lambda_3\lambda_2\lambda_3)^{1/2}$ in (4.126). We normally will choose the negative value for d, because this corresponds to a westerly current in either (4.109) or (4.110).

A similar calculation shows that the Charney and DeVore model (4.97)-(4.99) has a butterfly point given by

$$(h_o/H)_b = \frac{\pm\, 2k(-\varepsilon_n\,\delta_{n2})^{1/2}}{\varepsilon_n\, c_2 + \delta_{n2}\, c_1} \tag{4.128}$$

$$\left.\begin{aligned} \hat{\psi}_K = \hat{\psi}_K^* &= \frac{(\varepsilon_n c_2 - \delta_{n2} c_1)}{2\,\varepsilon_n \delta_{n2}} (h_o/H)_b \\ \hat{\psi}_c &= 0 \\ \hat{\psi}_N &= 0 \end{aligned}\right\} \tag{4.129}$$

$$c_1 = 8\sqrt{2}n/(15\pi) \tag{4.130}$$

$$c_2 = 32\sqrt{2}n/[15\pi(n^2+4)] \tag{4.131}$$

Upon comparison of (4.101)-(4.102) and (4.128), we observe that this butterfly point exists only for $n = 1$, which physically corresponds to a mountain peak at the middle of one boundary and a valley at the middle of the other.

Before proceeding to determine the type and unfolding of the stationary point, we make some estimates to see whether the values of the singularities in the two models represent realistic situations. For the modified Vickroy and Dutton model we may combine (4.127) and either (4.115)-(4.116) or (4.117)-(4.118) to find the critical values of the amplitudes $|U|$ of the quadratic or quartic forms for the zonal current U:

$$|\hat{U}|_1 = \frac{16\lambda_2\lambda_3 d}{3\ell(\lambda_3^2\Lambda_3 + \lambda_2^2\Lambda_2)} \tag{4.132}$$

$$|\hat{U}|_2 = 2d/c \tag{4.133}$$

in which $c = \ell[\frac{9}{2}(\lambda_3\Lambda_3 + \lambda_2\Lambda_2) + (\frac{3}{16}\pi^2 - \frac{45}{32})(\Lambda_3\lambda_3^2 + \Lambda_2\lambda_2^2)]\lambda_2^{-1}\lambda_3^{-1}$.

For the basic wind profiles (4.109) and (4.110) we may use the values (Vickroy and Dutton, 1979) $\nu = 0.01$, $\ell = 1$, $\lambda_1 = 1$ and $\lambda_2 = 2$ and $\lambda_3 = 10$ in (4.6)-(4.8), (4.127), and (4.132)-(4.133) to calculate $\Lambda_1 = 0.4323$, $\Lambda_2 = -0.2432$, $\Lambda_3 = 0.005404$, $d = -0.001621$, and $|\hat{U}|_1 = 0.400$ and $|\hat{U}|_2 = 0.0303$. For characteristic length and time scales of 3.6×10^6 m and one day, these amplitudes correspond to dimensional maximum velocities of 41 ms^{-1} for the quadratic wind profile but only 7.7 ms^{-1} for the quartic wind profile. The critical value for the quartic profile is well within observable values, and even those for the quadratic profile are possible.

In the Charney and DeVore (1979) system (4.97)-(4.99), we noted above that a butterfly point exists only when the longitudinal wavenumber $n = 1$; from (4.101)-(4.102) and (4.128)-(4.131) we find that in this case the critical value is given by $(h_o/H)_b = -0.229$ and $\hat{\psi}_K^* = 0.048$ for the choice $k = 0.01$. For the topography, this corresponds to peaks of 2.3 km in the middle of the right boundary of a domain whose height is 10 km; for the momentum forcing, this corresponds to a driving velocity amplitude of 17 ms^{-1} with use of the length scale 3.6×10^6 m. The magnitude of the butterfly point here is in the upper range of realistic values.

Substitution of (4.125) and (4.126) into (4.121)-(4.123) gives the form of the spectral system that we unfold with our contact catastrophe procedure:

$$\dot{\alpha}_1 = -\nu\lambda_1\alpha_1 + \Lambda_1\alpha_2\alpha_3 \tag{4.134}$$

$$\dot{\alpha}_2 = -\nu\lambda_2\alpha_2 + d\Lambda_3^{-1}\alpha_3 + \Lambda_2\alpha_1\alpha_3 \tag{4.135}$$

$$\dot{\alpha}_3 = -\nu\lambda_3\alpha_3 - d\Lambda_2^{-1}\alpha_2 + \Lambda_3\alpha_1\alpha_2 \tag{4.136}$$

We may write (4.134)-(4.136) in the form (2.247) as

$$\underset{\sim}{v} = \begin{bmatrix} \alpha_1 \\ \alpha_2 \end{bmatrix} \tag{4.137}$$

$$\underset{\sim}{w} = [\alpha_3] \tag{4.138}$$

$$A(\underset{\sim}{v},\underset{\sim}{w}) = \begin{bmatrix} -\nu\lambda_1 & 0 \\ 0 & -\nu\lambda_2 \end{bmatrix} \tag{4.139}$$

$$B(\underset{\sim}{v},\underset{\sim}{w}) = \begin{bmatrix} \Lambda_1 v_2 \\ \Lambda_2 v_1 + d\Lambda_3^{-1} \end{bmatrix} \tag{4.140}$$

$$C(\underset{\sim}{v},\underset{\sim}{w}) = [0,\ \Lambda_3 v_1 - d\Lambda_2^{-1}] \tag{4.141}$$

$$D(\underset{\sim}{v},\underset{\sim}{w}) = [-\nu\lambda_3] \tag{4.142}$$

By combining (2.248), (4.127) and (4.139)-(4.142) we find that $\Delta(0,0) = 0$ so that the singularity is of corank 1.

We insert (4.139) and (4.140) into (2.249) to obtain the solutions

$$v_1{}^*(w^*) = \Lambda_1 d w^{*2}[\Lambda_3(\nu^2\lambda_1\lambda_2 - \Lambda_1\Lambda_2 w^{*2})]^{-1} \tag{4.143}$$

$$v_2{}^*(w^*) = \nu\lambda_1 d w^*[\Lambda_3(\nu^2\lambda_1\lambda_2 - \Lambda_1\Lambda_2 w^{*2})]^{-1} \tag{4.144}$$

The function $p(w^*)$ governing the steady states of (4.134)-(4.136) is found by combining (2.250) and (4.141)-(4.144):

$$p(w^*) = -\nu\lambda_3\Lambda_1{}^2\Lambda_2{}^2 w^{*5}(\nu^2\lambda_1\lambda_2 - \Lambda_1\Lambda_2 w^{*2})^{-2} \tag{4.145}$$

Because $\Lambda_1\Lambda_2 < 0$, the denominator of (4.145) cannot vanish for any values of w^*, and the singular behavior of $p(w^*)$ is contained in its numerator

$$q(w^*) = -\nu\lambda_3\Lambda_1{}^2\Lambda_2{}^2 w^{*5} \qquad (4.146)$$

Because $q(w^*)$ contains only the quintic term, the singularity is of butterfly type, and we must add four lower order terms to (4.146) to unfold about it.

Just as in the unfolding about the cusp point $H_2 = \hat{H}_2$, we must clear the denominator of $p(w^*)$ by using a contact transformation similar in form to (4.30). After this has been done, we may write the versal unfolding of (4.134)-(4.136) about the singularity given by (4.125)-(4.127) as

$$\dot{\alpha}_1 = -\nu\lambda_1\alpha_1 + \Lambda_1\alpha_2\alpha_3 \qquad (4.147)$$

$$\dot{\alpha}_2 = -\nu\lambda_2\alpha_2 + \alpha_3 d\Lambda_3{}^{-1} + \Lambda_2\alpha_1\alpha_3 \qquad (4.148)$$

$$\dot{\alpha}_3 = -\nu\lambda_3\alpha_3 - \alpha_2 d\Lambda_2{}^{-1} + \Lambda_3\alpha_1\alpha_2 + \gamma_4\alpha_3{}^3 + \gamma_3\alpha_3{}^2 + \gamma_2\alpha_3 + \gamma_1 \qquad (4.149)$$

Clearly the unfolded system (4.147)-(4.149) is not usable in its present form because the quadratic and cubic terms cannot be obtained from substitution of the spectral expansion into a modified form of the governing partial differential equation.

We show below that the unfolding parameters $\gamma_1 - \gamma_4$ are equivalent to the parameters H_1, H_2, H_3 and $|U|$. The calculations must be performed separately for the Charney and DeVore model and for the two wind profiles (4.109) and (4.110) in the modified Vickroy and Dutton model; here we give the results only for the quadratic form in the latter model. The parameter γ_1 can be interpreted immediately as the forcing term $-H_3/\lambda_3$; we will find, however, that γ_1 also depends on H_2.

To apply Mather's Theorem I to move the unfolding parameters, we first substitute (4.134)-(4.136) into (2.261) to write

$$\tilde{N}_i(\alpha_1,\alpha_2,\alpha_3) = \begin{bmatrix} -\nu\lambda_1 & \Lambda_1\alpha_3 & \Lambda_1\alpha_2 \\ \Lambda_2\alpha_3 & -\nu\lambda_2 & \Lambda_2\alpha_1 + d\Lambda_3^{-1} \\ \Lambda_3\alpha_2 & \Lambda_3\alpha_1 - d\Lambda_2^{-1} & -\nu\lambda_3 \end{bmatrix} \begin{bmatrix} g_1 \\ g_2 \\ g_3 \end{bmatrix} \tag{4.150}$$

$$+ \begin{bmatrix} h_{11} & h_{12} & h_{13} \\ h_{21} & h_{22} & h_{23} \\ h_{31} & h_{32} & h_{33} \end{bmatrix} \begin{bmatrix} -\nu\lambda_1\alpha_1 + \Lambda_1\alpha_2\alpha_3 \\ -\nu\lambda_2\alpha_2 + \alpha_3(\Lambda_2\alpha_1 + d\Lambda_3^{-1}) \\ -\nu\lambda_3\alpha_3 + \alpha_2(\Lambda_3\alpha_1 - d\Lambda_2^{-1}) \end{bmatrix}$$

$$+ s_1 \begin{bmatrix} 1 \\ 0 \\ 0 \end{bmatrix} + s_2 \begin{bmatrix} 0 \\ 1 \\ 0 \end{bmatrix} + s_3 \begin{bmatrix} 0 \\ 0 \\ 1 \end{bmatrix} + s_4 \begin{bmatrix} 0 \\ \lambda_3^2\alpha_3 \\ -\lambda_2^2\alpha_2 \end{bmatrix}$$

in which $\tilde{N}_i(\alpha_1,\alpha_2,\alpha_3)$ are the original unfolding functions in (4.149) and the new unfolding functions are multiplied by $s_1 - s_4$ in (4.150). The first three functions correspond to the three inhomogeneous heating coefficients, and the ratio of the two components of the last function is the same as the magnitude of the ratio of the Fourier coefficients (4.115)-(4.116) of the quadratic zonal wind profile in the truncated system (4.112)-(4.114). In order to perform this calculation for the quartic wind profile, we would base the last unfolding function of (4.150) on the ratio of (4.117) and (4.118).

Upon choosing $N_1 = [0,0,\alpha_3]^T$, we find that the linear unfolding function can be written in the form (4.150) if we set

$$g_1 = e_1 \nu^{-1} \lambda_1^{-1} + \alpha_1(2\nu\lambda_3)^{-1} \tag{4.151}$$

$$g_2 = \alpha_2(2\nu\lambda_3)^{-1} \tag{4.152}$$

$$g_3 = 0 \tag{4.153}$$

$$h_{11} = h_{22} = h_{33}/2 = -(2\nu\lambda_3)^{-1} \tag{4.154}$$

$$h_{12} = h_{13} = h_{21} = h_{23} = h_{31} = h_{32} = 0 \tag{4.155}$$

$$s_1 = e_1 \tag{4.156}$$

$$s_4 = e_7 \tag{4.157}$$

$$s_2 = s_3 = 0 \tag{4.158}$$

in which e_1 and e_7 are defined in (4.185) and (4.191), respectively.

For the quadratic form $N_2 = [0,0,\alpha_3^2]^T$, we have

$$g_1 = -2d(5\nu\lambda_3\Lambda_2\Lambda_3)^{-1}\alpha_3 + 3\lambda_2(5\lambda_3\Lambda_2)^{-1}\alpha_2 \tag{4.159}$$

$$g_2 = -2\lambda_1 d(5\lambda_3\Lambda_1\Lambda_2\Lambda_3)^{-1} + 2\lambda_1(5\Lambda_1\lambda_3)^{-1}\alpha_1 \tag{4.160}$$

$$g_3 = 3\nu\lambda_1\lambda_2(5\lambda_3\Lambda_1\Lambda_2)^{-1} + 2d\,\lambda_1(5\nu\lambda_3^2\Lambda_1\Lambda_2)^{-1}\alpha_1 \tag{4.161}$$

$$h_{11} = 0 \tag{4.162}$$

$$h_{12} = -2\lambda_1(5\lambda_3\Lambda_2)^{-1} \tag{4.163}$$

$$h_{13} = -2d\lambda_1(5\nu\lambda_3^2\Lambda_2\Lambda_3)^{-1} \tag{4.164}$$

$$h_{21} = -\lambda_2(5\lambda_3\Lambda_1)^{-1} + 2d(5\nu^2\lambda_3^2\Lambda_1)^{-1}\alpha_1 \tag{4.165}$$

$$h_{22} = 0 \tag{4.166}$$

$$h_{23} = -2d(5\nu^2\lambda_3^2\Lambda_3)^{-1}\alpha_3 \tag{4.167}$$

$$h_{31} = -6d(5\nu\lambda_3\Lambda_1\Lambda_2)^{-1} + 2\Lambda_3(5\nu\Lambda_1\lambda_3)^{-1}\alpha_1 \tag{4.168}$$

$$h_{32} = 3\Lambda_3(5\nu\Lambda_2\lambda_3)^{-1}\alpha_2 \tag{4.169}$$

$$h_{33} = -(\nu\lambda_3)^{-1}\alpha_3 \tag{4.170}$$

$$s_1 = s_4 = 0 \tag{4.171}$$

$$s_2 = e_5 \tag{4.172}$$

$$s_3 = e_6 \tag{4.173}$$

in which e_5 and e_6 are defined in (4.189) and (4.190).

Finally we rewrite the cubic term $N_3 = [0,0,\alpha_3^3]^T$ with the aid of

$$g_1 = -d\nu\lambda_1\lambda_2(2\Lambda_1\Lambda_2^2\Lambda_3\lambda_3)^{-1} - \nu\lambda_1\lambda_2(2\Lambda_1\Lambda_2\lambda_3)^{-1}\alpha_1 \tag{4.174}$$

$$g_2 = -\nu\lambda_1\lambda_2(2\Lambda_1\Lambda_2\lambda_3)^{-1}\alpha_2 \tag{4.175}$$

$$g_3 = 0 \tag{4.176}$$

$$h_{11} = \nu\lambda_1\lambda_2(2\Lambda_1\Lambda_2\lambda_3)^{-1} \tag{4.177}$$

$$h_{22} = \nu\lambda_1\lambda_2(2\Lambda_1\Lambda_2\lambda_3)^{-1} \tag{4.178}$$

$$h_{31} = -\, d(\nu\lambda_3\Lambda_1\Lambda_2)^{-1}\,\alpha_3 + \Lambda_3(\nu\lambda_3\Lambda_1)^{-1}\,\alpha_1\alpha_3 \tag{4.179}$$

$$h_{32} = -\, 2\lambda_1 d(\Lambda_1\Lambda_2{}^2\lambda_3)^{-1} + \lambda_1\Lambda_3(\lambda_3\Lambda_1\Lambda_2)^{-1}\,\alpha_1 \tag{4.180}$$

$$h_{33} = -\,(\nu\lambda_3)^{-1}\,\alpha_3{}^2 + 2\nu\lambda_1\lambda_2(\lambda_3\Lambda_1\Lambda_2)^{-1} \tag{4.181}$$

$$h_{12} = h_{13} = h_{21} = h_{23} = 0 \tag{4.182}$$

$$s_1 = e_4 \tag{4.183}$$

$$s_2 = s_3 = s_4 = 0 \tag{4.184}$$

in which

$$e_1 = \lambda_1 d(\lambda_2{}^2\Lambda_2 - \lambda_3{}^2\Lambda_3)\;[2\Lambda_2\Lambda_3\lambda_3(\lambda_3{}^2\Lambda_3 + \lambda_2{}^2\Lambda_2)]^{-1} \tag{4.185}$$

$$e_2 = d\lambda_3[\nu(\lambda_3{}^2\Lambda_3 + \lambda_2{}^2\Lambda_2)]^{-1} \tag{4.186}$$

$$e_3 = -\, d\lambda_2{}^2[\nu\lambda_3(\lambda_3{}^2\Lambda_3 + \lambda_2{}^2\Lambda_2)]^{-1} \tag{4.187}$$

$$e_4 = -\, d\nu^2\lambda_1{}^2\lambda_2(2\Lambda_1\Lambda_2{}^2\Lambda_3\lambda_3)^{-1} \tag{4.188}$$

$$e_5 = -\,\nu\lambda_1\lambda_2 d(\lambda_3\Lambda_1\Lambda_2\Lambda_3)^{-1} \tag{4.189}$$

$$e_6 = \nu^2\lambda_1\lambda_2(\Lambda_1\Lambda_2)^{-1} \tag{4.190}$$

$$e_7 = e_2\lambda_3{}^{-2} = -\, e_3\lambda_2{}^{-2} \tag{4.191}$$

Thus (4.147)-(4.149) is contact equivalent to the system

$$\dot\alpha_1 = -\,\nu\lambda_1\alpha_1 + \Lambda_1\alpha_2\alpha_3 + e_1\gamma_2 + e_4\gamma_4 \tag{4.192}$$

$$\dot\alpha_2 = -\,\nu\lambda_2\alpha_2 + \Lambda_2\alpha_1\alpha_3 + \alpha_3(d\Lambda_3{}^{-1} + \gamma_2 e_2) + \gamma_3 e_5 \tag{4.193}$$

$$\dot\alpha_3 = -\,\nu\lambda_3\alpha_3 + \Lambda_3\alpha_1\alpha_2 + \alpha_2(-\, d\Lambda_2{}^{-1} + \gamma_2 e_3) + \gamma_3 e_6 + \gamma_1 \tag{4.194}$$

We may write (4.192)-(4.194) in the original variables a_i via (4.120):

$$\left.\begin{aligned} a_1 &= \alpha_1 + \hat{A}_1 \\ a_2 &= \alpha_2 \\ a_3 &= \alpha_3 \end{aligned}\right\} \tag{4.195}$$

We have set $A_1 = \hat{A}_1$ in (4.195) so that $\gamma_1 = \gamma_2 = \gamma_3 = \gamma_4 = 0$ still gives the butterfly point in the transformed system. Thus we replace (4.192)-(4.194) with

$$\dot{a}_1 = -\nu\lambda_1 a_1 + \Lambda_1 a_2a_3 - \hat{H}_1/\lambda_1 + e_1\gamma_2 + e_4\gamma_4 \tag{4.196}$$

$$\dot{a}_2 = -\nu\lambda_2 a_2 + \Lambda_2 a_1a_3 + a_3(\hat{\Gamma}_1 + \gamma_2e_2) + \gamma_3 e_5 \tag{4.197}$$

$$\dot{a}_3 = -\nu\lambda_3 a_3 + \Lambda_3 a_1a_2 - a_2(\hat{\Gamma}_2 - \gamma_2e_3) + \gamma_3 e_6 + \gamma_1 \tag{4.198}$$

in which we have used (4.125)-(4.126). Upon comparison of (4.112)-(4.114) and (4.196)-(4.198), we conclude that the original parameters and the independent unfolding parameters are related by (for the quadratic zonal wind profile)

$$\gamma_1 = -H_3\lambda_3^{-1} + H_2e_6(\lambda_2e_5)^{-1} \tag{4.199}$$

$$\gamma_2 = 3\ell\lambda_3(|U|_1 - |\hat{U}|_1)(8\lambda_2e_2)^{-1} \tag{4.200}$$

$$\gamma_3 = -H_2(\lambda_2e_5)^{-1} \tag{4.201}$$

$$\gamma_4 = -(H_1 - \hat{H}_1)(\lambda_1e_4)^{-1} - 3\ell\lambda_3e_1(|U|_1 - |\hat{U}|_1)(8\lambda_2e_2e_4)^{-1} \tag{4.202}$$

in which $|\hat{U}|_1$ is given by (4.132) and $\hat{H}_1$ is given by

$$\hat{H}_1 = d\nu\lambda_1^{\,2}(\lambda_3^{\,2}\Lambda_3 - \lambda_2^{\,2}\Lambda_2)\,[\Lambda_2\Lambda_3(\lambda_3^{\,2}\Lambda_3 + \lambda_2^{\,2}\Lambda_2)]^{-1} \tag{4.203}$$

The singularity set for (4.192)-(4.194) giving the critical parameter values at which transitions occur will be of the standard butterfly type only when it is displayed as a function of the parameters $\gamma_1 - \gamma_4$. However, the singularity set is of more value when shown as a function of the parameters $H_1' = H_1 - \hat{H}_1$, H_2, H_3 and $U' = |U| - |\hat{U}|$, even though when displayed in this way it will not have the standard appearance.

To find this set, we begin by writing the steady state polynomial as

$$y_1\alpha_3^{\,5} + y_2\alpha_3^{\,4} + y_3\alpha_3^{\,3} + y_4\alpha_3^{\,2} + y_5\alpha_3 + y_6 = 0 \tag{4.204}$$

in which we have substituted (4.199)-(4.202) into (4.192)-(4.194) and defined

$$y_1 = 1 \tag{4.205}$$

$$y_2 = b_1 H_3 \tag{4.206}$$

$$y_3 = b_2 U'^2 + b_3 H_1' U' + b_4 U' + b_5 H_1' \tag{4.207}$$

$$y_4 = b_6 H_2 U' + b_7 H_1' H_2 + b_8 H_2 + b_9 H_3 \tag{4.208}$$

$$y_5 = b_{10} U'^2 + b_{11} H_1' U' + b_{12} U' + b_{13} H_2^2 + b_{14} H_1'^2 \tag{4.209}$$

$$y_6 = b_{15} H_2 U' + b_{16} H_2 + b_{17} H_1' H_2 + b_{18} H_3 \tag{4.210}$$

and

$$b_1 = (\nu \lambda_3^2)^{-1} \tag{4.211}$$

$$b_2 = -9\ell^2 \lambda_1 (\lambda_2^2 \Lambda_2 + \lambda_3^2 \Lambda_3)(64\lambda_2^2 \lambda_3 \Lambda_1 \Lambda_2^2)^{-1} \tag{4.212}$$

$$b_3 = 3\ell (\lambda_2^2 \Lambda_2 + \lambda_3^2 \Lambda_3)(8\nu \lambda_1 \lambda_2 \lambda_3^2 \Lambda_1 \Lambda_2)^{-1} \tag{4.213}$$

$$b_4 = -3\ell \lambda_1 d(3\lambda_3^2 \Lambda_3 + \lambda_2^2 \Lambda_2)(8\lambda_2 \lambda_3^2 \Lambda_1 \Lambda_2^2 \Lambda_3)^{-1} \tag{4.214}$$

$$b_5 = 2d(\nu \lambda_3 \lambda_1 \Lambda_1 \Lambda_2)^{-1} \tag{4.215}$$

$$b_6 = 3\ell \lambda_1 (2\lambda_3^2 \Lambda_3 + \lambda_2^2 \Lambda_2)(8\lambda_2^2 \lambda_3^2 \Lambda_1 \Lambda_2^2)^{-1} \tag{4.216}$$

$$b_7 = -\Lambda_3 (\nu \lambda_1 \lambda_2 \lambda_3 \Lambda_1 \Lambda_2)^{-1} \tag{4.217}$$

$$b_8 = 3\lambda_1 d(\lambda_2 \lambda_3 \Lambda_1 \Lambda_2^2)^{-1} \tag{4.218}$$

$$b_9 = -2\nu \lambda_1 \lambda_2 (\lambda_3^2 \Lambda_1 \Lambda_2)^{-1} \tag{4.219}$$

$$b_{10} = -9\ell^2 \nu^2 \lambda_1^2 \lambda_2 (64\lambda_3 \Lambda_1^2 \Lambda_2^2)^{-1} \tag{4.220}$$

$$b_{11} = 3\ell \nu (\lambda_3^2 \Lambda_3 - \lambda_2^2 \Lambda_2)(8\lambda_3^2 \Lambda_1^2 \Lambda_2^2)^{-1} \tag{4.221}$$

$$b_{12} = -3\ell \nu^2 \lambda_1^2 d(\lambda_3^2 \Lambda_3 + \lambda_2^2 \Lambda_2)(8\lambda_3^2 \Lambda_1^2 \Lambda_2^3 \Lambda_3)^{-1} \tag{4.222}$$

$$b_{13} = -\lambda_1 \Lambda_3 (\lambda_2^2 \lambda_3 \Lambda_1 \Lambda_2^2)^{-1} \tag{4.223}$$

$$b_{14} = -\lambda_2 \Lambda_3 (\lambda_1^2 \lambda_3 \Lambda_1^2 \Lambda_2)^{-1} \tag{4.224}$$

$$b_{15} = - 3\ell\nu^2\lambda_1{}^2\lambda_2(8\lambda_3{}^2\Lambda_1{}^2\Lambda_2{}^2)^{-1} \quad (4.225)$$

$$b_{16} = - \nu^2\lambda_1{}^2 d(\lambda_3\Lambda_1{}^2\Lambda_2{}^3)^{-1} \quad (4.226)$$

$$b_{17} = - \nu\Lambda_3(\lambda_3\Lambda_1{}^2\Lambda_2{}^2)^{-1} \quad (4.227)$$

$$b_{18} = \nu^3\lambda_1{}^2\lambda_2{}^2(\lambda_3{}^2\Lambda_1{}^2\Lambda_2{}^2)^{-1} \quad (4.228)$$

We note that at the butterfly point $H_1' = H_2 = H_3 = U' = 0$ (4.204) reduces to (4.146).

The singularity set of (4.204) can be obtained by finding the fold points given by the values of the parameters at which both the steady state polynomial (4.204) and its first derivative

$$5y_1\alpha_3{}^4 + 4y_2\alpha_3{}^3 + 3y_3\alpha_3{}^2 + 2y_4\alpha_3 + y_5 = 0 \quad (4.229)$$

have common roots. After noting that H_3 appears as a linear term in y_2, y_4 and y_6, we find that these fold points are given by

$$H_3 = - (5y_1\alpha_3{}^4 + 3y_3\alpha_3{}^2 + 2q_1\alpha_3 + y_5)\ (4b_1\alpha_3{}^3 + 2b_9\alpha_3)^{-1} \quad (4.230)$$

in which α_3 is a root of

$$q_3\alpha_3{}^8 + q_4\alpha_3{}^6 + q_5\alpha_3{}^5 + q_6\alpha_3{}^4 + q_7\alpha_3{}^3 + q_8\alpha_3{}^2 + q_9\alpha_3 + q_{10} = 0 \quad (4.231)$$

Here we have used

$$q_1 = b_6H_2U' + b_7H_1'H_2 + b_8H_2 \quad (4.232)$$

$$q_2 = b_{15}H_2U' + b_{16}H_2 + b_{17}H_1'H_2 \quad (4.233)$$

$$q_3 = - b_1y_1 \quad (4.234)$$

$$q_4 = b_1y_3 - 3b_9y_1 \quad (4.235)$$

$$q_5 = 2b_1q_1 \quad (4.236)$$

$$q_6 = 3b_1y_5 - b_9y_3 - 5y_1b_{18} \quad (4.237)$$

$$q_7 = 4b_1q_2 \quad (4.238)$$

$$q_8 = b_9y_5 - 3y_3b_{18} \quad (4.239)$$

$$q_9 = 2(b_9 q_2 - b_{18} q_1) \tag{4.240}$$

$$q_{10} = - b_{18} y_5 \tag{4.241}$$

In practice, we choose values for H_1', H_2 and U', solve (4.231) for α_3 and then substitute the result into (4.230) to obtain H_3.

Two orthogonal three-dimensional sections through the four-dimensional butterfly singularity set are shown in Figs. 4.2a-b. These sections pass near to, but not through, the butterfly point B given by $H_1' = H_2 = H_3 = U' = 0$. In Fig. 4.2a, $U' = 0.17$, and in Fig. 4.2b, $H_1' = -0.02$; in both examples seven two-dimensional cross sections for $H_2 = 0, \pm 0.02, \pm 0.04$, and ± 0.06 are given with the value of H_2 increasing from left to right. Parameter values at which five distinct real solutions of (4.204) exist are indicated by the dark regions, at which three distinct real solutions exist by the shaded regions and at which only one exists by the unshaded areas.

The changes in appearance of the singularity set cross sections depicted in Fig. 4.2 indicate changes in the possible transitions among the five stationary solutions. To illustrate some of these changes we show in Fig. 4.3 several plots of the magnitude of α_3 as functions of either H_1' (Fig. 4.3a-c, corresponding to Fig. 4.2a) or of the new parameter U' (Fig. 4.3d-f, corresponding to Fig. 4.2b).

In Fig. 4.3a, the parameters $H_2 = H_3 = 0$, which is a line passing through the cusps in the middle cross section of Fig. 4.2a. Figures 4.3a and 3.1b are similar in appearance, with each parabola branching from the trivial solution at a cusp point. However, once the value of H_2 becomes nonzero (Fig. 4.3b), then the branching diagram is altered in a form similar to that found in Fig. 3.1c. The branching diagram changes character again after the value of H_2 has been increased from 0.02 to 0.06 (Fig. 4.3c). The two neighboring parabolas in Fig. 4.3b change to nearly concentric ones as a swallowtail point S (Fig. 4.2a) is crossed.

The cross section in Fig. 4.3d is perpendicular to the one given in Fig. 4.3a and it corresponds to the line $H_2 = H_3 = 0$ through the two cusps in the middle cross section of Fig. 4.2b. Ignoring the possibility of Hopf bifurcations for the moment, we expect that the stability of a steady solution will change whenever two or more solutions meet. Branching in the same sense as that of the parameter variation will lead to the new solution being stable, and branching in the opposite sense to the new solution being unstable (Iooss and Joseph, 1980). Thus, as the value of U' is decreased we expect a transition to occur at C_1 from the trivial solution to either the upper or lower branch and then a sudden transition back to the trivial solution at C_3. Hysteresis is indicated, however, because as the value of U' is increased, a sudden transition will occur at the different point C_2.

Once the value of H_2 is different from zero, more complicated hysteresis effects are indicated (Figs. 4.3e-f); these two figures resemble the form shown in Fig. 2.3. One sudden change is expected in Fig. 4.3e as the value of U' is

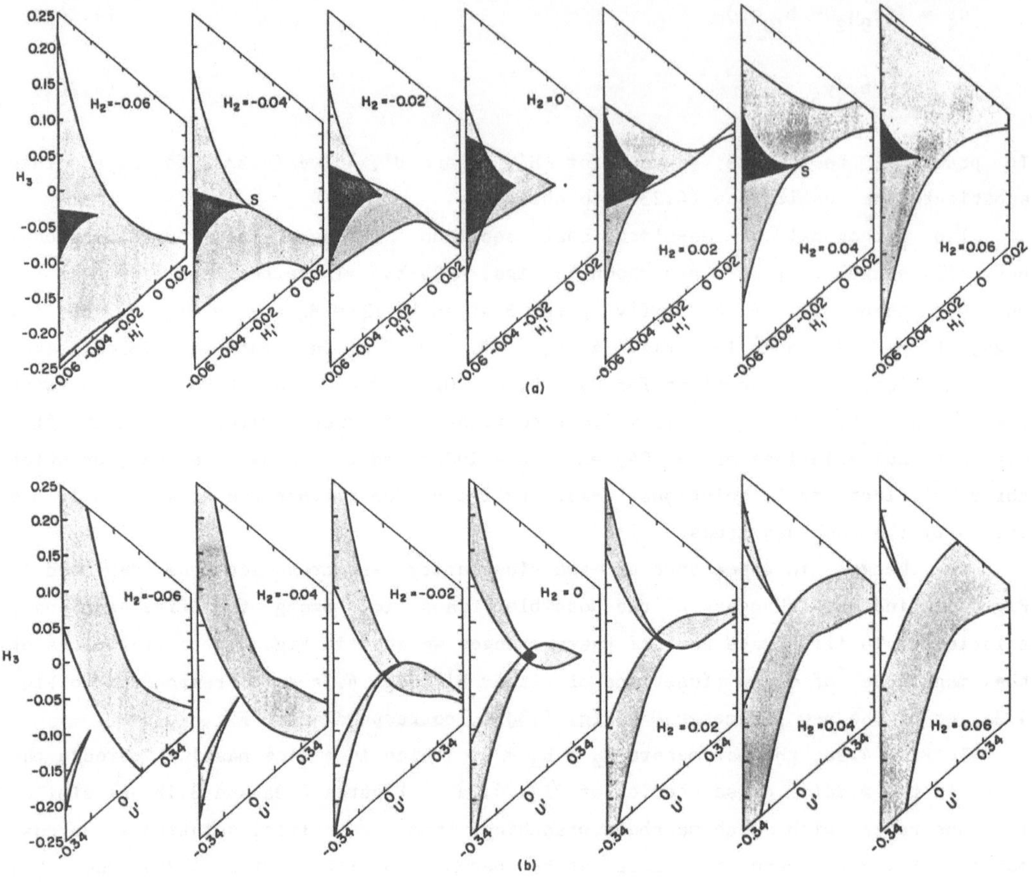

Figure 4.2 Two three-dimensional sections through the butterfly singularity set of steady solutions to (4.204). Both sections are displayed by showing seven constant - H_2 cross sections; here the magnitude of H_2 increases from -0.06 in the left most plane to 0.06 in the right most one. In (a), $U' = |U| - |\hat{U}| = 0.17$ and in (b) $H_1' = H_1 - \hat{H}_1 = -0.02$. The dark regions denote parameter values for which five real-valued solutions exist, the shaded portions three, and the remaining regions one. In (a) S denotes a swallowtail point.

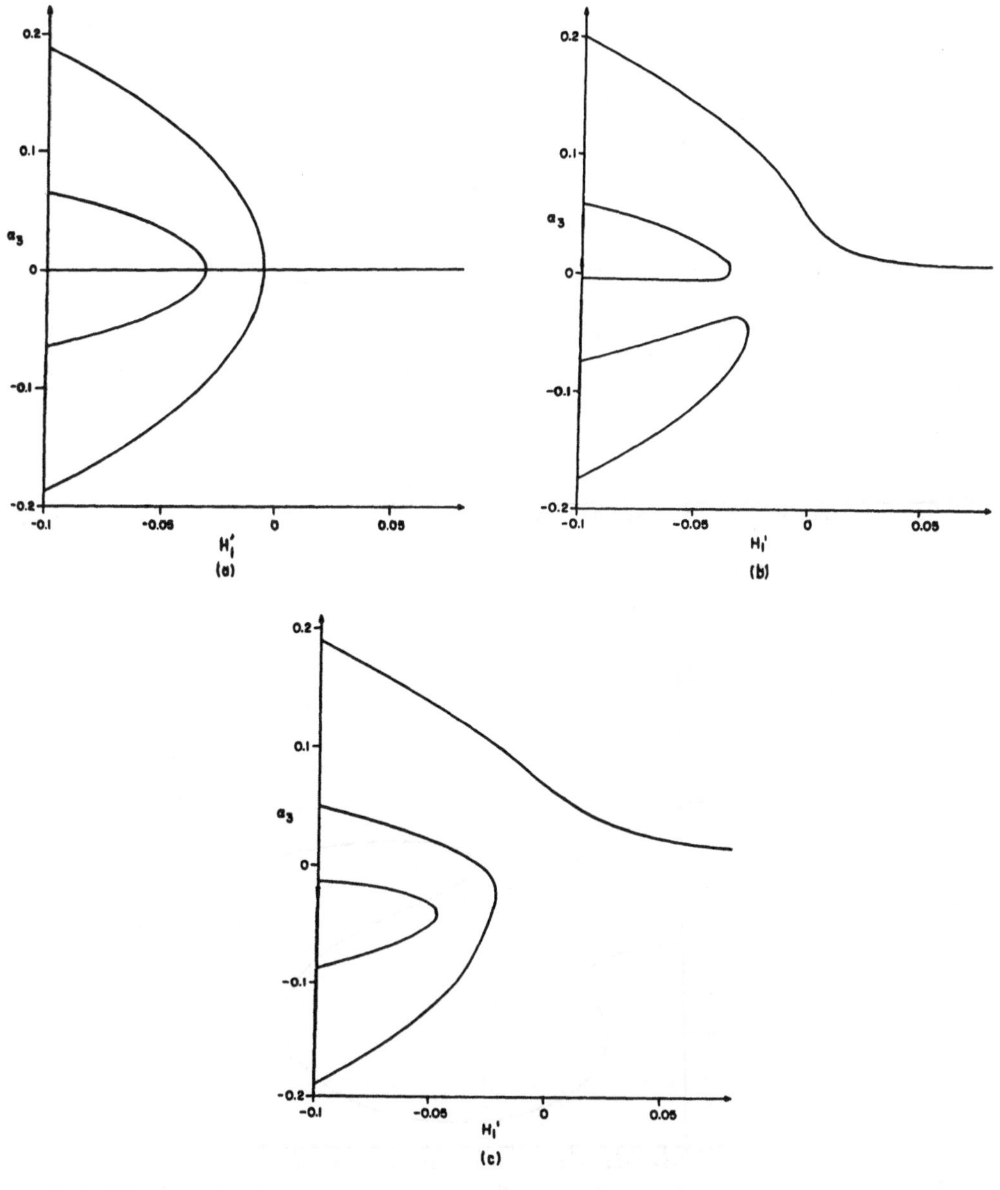

Figure 4.3 The amplitude of the spectral coefficient α_3 as functions of H_1' (a)-(c) and U' (d)-(f). Some of the branching behavior described by Fig. 4.2a is shown in (a)-(c) and some by Fig. 4.2b in (d)-(f). In (a), $U' = 0.17$, $H_2 = 0$, $H_3 = 0$; in (b), $U' = 0.17$, $H_2 = 0.02$, $H_3 = 0$; in (c), $U' = 0.17$, $H_2 = 0.06$, $H_3 = 0.05$; in (d), $H_1' = -0.02$, $H_2 = 0$, $H_3 = 0$; in (e), $H_1' = -0.02$, $H_2 = 0.02$, $H_3 = 0.025$; in (f), $H_1' = -0.02$, $H_2 = 0.06$, $H_3 = 0.15$.

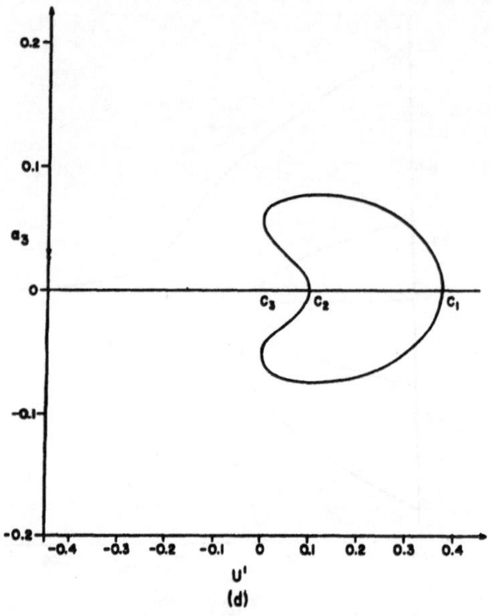
0.2
0.1
a3
0
-0.1
-0.2
C3
C2
C1
-0.4
-0.3
-0.2
-0.1
0
0.1
0.2
0.3
0.4
U'
(d)

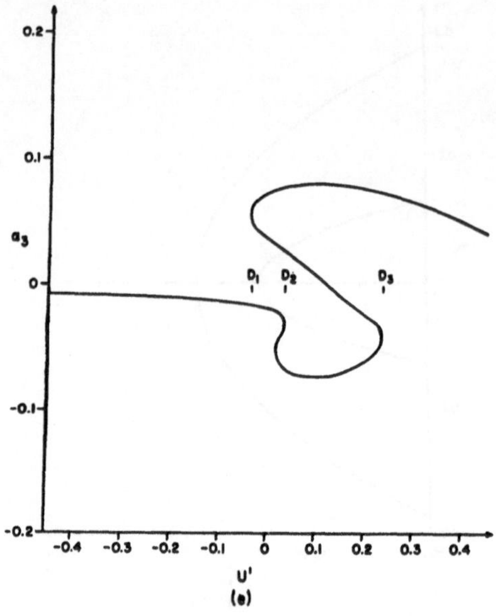
0.2
0.1
a3
0
-0.1
-0.2
D1
D2
D3
-0.4
-0.3
-0.2
-0.1
0
0.1
0.2
0.3
0.4
U'
(e)

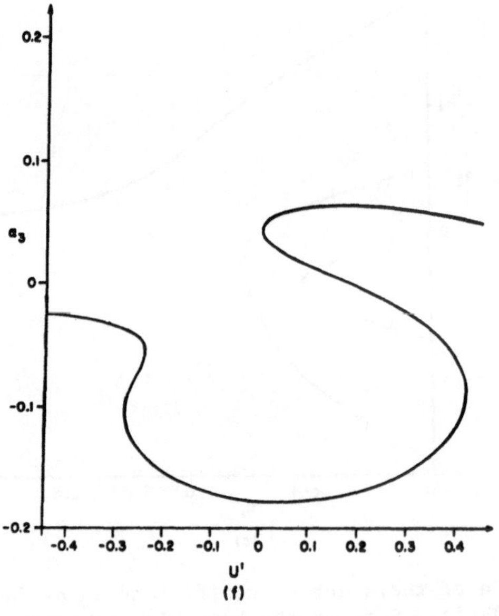
0.2
0.1
a3
0
-0.1
-0.2
-0.4
-0.3
-0.2
-0.1
0
0.1
0.2
0.3
0.4
U'
(f)

Fig. 4.3 (con't)

decreased past D_1, but two sudden changes involving the lower curve are possible as the value of U' is increased, one at D_2 and the other at D_3. After the cusp in Fig. 4.2b has withdrawn from the other shaded area (Fig. 4.3f), two sudden changes in the magnitude of α_3 occur for both the cases of increasing and decreasing values of U', and two hysteresis regions are likely.

Thus, by knowing the form of the singularity set as shown in Fig. 4.2 we find many different types of transitions possible, and the existence of several different transition classes has important physical implications. The complicated hysteresis possibilities shown in Figs. 4.3d-f are introduced only when the fourth parameter, $|U|$, is added to the differential system. Thus, we must consider the Vickroy and Dutton (1979) model (4.3)-(4.5) of quasi-geostrophic flow to be a special case (U=0) of the more general system (4.112)-(4.114) in which $|U| \neq 0$; an alternative view is that the height of the bottom topography varies in the more general system, as in the model (4.97)-(4.99) of Charney and DeVore (1979). Additional new parameters will not introduce more complicated transitions between the steady states because they are governed by a quintic polynomial. More general branching behavior may occur when there are more spectral equations in which case the steady states are governed by higher order polynomials; this might be the case, for example, in the five-component model of Wiin-Nielsen (1979) in which the steady states are governed by a ninth degree polynomial.

In order to find the most general branching behavior, then, we try to cause all but the highest order term in the steady state polynomial to vanish by finding the necessary critical values of the external parameters. In order to accomplish this, sometimes we must consider additional physical effects that introduce new parameters into the differential system. But once this is done successfully, important new stability results may be obtained as we saw here with the Vickroy and Dutton (1979) model. Whether there is an upper bound to the number of lower order terms that can be caused to vanish in this way is an open question that has important physical implications.

CHAPTER 5

ROTATING AXISYMMETRIC FLOW

As we saw in Chapter 3, axisymmetric flow, which is produced by fluid circulating in a vertical plane, must be viewed as being driven thermally in both the horizontal and the vertical. But for application of these results to the atmosphere, we must consider the effects of rotation in many cases; in this chapter we shall find that the singularities in this rotating axisymmetric flow are of butterfly type.

The atmospheric examples of axisymmetric flow divide into two classes according to the sign of the external temperature difference in the vertical, or of the Rayleigh number r (cf. (3.9), (3.13)). If $r > 0$ then the statically unstable fluid is being forced in the vertical and we obtain the rotating Rayleigh-Bénard problem whose atmospheric prototype is cloud streets in the planetary boundary layer; in this case, as we saw in Chapter 3, both direct and indirect flows are possible. If $r < 0$ and there is heating in the horizontal, then the statically stable fluid is being forced in the horizontal and we obtain the Hadley problem, whose atmospheric prototype is the long-term average meridional circulation in the tropics; here only direct circulations can occur.

The set of partial differential equations that govern rotating axisymmetric flow are obtained via introduction of the Coriolis parameter f and a v* equation to the shallow Boussinesq system (3.55)-(3.56); they may be written in nondimensional form as (cf. (3.2)-(3.3))

$$\frac{\partial}{\partial t^*} \tilde{\nabla}^2\psi^* = - K(\psi^*, \tilde{\nabla}^2\psi^*) - f^* \frac{\partial v^*}{\partial z^*} + \sigma(1 + a^2)^{-1} \tilde{\nabla}^4\psi^* + \sigma(1 + a^2) \frac{\partial \theta^*}{\partial x^*} + \sigma(1 + a^2)h \tag{5.1}$$

$$\frac{\partial v^*}{\partial t^*} = - K(\psi^*, v^*) + f^* \frac{\partial \psi^*}{\partial z^*} + \sigma(1 + a^2)^{-1} \tilde{\nabla}^2 v^* \tag{5.2}$$

$$\frac{\partial \theta^*}{\partial t^*} = - K(\psi^*, \theta^*) + r \frac{\partial \psi^*}{\partial x^*} + h \frac{\partial \psi^*}{\partial z^*} + (1 + a^2)^{-1} \tilde{\nabla}^2\theta^* \tag{5.3}$$

in which

$$f^* = fH^2 \pi^{-2} \kappa^{-1} (1 + a^2)^{-1} \tag{5.4}$$

is a nondimensional form of f, and we have used the following nondimensionalization for v (cf. (3.4)-(3.7))

$$v = (1 + a^2) \kappa \pi a^{-1} H^{-1} v^* \tag{5.5}$$

We recall that σ is the Prandtl number (3.11), a is the aspect ratio (3.12), r is the Rayleigh number (3.13) and h is the Hadley number (3,57).

Veronis (1966) studied the rotating Rayleigh-Bénard problem with a five-component spectral model. It is the smallest system in which each variable ψ^*, v^*, θ^* occurs somewhere in a nonlinear term; with this truncation, meridional transports of both heat and momentum can be represented. His five-coefficient model can be specified on the domain $0 \leq x^* \leq \pi$, $0 \leq z^* \leq \pi$ by

$$\psi^* = \sqrt{2}\, x_1 \sin x^* \sin z^* \tag{5.6}$$

$$\theta^* = \sqrt{2}\, x_2 \cos x^* \sin z^* - x_3 \sin 2z^* \tag{5.7}$$

$$v^* = -\sqrt{2}\, x_4 \sin x^* \cos z^* + x_5 \sin 2x^* \tag{5.8}$$

in which for simplicity we have adopted the coordinate system used in Chapter 3. This system is best suited for flow in a laboratory vessel; for application to the Hadley regime in the atmosphere we would interchange the roles of x^* and y^*, and of u^* and v^*, to maintain a right-handed coordinate system.

Upon substitution of (5.6)-(5.8) into (5.1)-(5.3) and integration of the result over the domain $0 \leq x^* \leq \pi$, $0 \leq z^* \leq \pi$, we obtain the system

$$\dot{x}_1 = -\sigma x_1 + \sigma x_2 + f^* (1 + a^2)^{-1} x_4 - 8\sqrt{2}\, \sigma h/\pi^2 \tag{5.9}$$

$$\dot{x}_2 = -x_1 x_3 + r x_1 - x_2 \tag{5.10}$$

$$\dot{x}_3 = x_1 x_2 - b x_3 - 16\sqrt{2}\, h(3\pi^2)^{-1} x_1 \tag{5.11}$$

$$\dot{x}_4 = -x_1 x_5 - f^* x_1 - \sigma x_4 \tag{5.12}$$

$$\dot{x}_5 = x_1 x_4 - \sigma b a^2 x_5 \tag{5.13}$$

in which $b = 4(1 + a^2)^{-1}$. We must choose the phasing (5.8) for v^* in order that f^* appears in the vorticity equation (5.9); as a result, $u^* = -\partial\psi^*/\partial z^*$ and v^* satisfy the same boundary conditions. For $f^* = 0$ we note that (5.9)-(5.13) reduces to the unfolded version (3.58)-(3.60) of the Lorenz model, with $x_4 = x_5 = 0$.

When Veronis (1966) studied (5.9)-(5.13) for $h = 0$, he noted that subcritical branching was possible for some ranges of the rotation rate f^* and Prandtl number σ, while supercritical branching was expected for other values of these secondary parameters. This is the signature of a singularity of higher order than the cusp present in the Lorenz model. We will show below that Veronis (1966) considered special cases of steady states described by a butterfly surface on which as many as five solutions may meet. In addition, we will find that the cusp points that occur for $r < 0$ in the modified Lorenz model discussed in Chapter 3 can become butterfly points once the effects of rotation are included.

5.1 The Butterfly Points

In order to find the highest order singularity in a spectral model, we examine the coefficients of the polynomial governing the steady states to see how many of these coefficients can be made to vanish for certain special values of the external parameters. These singularities do not occur necessarily on the trivial solution, and so we must examine the coefficients of the polynomials obtained from arbitrary translation of the original variable; the amount of translation is then itself a parameter of the problem. We used this approach in Section 3.2 to find the cusp points of the modified Lorenz model (3.58)-(3.60) and in Section 4.2 in an attempt to find swallowtail points in the Vickroy and Dutton model (4.3)-(4.5). Here we will be able to find three sets of butterfly points in the modified Veronis model (5.9)-(5.13), each one corresponding to one of the cusp points $r = 1$ or $r = -26$ of the modified Lorenz model (3.58)-(3.60).

We may use the Lyapunov-Schmidt process to determine the polynomial governing the steady states of (5.9)-(5.13) (see Section 5.3). Alternatively, we may solve (5.10)-(5.13) directly for the spectral components $x_2 - x_5$ as functions of x_1 to obtain

$$x_2 = [r\, b\, x_1 + 16\sqrt{2}\, h\, (3\pi^2)^{-1}\, x_1^{\,2}]\, (x_1^{\,2} + b)^{-1} \tag{5.14}$$

$$x_3 = [r\, x_1^{\,2} - 16\sqrt{2}\, h\, (3\pi^2)^{-1}\, x_1]\, (x_1^{\,2} + b)^{-1} \tag{5.15}$$

$$x_4 = -\, f^*\, \sigma\, b\, a^2\, x_1\, (x_1^{\,2} + \sigma^2\, a^2\, b)^{-1} \tag{5.16}$$

$$x_5 = -\, f^*\, x_1^{\,2}\, (x_1^{\,2} + \sigma^2\, a^2\, b)^{-1} \tag{5.17}$$

Substitution of (5.14)-(5.17) into (5.9) yields the quintic polynomial

$$\begin{aligned} &x_1^{\,5} + g_1\, h\, x_1^{\,4} + (g_2 + g_3\, r + g_4\, f^{*2})\, x_1^{\,3} + g_5\, h\, x_1^{\,2} \\ &\quad + (g_6 + g_7\, r + g_8\, f^{*2})\, x_1 + g_9\, h = 0 \end{aligned} \tag{5.18}$$

in which

$$g_1 = 8\sqrt{2}\, (3\pi^2)^{-1} \tag{5.19}$$

$$g_2 = b\, (1 + \sigma^2\, a^2) \tag{5.20}$$

$$g_3 = -\, b \tag{5.21}$$

$$g_4 = b\, a^2 \,(1 + a^2)^{-1} \tag{5.22}$$

$$g_5 = 8\sqrt{2}\, b\, (3 + \sigma^2 a^2)\, (3\pi^2)^{-1} \tag{5.23}$$

$$g_6 = \sigma^2 a^2 b^2 \tag{5.24}$$

$$g_7 = -\sigma^2 a^2 b^2 \tag{5.25}$$

$$g_8 = a^2 b^2 (1 + a^2)^{-1} \tag{5.26}$$

$$g_9 = 8\sqrt{2}\, \sigma^2 a^2 b^2 \pi^{-2} \tag{5.27}$$

Because the butterfly point may occur for some nonzero value of x_1, we change coordinates via

$$x_1 = z + y \tag{5.28}$$

For $y \neq 0$, trivial solutions of the new polynomial in the variable z will correspond to nontrivial ones $x_1 = y$. Upon substitution of (5.28) into (5.18), we obtain the generalized polynomial governing the steady states of (5.9)-(5.13):

$$z^5 + k_4 z^4 + k_3 z^3 + k_2 z^2 + k_1 z + k_o \tag{5.29}$$

in which

$$k_4 = 5y + g_1 h \tag{5.30}$$

$$k_3 = 10y^2 + 4g_1 hy + g_2 + g_3 r + g_4 f*^2 \tag{5.31}$$

$$k_2 = 10y^3 + 6g_1 hy^2 + 3(g_2 + g_3 r + g_4 f*^2)y + g_5 h \tag{5.32}$$

$$k_1 = 5y^4 + 4g_1 hy^3 + 3(g_2 + g_3 r + g_4 f*^2)y^2 + 2g_5 hy + g_6 + g_7 r + g_8 f*^2 \tag{5.33}$$

$$k_o = y^5 + g_1 hy^4 + (g_2 + g_3 r + g_4 f*^2)y^3 + g_5 hy^2 + (g_6 + g_7 r + g_8 f*^2)y + g_9 h \tag{5.34}$$

Butterfly points correspond to the vanishing of each coefficient k_i individually. The quartic term vanishes if

$$h = -5\, y\, g_1^{-1} \tag{5.35}$$

the cubic term if

$$f^{*2} = -(g_2 + g_3\, r - 10y^2)\, g_4^{-1} \tag{5.36}$$

and the quadratic term if either

$$y = 0 \tag{5.37}$$

or

$$y^2 = g_5\, (2g_1)^{-1} \tag{5.38}$$

Thus, because there are three possible values for the translation distance y, there are three potential butterfly points. Finally, the linear term vanishes provided that

$$r = \left[15g_1\, g_4\, y^4 + 10(g_1\, g_8 - g_4\, g_5)y^2 + g_1\, (g_4\, g_6 - g_2\, g_8)\right] \times \left[g_1\, (g_3\, g_8 - g_4\, g_7)\right]^{-1} \tag{5.39}$$

and the constant term vanishes provided that

$$yr = \left\{6g_1\, g_4\, y^5 + (10g_1\, g_8 - 5g_4\, g_5)y^3 + \left[g_1\, (g_4\, g_6 - g_2\, g_8) - 5g_4\, g_9\right]y\right\} \times \left[g_1\, (g_3\, g_8 - g_4\, g_7)\right]^{-1} \tag{5.40}$$

For the first point $y = 0$, we see that (5.40) is satisfied automatically; hence, we can determine the critical values of r, f^* and h from (5.35), (5.36), and (5.39) to be

$$r = (g_4\, g_6 - g_2\, g_8)\, (g_3\, g_8 - g_4\, g_7)^{-1} \tag{5.41}$$

$$f^{*2} = (g_2\, g_7 - g_3\, g_6)\, (g_3\, g_8 - g_4\, g_7)^{-1} \tag{5.42}$$

$$h = 0 \tag{5.43}$$

Upon use of the definitions (5.19)-(5.27) of the variables g_i, we may rewrite the butterfly point (5.41)-(5.43) as

$$r = -(\sigma^2 a^2 - 1)^{-1} \tag{5.44}$$

$$f*^2 = -\sigma^4 a^2 (1 + a^2)(\sigma^2 a^2 - 1)^{-1} \tag{5.45}$$

$$h = 0 \tag{5.46}$$

From (5.45) we discover that this butterfly point exists only when $\sigma^2 a^2 < 1$ so that $f*^2 > 0$; in this case, we observe from (5.44) that $r > 0$. Thus, this is a butterfly point for the rotating Rayleigh-Bénard problem because the required thermal stratification is unstable and because the required horizontal heating rate h is zero. Here we have set only two parameters to nonzero critical values and so we must find two other parameters to complete the unfolding of the butterfly point. Determination of these parameters and their physical interpretation will serve as topics for Sections 5.3 and 5.4.

There are two other potential butterfly points given by (5.38) as $y^2 = g_5(2g_1)^{-1}$. Inserting this value of y^2 into (5.39) and (5.40) and equating the two resulting expressions produces the relationship

$$g_5^2 = 20g_1 g_9 \tag{5.47}$$

In this case we conclude that the butterfly points correspond to

$$r = [5g_5 g_8 - 25g_4 g_9 + g_1(g_4 g_6 - g_2 g_8)][g_1(g_3 g_8 - g_4 g_7)]^{-1} \tag{5.48}$$

$$f*^2 = [g_1(g_2 g_7 - g_3 g_6) + 5(5g_3 g_9 - g_5 g_7)][g_1(g_3 g_8 - g_4 g_7)]^{-1} \tag{5.49}$$

$$h = \mp 5[g_5 (2g_1^3)^{-1}]^{1/2} \tag{5.50}$$

These butterfly points can be written with the aid of (5.19)-(5.27) and $b = 4(1 + a^2)^{-1}$ as

$$\sigma^2 a^2 = 27 \pm 3\sqrt{80} \tag{5.51}$$

$$r = -14(5\sigma^2 a^2 - 1)(\sigma^2 a^2 - 1)^{-1} \tag{5.52}$$

$$f*^2 = (1 + a^2)\sigma^2 a^2 (4\sigma^2 a^2 - 60)[a^2(\sigma^2 a^2 - 1)]^{-1} \tag{5.53}$$

$$y^2 = 2(3 + \sigma^2 a^2)(1 + a^2)^{-1} \tag{5.54}$$

$$h = \mp 15\pi^2 y (8\sqrt{2})^{-1} \tag{5.55}$$

Because both choices for $\sigma^2 a^2$ are positive, (5.51)-(5.55) correspond to two pairs of butterfly points, which are (cf. (3.69)-(3.70) for the modified Lorenz model)

$$\left.\begin{aligned} \sigma &= 0.409\ a^{-1} \\ r &= -\ 2.76 \\ f^{*2} &= 11.91\ (1 + a^2)\ a^{-2} \\ y &= \pm\ 2.52\ (1 + a^2)^{-1/2} \\ h &= \mp\ 32.93\ (1 + a^2)^{-1/2} \end{aligned}\right\} \tag{5.56}$$

and

$$\left.\begin{aligned} \sigma &= 7.34\ a^{-1} \\ r &= -\ 71.06 \\ f^{*2} &= 158.3\ (1 + a^2)\ a^{-2} \\ y &= \pm\ 10.66\ (1 + a^2)^{-1/2} \\ h &= \mp\ 139.5\ (1 + a^2)^{-1/2} \end{aligned}\right\} \tag{5.57}$$

or

$$\left.\begin{aligned} x_1 &= \pm\ 2.52\ (1 + a^2)^{-1/2} \\ x_2 &= \mp\ 18.13\ (1 + a^2)^{-1/2} \\ x_3 &= 4.43 \\ x_4 &= \mp\ 2.02 \\ x_5 &= -\ 3.12\ (1 + a^2)^{1/2}\ a^{-1} \end{aligned}\right\} \tag{5.58}$$

and

$$\left.\begin{aligned} x_1 &= \pm\ 10.66\ (1 + a^2)^{-1/2} \\ x_2 &= \mp\ 128.7\ (1 + a^2)^{-1/2} \\ x_3 &= -\ 58.98 \\ x_4 &= \mp\ 11.96 \\ x_5 &= -\ 4.34\ (1 + a^2)^{1/2}\ a^{-1} \end{aligned}\right\} \tag{5.59}$$

These butterfly points evidently apply to the Hadley regime because the critical values of r are negative, corresponding to stable thermal stratification, and because the critical values of h are nonzero, corresponding to heating rates that vary in the horizontal. Moreover, the circulation can be seen to be a direct one, as is necessary in the Hadley regime. When $r < 0$ and $h < 0$, we see from (3.57) that $\Delta_x T < 0$ corresponding to a warm left wall and a cool right one; a direct circulation is one in which fluid rises near the warm wall, which from (5.6) and Fig. 3.2 is seen to occur in this case if $x_1 > 0$. Consequently, direct circulations are ones in which $x_1\, h < 0$ (a similar argument holds for $r < 0$, $h > 0$ leading to $x_1 < 0$). Thus, at the butterfly point (5.56) or (5.57) we have that $x_1 = y$ and that $x_1\, h < 0$ as required. Of course, these values apply strictly to the Hadley regime of a rotating annulus within which the flow is described by the shallow Boussinesq equations (5.1)-(5.3); a similar analysis applied to a truncated spectral model of a deep Boussinesq fluid would produce the critical values for the atmosphere.

5.2 Unfolding about the Butterfly Point: The Hadley Problem

In order to locate the butterfly points in the Hadley regime, we set four parameters to nonzero critical values in the previous section. Because four independent parameters are needed in order to unfold about a butterfly singularity, it is clear that these four parameters, r, h, f*, and σ, determine a complete unfolding.

However, to verify this contention, we would need to use the procedure outlined in Chapter 2 first to provide the canonical unfolding and then to move the parameters to the physically interpretable locations in the governing differential system. To accomplish this, we first translate the variables x_i to the butterfly point (5.58) or (5.59) by writing

$$x_i = X_i + \chi_i \quad , \quad i = 1, \ldots, 5 \tag{5.60}$$

and substitute (5.60) into (5.9)-(5.13). The unfolding produced by application of (2.252) is therefore

$$- \sigma_b\, \chi_1 + \sigma_b\, \chi_2 + f_b^* \,(1 + a^2)^{-1}\, \chi_4 + \gamma_4\, \chi_1^{\,3} + \gamma_3\, \chi_1^{\,2} + \gamma_2\, \chi_1 + \gamma_1 = 0 \tag{5.61}$$

$$(r_b - X_3)\, \chi_1 - \chi_2 - X_1\, \chi_3 - \chi_1\, \chi_3 = 0 \tag{5.62}$$

$$[X_2 - 16\sqrt{2}\, h_b\, (3\pi^2)^{-1}]\, \chi_1 + X_1\, \chi_2 - b\, \chi_3 + \chi_1\, \chi_2 = 0 \tag{5.63}$$

$$- (f_b^* + X_5)\, \chi_1 - \sigma_b\, \chi_4 - X_1\, \chi_5 - \chi_1\, \chi_5 = 0 \tag{5.64}$$

$$X_4\, \chi_1 + X_1\, \chi_4 - \sigma_b\, b\, a^2\, \chi_5 + \chi_1\, \chi_4 = 0 \tag{5.65}$$

in which the subscript b refers to the butterfly point (5.56) or (5.57).

In order to relate the canonical parameters to the physically interpretable ones, in Chapters 3 and 4 we used as new unfolding functions the ones that the parameters of interest multiplied in the original model because these parameters appeared in separate terms. However, here we wish to relate $\gamma_1 - \gamma_4$ in (5.61) to

$$\left.\begin{aligned} r' &= r - r_b \\ h' &= h - h_b \\ \sigma' &= \sigma - \sigma_b \\ f^{*\prime} &= f^* - f^*_b \end{aligned}\right\} \qquad (5.66)$$

Inspection of (5.9) reveals that two of these parameters appear in the product σh. As a consequence, we cannot use (2.261) to replace the unfolding functions in (5.61)-(5.65) with ones that reproduce entirely the product σh.

However, we may apply the first corollary of Mather's Theorem I to note that the product

$$\sigma h = \sigma_b h_b + \sigma_b h' + h_b \sigma' + \sigma' h' \qquad (5.67)$$

is part of a versal unfolding if and only if the linearized form $\sigma_b h' + h_b \sigma'$ is part of a versal unfolding. Accordingly, we may use (2.261) to replace the unfolding functions

$$\begin{bmatrix} X_1^{\,3} \\ 0 \\ 0 \\ 0 \\ 0 \end{bmatrix}, \quad \begin{bmatrix} X_1^{\,2} \\ 0 \\ 0 \\ 0 \\ 0 \end{bmatrix}, \quad \begin{bmatrix} X_1 \\ 0 \\ 0 \\ 0 \\ 0 \end{bmatrix}, \quad \begin{bmatrix} 1 \\ 0 \\ 0 \\ 0 \\ 0 \end{bmatrix} \qquad (5.68)$$

with the unfolding functions

$$\begin{bmatrix} \chi_2 - \chi_1 - \frac{8\sqrt{2}}{\pi^2} h_b \\ 0 \\ 0 \\ -\chi_4 \\ -b\,a^2\,\chi_5 \end{bmatrix}, \quad \begin{bmatrix} -\frac{8\sqrt{2}}{\pi^2}\sigma_b \\ 0 \\ -\frac{16\sqrt{2}}{3\pi^2}\chi_1 \\ 0 \\ 0 \end{bmatrix}, \quad \begin{bmatrix} \frac{\chi_4}{1+a^2} \\ 0 \\ 0 \\ -\chi_1 \\ 0 \end{bmatrix}, \quad \begin{bmatrix} 0 \\ \chi_1 \\ 0 \\ 0 \\ 0 \end{bmatrix} \qquad (5.69)$$

The first function in (5.69) represents σ', the second h', the third $f^{*\prime}$ and the last r'. Because the calculation is extremely tedious and because the four physically acceptable unfolding parameters are known, we omit the details of the calculation verifying that we can replace (5.68) with (5.69) and consequently $\gamma_1 - \gamma_4$ in (5.61) with r', h', σ' and $f^{*\prime}$.

In the remainder of this chapter we discuss the unfolding of the first butterfly point, (5.44)-(5.46), found in Section 5.1; for this unfolding, two new parameters must be identified.

5.3 Unfolding about the Butterfly Point: The Rotating Rayleigh-Bénard Problem

Although we have already found the appropriate butterfly point (5.44)-(5.46) previously, we illustrate here how to use the Lyapunov-Schmidt procedure to find it. At this singularity, the trivial solution is neutrally stable, so we may determine the critical values of the external parameters by performing a linear stability analysis.

Upon linearizing (5.9)-(5.13) about the zero solution and substituting the form $x_i = \tilde{x}_i \exp(\lambda t)$ into the result, we obtain the characteristic polynomial

$$(\lambda + b)\,(\lambda + \sigma\, b\, a^2)\,\{\lambda^3 + (2\sigma + 1)\,\lambda^2 + [\sigma\,(\sigma + 2 - r) + f^{*2}\,(1 + a^2)^{-1}]\,\lambda + \sigma^2\,[-r + 1 + f^{*2}\,\sigma^{-2}\,(1 + a^2)^{-1}]\} = 0 \qquad (5.70)$$

Bifurcation is signaled by $\lambda = 0$; from (5.70) we see that this occurs when r and f^{*2} are related by

$$r_s = 1 + f^{*2}\,\sigma^{-2}\,(1 + a^2)^{-1} \qquad (5.71)$$

We may now write (5.9)-(5.13) in the form (2.247) as (for $r = r_s$, $h = 0$)

$$\underset{\sim}{v} = \begin{bmatrix} x_2 \\ x_3 \\ x_4 \\ x_5 \end{bmatrix} \tag{5.72}$$

$$\underset{\sim}{w} = [x_1] \tag{5.73}$$

$$A(\underset{\sim}{v}, \underset{\sim}{w}) = \begin{bmatrix} -1 & 0 & 0 & 0 \\ 0 & -b & 0 & 0 \\ 0 & 0 & -\sigma & 0 \\ 0 & 0 & 0 & -\sigma b a^2 \end{bmatrix} \tag{5.74}$$

$$B(\underset{\sim}{v}, \underset{\sim}{w}) = \begin{bmatrix} r_s - v_2 \\ v_1 \\ -f^* - v_4 \\ v_3 \end{bmatrix} \tag{5.75}$$

$$C(\underset{\sim}{v}, \underset{\sim}{w}) = [\sigma, 0, f^*(1 + a^2)^{-1}, 0] \tag{5.76}$$

$$D(\underset{\sim}{v}, \underset{\sim}{w}) = [-\sigma] \tag{5.77}$$

By combining (2.248) and (5.74)-(5.77) we find that $\Delta(0,0) = 0$ so that the singularity r_s is of corank 1.

We insert (5.74) and (5.75) into (2.249) to obtain the solutions

$$v_1^*(w^*) = r_s\, b\, w^*(w^{*2} + b)^{-1} \tag{5.78}$$

$$v_2^*(w^*) = r_s\, w^{*2}(w^{*2} + b)^{-1} \tag{5.79}$$

$$v_3^*(w^*) = -f^*\, \sigma\, b\, a^2\, w^*(w^{*2} + \sigma^2 a^2 b)^{-1} \tag{5.80}$$

$$v_4^*(w^*) = -f^*\, w^{*2}(w^{*2} + \sigma^2 a^2 b)^{-1} \tag{5.81}$$

After combining (2.250) and (5.76)-(5.77) we find that $p(w^*)$ is

$$p(w^*) = \{-\sigma\, w^{*5} + b\,\sigma^{-1}(1+a^2)^{-1}\left[f^{*2}(1-a^2\sigma^2) - \sigma^4 a^2(1+a^2)\right]w^{*3}\}$$

$$\times\left[(w^{*2}+b)\,(w^{*2}+\sigma^2 a^2 b)\right]^{-1} \qquad (5.82)$$

Clearly the denominator of (5.82) cannot vanish for any values of w^*, so all the singular behavior is contained in the numerator of $p(w^*)$.

We see from inspection of (5.82) that the singularity r_s (5.71) is of cusp type (i.e. the cubic term is nonzero) except when

$$f^{*2}_c = -\sigma^4 a^2 (1+a^2)/(\sigma^2 a^2 - 1) \qquad (5.83)$$

in which case the critical Rayleigh number is given by

$$r_c = -1/(\sigma^2 a^2 - 1) \qquad (5.84)$$

Now only the quintic term of $p(w^*)$ remains, so the singularity (r_c, f^*_c) is of butterfly type; from (5.83) we see that this point exists only when $\sigma a < 1$. After adding the required lower order terms to the numerator of $p(w^*)$ evaluated at (r_c, f^*_c), we find that the unfolding in the original spectral model is

$$\dot{x}_1 = -\sigma x_1 + \sigma x_2 + f^*_c(1+a^2)^{-1} x_4 + \mu_4 x_1^3 + \mu_3 x_1^2 + \mu_2 x_1 + \mu_1 \qquad (5.85)$$

$$\dot{x}_2 = -x_1 x_3 + r_c x_1 - x_2 \qquad (5.86)$$

$$\dot{x}_3 = x_1 x_2 - b x_3 \qquad (5.87)$$

$$\dot{x}_4 = -x_1 x_5 - f^*_c x_1 - \sigma x_4 \qquad (5.88)$$

$$\dot{x}_5 = x_1 x_4 - \sigma b a^2 x_5 \qquad (5.89)$$

We must find now physical interpretations for the coefficients $\mu_1 - \mu_4$ in (5.85). Our goal is to find suitable linear or constant terms for the parameters $\mu_1 - \mu_4$ to multiply, because these correspond to linear or constant terms in a (modified) version of (5.1)-(5.3). We discuss this in the following section.

5.4 Dynamic Similarity

We found in the previous section that the singularity in the Veronis (1966) model is of butterfly type so that four independent parameters are needed to unfold about the singularity $r = r_c$, $f^* = f^*_c$. Veronis (1966), however, only included two of these; as a result, we must conclude that some transitions not represented in his model are possible in a rotating convective system.

There are several candidates for the two missing parameters associated with μ_1 and μ_3 that multiply the constant and the quadratic terms, respectively, in (5.85). After using (2.261) to find possible locations for these two parameters, we discover that there are 3 constant and 12 linear terms that serve as possibilities for replacing the pair $[1, 0, 0, 0, 0]^T$ and $[x_1^2, 0, 0, 0, 0]^T$ in a new versal unfolding. Certainly, combinations of many of these fifteen terms will be associated with different physical effects, but the branching behavior created by any of them will be equivalent. We may use Mather's Theorem I to identify which combinations are actually independent.

The above observation suggests the following definition for dynamically similar physical systems: if the stationary phase portraits of the unfolded versions of two physical systems are the same, then these systems are dynamically similar for parameter values near their singular points. For corank 1 singularities, we saw in Chapter 2 that the steady state structure is determined by the number of parameters in a minimal versal unfolding. As a result, we may label two corank 1 systems as dynamically similar if, for parameter values in the neighborhood of their singularities, the number of parameters in the minimal versal unfolding is the same, even though these parameters may be associated with different physical effects in the two systems.

In this section we show that two equivalent effects are the Hadley number h representing horizontal heating rates and the tilt angle α of a vertically heated domain. Physically, fluid in a level vessel forced by an externally imposed horizontal temperature difference exhibits the same type of transitions as fluid in a tilted vessel forced by an externally imposed vertical temperature difference. By the definition given above, these two systems are dynamically similar. The effect present in both unfoldings is the Fourier coefficient q of a horizontally varying diabatic heating rate $Q(x^*)$ of Newtonian type. In Shirer and Wells (1982), it was noted incorrectly that α and q were equivalent; but we will demonstrate below that the unfolding functions associated with α and h are not independent, showing that α and h are equivalent.

Introduction of α into the governing Boussinesq equations leads to buoyancy terms in both the horizontal and vertical equations of motion. To see this, we observe that the basic state hydrostatic pressure field $p_0(x, z)$, constant density ρ_0, and α are related by (Fig. 5.1)

$$p_0(x, z) = p_{00} - \rho_0\, g\, (\cos \alpha)\, z - \rho_0\, g\, (\sin \alpha)\, x \qquad (5.90)$$

in which we have assumed that the domain is not tilted in the y-direction. If we assume that the angle α is small, then we may use the approximate form

$$p_0(x, z) = p_{00} - \rho_0\, g\, z - \rho_0\, g\, \alpha\, x \qquad (5.91)$$

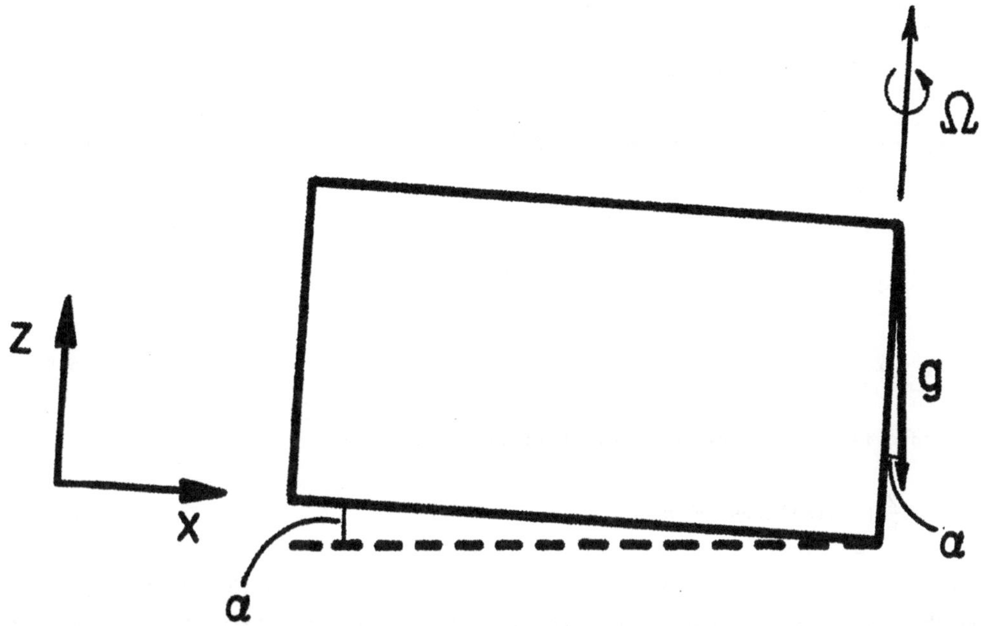

Fig. 5.1 Relationship between a rotating laboratory tank and the true horizontal, denoted by the dashed line. Here α is either the angle between the base (x-direction) and the horizontal or the angle between the side (z-direction) and the gravity vector $\underline{g}$. Note that rotation occurs about the tilted z-axis.

After we have used the basic state pressure field (5.91) and after we have introduced an inhomogeneous horizontally varying Newtonian heating rate $Q(x^*)$ into the thermodynamic equation, we may write the modified version of (5.1)-(5.3) as

$$\frac{\partial}{\partial t^*} \tilde{\nabla}^2 \psi^* = - K(\psi^*, \tilde{\nabla}^2 \psi^*) - f^* \frac{\partial v^*}{\partial z^*} + \sigma(1 + a^2)^{-1} \tilde{\nabla}^4 \psi^* \tag{5.92}$$

$$+ \sigma(1 + a^2) \frac{\partial \theta^*}{\partial x^*} - \sigma\, \alpha(1 + a^2)\, a^{-1} \frac{\partial \theta^*}{\partial z^*} + \sigma(1 + a^2)\, (h + \alpha\, r\, a^{-1})$$

$$\frac{\partial v^*}{\partial t^*} = - K(\psi^*, v^*) + f^* \frac{\partial \psi^*}{\partial z^*} + \sigma(1 + a^2)^{-1} \tilde{\nabla}^2 v^* \tag{5.93}$$

$$\frac{\partial \theta^*}{\partial t^*} = - K(\psi^*, \theta^*) + r \frac{\partial \psi^*}{\partial x^*} + h \frac{\partial \psi^*}{\partial z^*} + (1 + a^2)^{-1} \tilde{\nabla}^2 \theta^* + Q(x^*) \tag{5.94}$$

Upon substitution of (5.6)-(5.8) into (5.92)-(5.94) and integration over the domain, we obtain the proposed, revised spectral model

$$\dot{x}_1 = -\sigma x_1 + \sigma x_2 + 16\sqrt{2}\,\sigma\,\alpha(3\pi^2 a)^{-1} x_3 + f^* (1 + a^2)^{-1} x_4 \tag{5.95}$$
$$- 8\sqrt{2}\,\sigma\,(h + \alpha r a^{-1})\,\pi^{-2}$$

$$\dot{x}_2 = -x_1 x_3 + r x_1 - x_2 + 2\sqrt{2}\,q\,\pi^{-1} \tag{5.96}$$

$$\dot{x}_3 = x_1 x_2 - b x_3 - 16\sqrt{2}\,h(3\pi^2)^{-1} x_1 \tag{5.97}$$

$$\dot{x}_4 = -x_1 x_5 - f^* x_1 - \sigma x_4 \tag{5.98}$$

$$\dot{x}_5 = x_1 x_4 - \sigma b a^2 x_5 \tag{5.99}$$

in which q is the Fourier coefficient of $Q(x^*)$ given by

$$q = \frac{2}{\pi}\int_0^{\pi} Q(x^*) \cos x^* \, dx^* \tag{5.100}$$

We next show that (5.95)-(5.99) is an unfolding equivalent to (5.85)-(5.89), and we find the relationship between the unfolding and physically interpretable parameters.

Upon comparison of (5.9) and (5.85) we see that μ_1 is proportional to h. However, as in the Lorenz model discussed in Chapter 3, we must account for the appearance of h in a linear term of (5.97) to obtain the precise relation. Similarly, we must allow for the occurrence of f* in the two equations (5.95) and (5.98). Because two parameters, α and r, appear in a product in (5.95), we must use the linearization of them as we discussed in Section 5.2.

To apply Mather's Theorem I, we first substitute (5.85)-(5.89) into (2.261) to write

$$
\underset{\sim}{N}_1(x_1, x_2, x_3, x_4, x_5) = \begin{bmatrix} -\sigma & \sigma & 0 & f_c^*(1+a^2)^{-1} & 0 \\ r_c - x_3 & -1 & -x_1 & 0 & 0 \\ x_2 & x_1 & -b & 0 & 0 \\ -(f_c^* + x_5) & 0 & 0 & -\sigma & -x_1 \\ x_4 & 0 & 0 & x_1 & -\sigma\, b\, a^2 \end{bmatrix} \begin{bmatrix} g_1 \\ g_2 \\ g_3 \\ g_4 \\ g_5 \end{bmatrix}
$$

$$
+ \begin{bmatrix} h_{11} & h_{12} & h_{13} & h_{14} & h_{15} \\ h_{21} & h_{22} & h_{23} & h_{24} & h_{25} \\ h_{31} & h_{32} & h_{33} & h_{34} & h_{35} \\ h_{41} & h_{42} & h_{43} & h_{44} & h_{45} \\ h_{51} & h_{52} & h_{53} & h_{54} & h_{55} \end{bmatrix} \begin{bmatrix} -\sigma x_1 + \sigma x_2 + f_c^*(1+a^2)^{-1} x_4 \\ -x_1 x_3 + r_c x_1 - x_2 \\ x_1 x_2 - b x_3 \\ -x_1 x_5 - f_c^* x_1 - \sigma x_4 \\ x_1 x_4 - \sigma b a^2\, x_5 \end{bmatrix}
$$

$$
+ S_1 \begin{bmatrix} 1 \\ 0 \\ 2\, x_1 (3\sigma)^{-1} \\ 0 \\ 0 \end{bmatrix} + S_2 \begin{bmatrix} x_4 (1+a^2)^{-1} \\ 0 \\ 0 \\ -x_1 \\ 0 \end{bmatrix} + S_3 \begin{bmatrix} 0 \\ x_1 \\ 0 \\ 0 \\ 0 \end{bmatrix}
$$

$$
+ S_4 \begin{bmatrix} \frac{2}{3} x_3 - r_c \\ 0 \\ 0 \\ 0 \\ 0 \end{bmatrix} + S_5 \begin{bmatrix} 0 \\ 1 \\ 0 \\ 0 \\ 0 \end{bmatrix} \qquad (5.101)
$$

The forms of the unfolding functions that serve as the factors of $S_1 - S_5$ in (5.101) were chosen to agree with those found in (5.95)-(5.99). The first and fourth unfolding functions are two of the candidates that together with the fifth one allow moving the constant and cubic coefficients μ_1 and μ_3 in (5.85); we will mention the others in Section 5.4.3.

It is a simple matter to show that the first and fourth unfolding functions in (5.101) are not independent and therefore cannot both be in a minimal versal unfolding. Accordingly, we have from (5.101) that

$$\begin{bmatrix} 1 \\ 0 \\ \frac{2x_1}{3\sigma} \\ 0 \\ 0 \end{bmatrix} = \cdots + S_o \begin{bmatrix} \frac{2x_3}{3} - r_c \\ 0 \\ 0 \\ 0 \\ 0 \end{bmatrix} \qquad (5.102)$$

if we choose

$$g_2 = 2(3\sigma\ r_c)^{-1}\ x_3 \qquad (5.103)$$

$$g_3 = -\ 2(3\sigma\ b\ r_c)\ x_2 \qquad (5.104)$$

$$g_1 = g_4 = g_5 = 0 \qquad (5.105)$$

$$h_{23} = -\ 2(3\sigma\ b\ r_c) \qquad (5.106)$$

$$h_{32} = 2(3\sigma\ r_c) \qquad (5.107)$$

$$\text{all other } h_{ij} = 0 \qquad (5.108)$$

$$S_o = -\ 1/r_c \qquad (5.109)$$

We note that the dependence of these two unfolding functions comes only from the presence of constant terms in each.

Thus, we conclude that we can relate μ_1 and μ_3 to either h and q or α and q; moreover, we suspect from the calculation of the butterfly points r_c and f^*_c that we can relate μ_2 and μ_4 to r and f*. We demonstrate this in the following subsections by finding solutions to (5.101).

5.4.1 Horizontal heating. We relate the constant term in (5.85) to h and q in (5.95)-(5.99) by choosing $N_1 = [1, 0, 0, 0, 0]^T$ and

$$g_2 = -2/\sigma \tag{5.110}$$

$$g_1 = g_3 = g_4 = g_5 = h_{ij} = 0 \text{ for all } i, j \tag{5.111}$$

$$S_1 = 3 \tag{5.112}$$

$$S_5 = -2/\sigma \tag{5.113}$$

$$S_2 = S_3 = S_4 = 0 \tag{5.114}$$

Upon setting $N_3 = [x_1^2, 0, 0, 0, 0]^T$, we find that the unfolding parameter μ_3 can be related to h and q; from (5.101) we have

$$g_1 = -\sigma\ b\ a^2/5 \tag{5.115}$$

$$\begin{aligned} g_2 = &- b(14\ \sigma^2 a^2 - 15)\ [5\ \sigma(\sigma^2 a^2 - 1)]^{-1} \\ &+ b(7\ \sigma^2 a^2 - 5)\ [5\ \sigma(\sigma^2 a^2 - 1)]^{-1}\ x_3 \\ &+ 2\ \sigma\ b\ a^4\ f_c^*[5\ (a^2 + 1)\ (\sigma^2 a^2 - 1)]^{-1}\ x_5 \end{aligned} \tag{5.116}$$

$$\begin{aligned} g_3 = &(\sigma^4 a^4 - 8\ \sigma^2 a^2 + 5)\ [5\ \sigma(\sigma^2 a^2 - 1)]^{-1}\ x_2 \\ &- 2\ a^2\ f_c^*[5(1 + a^2)\ (\sigma^2 a^2 - 1)]^{-1}\ x_4 \end{aligned} \tag{5.117}$$

$$g_4 = f_c^*\ b\ a^2/5 + 2\ b\ f_c^*(5\ \sigma^2)^{-1}\ x_3 + b\ a^2(3\sigma^2 a^2 - 5)\ [5(\sigma^2 a^2 - 1)]^{-1}\ x_5 \tag{5.118}$$

$$g_5 = -2\ f_c^*(5\ \sigma)^{-1}\ x_2 - 2(\sigma^2 a^2 - 2)\ [5(\sigma^2 a^2 - 1)]^{-1}\ x_4 \tag{5.119}$$

$$h_{11} = -x_1/\sigma \tag{5.120}$$

$$h_{13} = 1 \tag{5.121}$$

$$h_{15} = f_c^*[\sigma(1 + a^2)]^{-1} \tag{5.122}$$

$$h_{23} = (\sigma^4 a^4 - 8\ \sigma^2 a^2 + 5)\ [5\ \sigma(\sigma^2 a^2 - 1)]^{-1} \tag{5.123}$$

$$h_{25} = -2\ a^2\ f_c^*[5(1 + a^2)\ (\sigma^2 a^2 - 1)]^{-1} \tag{5.124}$$

$$h_{31} = 2\ b\ a^2/5 \tag{5.125}$$

$$h_{32} = b(7\ \sigma^2\ a^2 - 5)\ [5\ \sigma(\sigma^2\ a^2 - 1)]^{-1} \tag{5.126}$$

$$h_{34} = 2\ \sigma\ b\ a^4\ f_c^*[5(1 + a^2)\ (\sigma^2\ a^2 - 1)]^{-1} \tag{5.127}$$

$$h_{43} = -\ 2\ f_c^*(5\ \sigma)^{-1} \tag{5.128}$$

$$h_{45} = -\ 2(\sigma^2\ a^2 - 2)\ [5(\sigma^2\ a^2 - 1)]^{-1} \tag{5.129}$$

$$h_{51} = -\ 2\ b\ f_c^*(\sigma^2\ a^2 - 1)\ (5\ \sigma^3)^{-1} \tag{5.130}$$

$$h_{52} = 2\ b\ f_c^*(5\ \sigma^2)^{-1} \tag{5.131}$$

$$h_{54} = b\ a^2(3\ \sigma^2\ a^2 - 5)\ [5(\sigma^2\ a^2 - 1)]^{-1} \tag{5.132}$$

$$h_{12} = h_{14} = h_{21} = h_{22} = h_{24} = h_{33} = h_{35} = h_{41} = h_{42} = h_{44} = h_{53} = h_{55} = 0 \tag{5.133}$$

$$S_1 = 3\ b \tag{5.134}$$

$$S_5 = -\ 3\ b\ \sigma^{-1} \tag{5.135}$$

$$S_2 = S_3 = S_4 = 0 \tag{5.136}$$

in which we have used (5.83) and (5.84).

We next set $N_2 = [x_1, 0, 0, 0, 0]^T$ to find

$$g_1 = 0 \tag{5.137}$$

$$g_2 = x_2/\sigma \tag{5.138}$$

$$g_3 = x_3/\sigma \tag{5.139}$$

$$g_4 = x_4(2\ \sigma)^{-1} \tag{5.140}$$

$$g_5 = x_5(2\ \sigma)^{-1} \tag{5.141}$$

$$h_{11} = h_{22} = h_{33} = 2\ h_{44} = 2\ h_{55} = -\ 1/\sigma \tag{5.142}$$

$$h_{ij} = 0 \text{ for } i \neq j \tag{5.143}$$

$$S_2 = f_c^*(2\ \sigma)^{-1} \tag{5.144}$$

$$S_3 = r_c/\sigma \tag{5.145}$$

$$S_1 = S_4 = S_5 = 0 \tag{5.146}$$

and we have again used (5.83) and (5.84).

Finally, the cubic unfolding function $N_4 = [x_1^3, 0, 0, 0, 0]^T$ can be moved to an equivalent position via

$$g_1 = 0 \tag{5.147}$$

$$g_2 = \sigma\ b\ a^2\ x_2 \tag{5.148}$$

$$g_3 = \sigma\ b\ a^2\ x_3 \tag{5.149}$$

$$g_4 = b(2\ \sigma)^{-1}\ x_4 \tag{5.150}$$

$$g_5 = b(2\ \sigma)^{-1}\ x_5 \tag{5.151}$$

$$h_{11} = -\ x_1^2/\sigma - b(\sigma^2\ a^2 + 1)/\sigma \tag{5.152}$$

$$h_{12} = -\ b \tag{5.153}$$

$$h_{13} = x_1 \tag{5.154}$$

$$h_{14} = -\ b\ a^2\ f_c^*(a^2 + 1)^{-1} \tag{5.155}$$

$$h_{15} = f_c^*[\sigma(a^2 + 1)]^{-1}\ x_1 \tag{5.156}$$

$$h_{22} = -\ \sigma\ b\ a^2 \tag{5.157}$$

$$h_{33} = -\ \sigma\ b\ a^2 \tag{5.158}$$

$$h_{44} = -\ b(2\ \sigma)^{-1} \tag{5.159}$$

$$h_{55} = -\ b(2\ \sigma)^{-1} \tag{5.160}$$

$$h_{21} = h_{23} = h_{24} = h_{25} = h_{31} = h_{32} = h_{34} = h_{35} = h_{41} = h_{42} = h_{43} = h_{45} = h_{51} = h_{52} = h_{53} = h_{54} = 0 \tag{5.161}$$

$$S_2 = f_c^* \, b(2\,\sigma)^{-1} \tag{5.162}$$

$$S_3 = r_c \, \sigma \, b \, a^2 \tag{5.163}$$

$$S_1 = S_4 = S_5 = 0 \tag{5.164}$$

We may use the above calculations to transform (5.85)-(5.89) to an equivalent versally unfolded system whose form resembles the physically derived one (5.95)-(5.99). Accordingly, we find that

$$\dot{x}_1 = -\sigma\, x_1 + \sigma\, x_2 + f_c^*(1 + a^2)^{-1}\,(2\,\sigma + \mu_2 + b\,\mu_4)\,(2\,\sigma)^{-1}\, x_4 + 3(\mu_1 + b\,\mu_3) \tag{5.165}$$

$$\dot{x}_2 = -x_1\, x_3 + r_c(1 + \mu_2\,\sigma^{-1} + \sigma\, b\, a^2\, \mu_4)\, x_1 - x_2 - (2\,\mu_1 + 3\, b\,\mu_3)\,\sigma^{-1} \tag{5.166}$$

$$\dot{x}_3 = x_1\, x_2 + 2(\mu_1 + b\,\mu_3)\,\sigma^{-1}\, x_1 - b\, x_3 \tag{5.167}$$

$$\dot{x}_4 = -x_1\, x_5 - f_c^*(2\,\sigma + \mu_2 + b\,\mu_4)\,(2\,\sigma)^{-1}\, x_1 - \sigma\, x_4 \tag{5.168}$$

$$\dot{x}_5 = x_1\, x_4 - \sigma\, b\, a^2\, x_5 \tag{5.169}$$

Upon comparison of (5.95)-(5.99) with (5.165)-(5.169) we find that (for $\alpha = 0$)

$$h = -3\pi^2(\mu_1 + b\,\mu_3)\,(8\sqrt{2}\,\sigma)^{-1} \tag{5.170}$$

$$q = -\pi(2\,\mu_1 + 3\, b\,\mu_3)\,(2\sqrt{2}\,\sigma)^{-1} \tag{5.171}$$

$$f^* = f_c^*\,[1 + (\mu_2 + b\,\mu_4)\,(2\,\sigma)^{-1}] \tag{5.172}$$

$$r = r_c\,[1 + (\mu_2 + b\,\sigma^2\, a^2\,\mu_4)/\sigma] \tag{5.173}$$

Thus, we have obtained one possible transformation (5.170)-(5.173) between the unfolding parameters μ_i, which yields the proper control variables that produce the standard butterfly surface, and the physically meaningful parameters h, r, f*, and q. As in the Vickroy and Dutton model discussed in Chapter 4, we find that the surface of steady states will not appear in the standard form when it is displayed as a function of the original control variables. We note that the factors in brackets in (5.172)-(5.173) are independent only when $\sigma\, a \neq 1$, which is a result consistent with the fact that the butterfly point exists only when $\sigma\, a < 1$.

5.4.2 Tilting domain. In order to show that the small tilt angle α of a vertically heated domain together with the Newtonian heating rate q provide a second choice for the unfolding parameters, μ_1 and μ_3, we must recalculate two of the unfolding functions, N_1 and N_3. Relationships (5.137)-(5.164) for N_2 and N_4 remain the same. The constant term μ_1 of the unfolding can be viewed as a function of α and q if we take $N_1 = [1, 0, 0, 0, 0]^T$ and

$$g_2 = -2\sigma^{-1} + 2(\sigma r_c)^{-1} x_3 \tag{5.174}$$

$$g_3 = -2(\sigma b r_c)^{-1} x_2 \tag{5.175}$$

$$g_1 = g_4 = g_5 = 0 \tag{5.176}$$

$$h_{23} = -2(\sigma b r_c)^{-1} \tag{5.177}$$

$$h_{32} = 2(\sigma r_c)^{-1} \tag{5.178}$$

$$\text{all other } h_{ij} = 0 \tag{5.179}$$

$$S_4 = -3 r_c^{-1} \tag{5.180}$$

$$S_5 = -2\sigma^{-1} \tag{5.181}$$

We find an alternate form for the quadratic unfolding function when we set $N_3 = [x_1^2, 0, 0, 0, 0]^T$ and

$$g_1 = -\sigma b a^2/5 \tag{5.182}$$

$$g_2 = [-b(14\sigma^2 a^2 - 15) + (-10\sigma^4 a^4 + 27\sigma^2 a^2 - 15)b x_3 + 2\sigma^2 b a^4 f_c^*(1 + a^2)^{-1} x_5][5\sigma(\sigma^2 a^2 - 1)]^{-1} \tag{5.183}$$

$$g_3 = -[(-11\sigma^4 a^4 + 28\sigma^2 a^2 - 15) x_2 + 2 a^2 \sigma f_c^*(1 + a^2)^{-1} x_4] \times [5\sigma(\sigma^2 a^2 - 1)]^{-1} \tag{5.184}$$

$$g_4 = f_c^* b a^2/5 + 2 b f_c^*(5\sigma^2)^{-1} x_3 + (3\sigma^2 a^2 - 5) b a^2[5(\sigma^2 a^2 - 1)]^{-1} x_5 \tag{5.185}$$

$$g_5 = -2 f_c^*(5\sigma)^{-1} x_2 - 2(\sigma^2 a^2 - 2)[5(\sigma^2 a^2 - 1)]^{-1} x_4 \tag{5.186}$$

$$h_{11} - x_1/\sigma \tag{5.187}$$

$$h_{13} = 1 \tag{5.188}$$

$$h_{15} = f_c^*[\sigma(1 + a^2)]^{-1} \tag{5.189}$$

$$h_{23} = (11\,\sigma^4\,a^4 - 28\,\sigma^2\,a^2 + 15)\,[5\,\sigma(\sigma^2\,a^2 - 1)]^{-1} \tag{5.190}$$

$$h_{25} = -\,2\,a^2\,f_c^*[5(\sigma^2\,a^2 - 1)\,(a^2 + 1)]^{-1} \tag{5.191}$$

$$h_{31} = 2\,b\,a^2/5 \tag{5.192}$$

$$h_{32} = -\,(10\,\sigma^4\,a^4 - 27\,\sigma^2\,a^2 + 15)b\,[5\,\sigma(\sigma^2\,a^2 - 1)]^{-1} \tag{5.193}$$

$$h_{34} = 2\,\sigma\,b\,a^4\,f_c^*[5(\sigma^2\,a^2 - 1)\,(a^2 + 1)]^{-1} \tag{5.194}$$

$$h_{43} = -\,2\,f_c^*(5\,\sigma)^{-1} \tag{5.195}$$

$$h_{45} = -\,2(\sigma^2\,a^2 - 2)\,[5(\sigma^2\,a^2 - 1)]^{-1} \tag{5.196}$$

$$h_{51} = -\,2\,b\,f_c^*(\sigma^2\,a^2 - 1)\,(5\,\sigma^3)^{-1} \tag{5.197}$$

$$h_{52} = 2\,b\,f_c^*(5\,\sigma^2)^{-1} \tag{5.198}$$

$$h_{54} = b\,a^2(3\,\sigma^2\,a^2 - 5)\,[5(\sigma^2\,a^2 - 1)]^{-1} \tag{5.199}$$

$$h_{12} = h_{14} = h_{21} = h_{22} = h_{24} = h_{33} = h_{35} = h_{41} = h_{42} = h_{44} = h_{53} = h_{55} = 0 \tag{5.200}$$

$$S_4 = 3\,b(\sigma^2\,a^2 - 1) \tag{5.201}$$

$$S_5 = -\,3\,b/\sigma \tag{5.202}$$

$$S_1 = S_2 = S_3 = 0 \tag{5.203}$$

As before, we use (5.137)-(5.164) and (5.174)-(5.203) to transform (5.85)-(5.89) to an equivalent versally unfolded system. Thus, we have

$$\dot{x}_1 = -\,\sigma\,x_1 + \sigma\,x_2 + (\sigma^2\,a^2 - 1)\,(2\,\mu_1 + 2\,b\,\mu_3)x_3 + f_c^*(1 + a^2)^{-1}\,(2\,\sigma + \mu_2 + b\,\mu_4)\,(2\,\sigma)^{-1}\,x_4 + 3\,\mu_1 + 3\,b\,\mu_3 \tag{5.204}$$

$$\dot{x}_2 = -\,x_1\,x_3 + r_c(1 + \mu_2\,\sigma^{-1} + \sigma\,b\,a^2\,\mu_4)x_1 - x_2 - (2\,\mu_1 + 3\,b\,\mu_3)\sigma^{-1} \tag{5.205}$$

$$\dot{x}_3 = x_1\,x_2 - b\,x_3 \tag{5.206}$$

$$\dot{x}_4 = - x_1 x_5 - f_c^*(2 \sigma + \mu_2 + b \mu_4) (2 \sigma)^{-1} x_1 - \sigma x_4 \tag{5.207}$$

$$\dot{x}_5 = x_1 x_4 - \sigma b a^2 x_5 \tag{5.208}$$

Upon comparison of (5.95)-(5.99) with (5.204)-(5.208) we find that (for $h = 0$)

$$f^* = f_c^*[1 + (\mu_2 + b \mu_4) (2 \sigma)^{-1}] \tag{5.209}$$

$$r = r_c[1 + (\mu_2 + b \sigma^2 a^2 \mu_4)/\sigma] \tag{5.210}$$

$$q = - \pi(2 \mu_1 + 3 b \mu_3) (2\sqrt{2} \sigma)^{-1} \tag{5.211}$$

as before, but that

$$\alpha = 3 \pi^2 a(8\sqrt{2} \sigma)^{-1} (\sigma^2 a^2 - 1) (\mu_1 + b \mu_3) \tag{5.212}$$

We recall that we have replaced αr with αr_c in the inhomogeneous term of (5.95) in order to calculate (5.212).

Indeed, f^*, r, α, and q are a second set of physically relevant parameters that give an unfolding about the singularity f_c^*, r_c. From (5.170) and (5.212) we see that α and h are both proportional to the same linear combination of μ_1 and μ_3. In general, however, we must consider the effects on all unfolding functions and parameters even when checking to see if a new parameter can be substituted for a given one.

5.4.3 Other candidates. As mentioned previously, there are a total of three constant and twelve linear unfolding functions that serve as candidates for physical interpretation of the pair of unfolding functions $N_1 = [1, 0, 0, 0, 0]^T$ and $N_3 = [x_1^2, 0, 0, 0, 0]^T$. Two pairs of them were discussed earlier in this section. We find all candidates here by noting that in our applications of (5.101) to the moving of N_3 we had to solve 26 linear equations for the components g_i, h_{ij} and S_i. If we set

$$g_1 = g_{10} \tag{5.213}$$

$$g_2 = g_{20} + g_{23} x_3 + g_{25} x_5 \tag{5.214}$$

$$g_3 = g_{32} x_2 + g_{34} x_4 \tag{5.215}$$

$$g_4 = g_{40} + g_{43} x_3 + g_{45} x_5 \tag{5.216}$$

$$g_5 = g_{52} x_2 + g_{54} x_4 \tag{5.217}$$

$$h_{11} = h_{111}\, x_1 \tag{5.218}$$

$$h_{13} = h_{130} \tag{5.219}$$

$$h_{15} = h_{150} \tag{5.220}$$

$$h_{23} = h_{230} \tag{5.221}$$

$$h_{25} = h_{250} \tag{5.222}$$

$$h_{31} = h_{310} \tag{5.223}$$

$$h_{32} = h_{320} \tag{5.224}$$

$$h_{34} = h_{340} \tag{5.225}$$

$$h_{43} = h_{430} \tag{5.226}$$

$$h_{45} = h_{450} \tag{5.227}$$

$$h_{51} = h_{510} \tag{5.228}$$

$$h_{52} = h_{520} \tag{5.229}$$

$$h_{54} = h_{540} \tag{5.230}$$

then these equations are found by first inserting (5.213)-(5.230) into (5.101) and then by requiring that the coefficients of the functions vanish separately. Accordingly, we must solve

Row 1

$$\text{const}\ [0 = -\sigma\, g_{10} + \sigma\, g_{20} + f_c^*(1 + a^2)^{-1}\, g_{40} + S_1 - S_4\, r_c] \tag{5.231}$$

$$x_3\ [0 = \sigma\, g_{23} + f_c^*(1 + a^2)^{-1}\, g_{43} - b\, h_{130} + 2\, S_4/3] \tag{5.232}$$

$$x_5\ [0 = \sigma\, g_{25} + f_c^*(1 + a^2)^{-1}\, g_{45} - \sigma\, b\, a^2\, h_{150}] \tag{5.233}$$

$$x_1{}^2\ [1 = -\sigma\, h_{111}] \tag{5.234}$$

$$x_1\, x_2\ [0 = \sigma\, h_{111} + h_{130}] \tag{5.235}$$

$$x_1\, x_4\ [0 = f_c^*(1 + a^2)^{-1}\, h_{111} + h_{150}] \tag{5.236}$$

Row 2

$$\text{const}\ [0 = r_c\ g_{10} - g_{20} + S_5] \tag{5.237}$$

$$x_3\ [0 = -\ g_{10} - g_{23} - b\ h_{230}] \tag{5.238}$$

$$x_5\ [0 = -\ g_{25} - \sigma\ b\ a^2\ h_{250}] \tag{5.239}$$

$$x_1\ x_2\ [0 = -\ g_{32} + h_{230}] \tag{5.240}$$

$$x_1\ x_4\ [0 = -\ g_{34} + h_{250}] \tag{5.241}$$

Row 3

$$x_1\ [0 = g_{20} - \sigma\ h_{310} + r_c\ h_{320} - f_c^*\ h_{340} + 2\ S_1(3\ \sigma)^{-1}] \tag{5.242}$$

$$x_2\ [0 = g_{10} - b\ g_{32} + \sigma\ h_{310} - h_{320}] \tag{5.243}$$

$$x_4\ [0 = -\ b\ g_{34} + f_c^*(1 + a^2)^{-1}\ h_{310} - \sigma\ h_{340}] \tag{5.244}$$

$$x_1\ x_3\ [0 = g_{23} - h_{230}] \tag{5.245}$$

$$x_1\ x_5\ [0 = g_{25} - h_{340}] \tag{5.246}$$

Row 4

$$\text{const}\ [0 = -\ f_c^*\ g_{10} - \sigma\ g_{40}] \tag{5.247}$$

$$x_3\ [0 = -\ \sigma\ g_{43} - b\ h_{430}] \tag{5.248}$$

$$x_5\ [0 = -\ g_{10} - \sigma\ g_{45} - \sigma\ b\ a^2\ h_{450}] \tag{5.249}$$

$$x_1\ x_2\ [0 = -\ g_{52} + h_{430}] \tag{5.250}$$

$$x_1\ x_4\ [0 = -\ g_{54} + h_{450}] \tag{5.251}$$

Row 5

$$x_1 \; [0 = g_{40} - \sigma \, h_{510} + r_c \, h_{520} - f_c^* \, h_{540}] \tag{5.252}$$

$$x_2 \; [0 = - \sigma \, b \, a^2 \, g_{52} + \sigma \, h_{510} - h_{520}] \tag{5.253}$$

$$x_4 \; [0 = g_{10} - \sigma \, b \, a^2 \, g_{54} + f_c^*(1 + a^2)^{-1} \, h_{510} - \sigma \, h_{540}] \tag{5.254}$$

$$x_1 \, x_3 \; [0 = g_{43} - h_{520}] \tag{5.255}$$

$$x_1 \, x_5 \; [0 = g_{45} - h_{540}] \tag{5.256}$$

When $S_4 = 0$, then the solution of (5.231)-(5.256) is (5.115)-(5.136) and we obtain the relations (5.170)-(5.171) between (μ_1, μ_3) and (h, q). For $S_1 = 0$, the solution of (5.231)-(5.256) is instead (5.182)-(5.203) and we have the relations (5.211)-(5.212) between (μ_1, μ_3) and (α, q). However, in principle we can set $S_1 = S_4 = S_5 = 0$ if we introduce two constants s_1 and s_2 into any two of the other equations in the set (5.231)-(5.256). After solving the new resulting set for s_1 and s_2 we find alternate locations for μ_1 and μ_3 in the spectral system. New constant and linear terms in the spectral system are more easily obtained than nonlinear terms by introduction of small terms into the governing partial differential equations. Thus, we restrict our attention to only these terms when we form our list of possible alternate unfolding functions for the pair $N_1 = [1, 0, 0, 0, 0]^T$ and $N_3 = [x_1^2, 0, 0, 0, 0]^T$:

$$\begin{bmatrix}1\\0\\0\\0\\0\end{bmatrix}, \begin{bmatrix}0\\1\\0\\0\\0\end{bmatrix}, \begin{bmatrix}0\\0\\0\\1\\0\end{bmatrix}, \begin{bmatrix}x_3\\0\\0\\0\\0\end{bmatrix}, \begin{bmatrix}x_5\\0\\0\\0\\0\end{bmatrix}, \begin{bmatrix}0\\x_3\\0\\0\\0\end{bmatrix}, \begin{bmatrix}0\\x_5\\0\\0\\0\end{bmatrix}, \begin{bmatrix}0\\0\\x_1\\0\\0\end{bmatrix}, \tag{5.257}$$

$$\begin{bmatrix}0\\0\\x_2\\0\\0\end{bmatrix}, \begin{bmatrix}0\\0\\x_4\\0\\0\end{bmatrix}, \begin{bmatrix}0\\0\\0\\x_3\\0\end{bmatrix}, \begin{bmatrix}0\\0\\0\\x_5\\0\end{bmatrix}, \begin{bmatrix}0\\0\\0\\0\\x_1\end{bmatrix}, \begin{bmatrix}0\\0\\0\\0\\x_2\end{bmatrix}, \begin{bmatrix}0\\0\\0\\0\\x_4\end{bmatrix}$$

These column vectors correspond to (5.231), (5.237), (5.247), (5.232), (5.233), (5.238), (5.239), (5.242), (5.243), (5.244), (5.248), (5.249), (5.252), (5.253), and (5.254), and so far we have found physical interpretations involving only the first, second, fourth and eighth ones in the list. As we have seen, in many cases physically meaningful unfolding functions involve combinations of the ones listed above. When we must replace pairs of canonical unfolding functions with pairs drawn from a list such as (5.257), not all combinations are independent, as we saw in Section 5.4.2. Thus calculations such as those displayed in this section are invaluable for identifying the physically independent effects.

We note that the coefficient equations that were solved to produce the equivalent forms for N_2 and N_4 in Section 5.4.1 originated from a set of functions different from those (5.231)-(5.256) needed above in the N_3 calculation. For example, with $N_4 = [x_1^3, 0, 0, 0, 0]^T$ and

$$g_2 = g_{22}\, x_2 \tag{5.258}$$

$$g_3 = g_{33}\, x_3 \tag{5.259}$$

$$g_4 = g_{44}\, x_4 \tag{5.260}$$

$$g_5 = g_{55}\, x_5 \tag{5.261}$$

$$h_{11} = h_{1111}\, x_1^2 + h_{110} \tag{5.262}$$

$$h_{12} = h_{120} \tag{5.263}$$

$$h_{13} = h_{131}\, x_1 \tag{5.264}$$

$$h_{14} = h_{140} \tag{5.265}$$

$$h_{15} = h_{151}\, x_1 \tag{5.266}$$

$$h_{22} = h_{220} \tag{5.267}$$

$$h_{33} = h_{330} \tag{5.268}$$

$$h_{44} = h_{440} \tag{5.269}$$

$$h_{55} = h_{550} \tag{5.270}$$

we found the solutions (5.147)-(5.164) of (5.101) from

Row 1

$$x_1\ [0 = r_c\, h_{120} - f_c^*\, h_{140} - \sigma\, h_{110}] \tag{5.271}$$

$$x_2\ [0 = -\, h_{120} + \sigma\, h_{110} + \sigma\, g_{22}] \tag{5.272}$$

$$x_4\ [0 = -\,\sigma\, h_{140} + f_c^*(1 + a^2)^{-1}\, h_{110} + f_c^*(1 + a^2)^{-1}\, g_{44} + S_2(1 + a^2)^{-1}] \tag{5.273}$$

$$x_1\, x_3\ [0 = -\, b\, h_{131} - h_{120}] \tag{5.274}$$

$$x_1\, x_5\ [0 = -\,\sigma\, b\, a^2\, h_{151} - h_{140}] \tag{5.275}$$

$$x_1^{\,3}\ [1 = -\,\sigma\, h_{1111}] \tag{5.276}$$

$$x_1^{\,2}\, x_2\ [0 = \sigma\, h_{1111} + h_{131}] \tag{5.277}$$

$$x_1^{\,2}\, x_4\ [0 = f_c^*(1 + a^2)^{-1}\, h_{1111} + h_{151}] \tag{5.278}$$

Row 2

$$x_1 \; [0 = r_c \, h_{220} + S_3] \tag{5.279}$$

$$x_2 \; [0 = - g_{22} - h_{220}] \tag{5.280}$$

$$x_1 \, x_3 \; [0 = - g_{33} - h_{220}] \tag{5.281}$$

Row 3

$$x_3 \; [0 = - b \, g_{33} - b \, h_{330}] \tag{5.282}$$

$$x_1 \, x_2 \; [0 = g_{22} + h_{330}] \tag{5.283}$$

Row 4

$$x_1 \; [0 = - f_c^* \, h_{440} - S_2] \tag{5.284}$$

$$x_4 \; [0 = - \sigma \, g_{44} - \sigma \, h_{440}] \tag{5.285}$$

$$x_1 \, x_5 \; [0 = - g_{55} - h_{440}] \tag{5.286}$$

Row 5

$$x_5 \; [0 = - \sigma \, b \, a^2 \, g_{55} - \sigma \, b \, a^2 \, h_{550}] \tag{5.287}$$

$$x_1 \, x_4 \; [0 = g_{44} + h_{550}] \tag{5.288}$$

The above partitioning of the coefficient functions into subsets yields different sets of candidate unfolding functions than those given in (5.257) as alternatives for the original unfolding function N_4; they are

$$\begin{bmatrix} x_1 \\ 0 \\ 0 \\ 0 \\ 0 \end{bmatrix}, \begin{bmatrix} x_2 \\ 0 \\ 0 \\ 0 \\ 0 \end{bmatrix}, \begin{bmatrix} x_4 \\ 0 \\ 0 \\ 0 \\ 0 \end{bmatrix}, \begin{bmatrix} 0 \\ x_1 \\ 0 \\ 0 \\ 0 \end{bmatrix}, \begin{bmatrix} 0 \\ x_2 \\ 0 \\ 0 \\ 0 \end{bmatrix}, \begin{bmatrix} 0 \\ 0 \\ x_3 \\ 0 \\ 0 \end{bmatrix}, \begin{bmatrix} 0 \\ 0 \\ 0 \\ x_1 \\ 0 \end{bmatrix}, \begin{bmatrix} 0 \\ 0 \\ 0 \\ x_4 \\ 0 \end{bmatrix}, \begin{bmatrix} 0 \\ 0 \\ 0 \\ 0 \\ x_5 \end{bmatrix} \tag{5.289}$$

This natural division of possibilities must occur because the original unfolding functions are themselves independent as a consequence of Mather's theory.

<u>5.4.4 Final comments.</u> After inspecting the form (5.83)-(5.84) of the singularity, we see that we must have $\sigma a < 1$ in order for it to exist. When $\sigma a > 1$, the singularity r_s is of cusp type only, and it is given by (5.71). In this case the branching behavior resembles that of the Lorenz model, with the critical Rayleigh number depending on the value of f^{*2}. However, the steady state polynomial is still a quintic. Thus, we would need to perform an analysis similar to that used in Section 4.3 to find a parameter that leads to a butterfly point for the case $\sigma a > 1$. Because introduction of either α or q does not alter the requirement $\sigma a < 1$ for the existence of a butterfly point, we conclude that a different parameter would be needed. These results suggest that the branching behavior within axisymmetric rotating convection is fundamentally different in the two cases $\sigma a < 1$ and $\sigma a > 1$ because different parameters are needed to produce butterfly points. This conclusion could be checked in a laboratory experiment.

CHAPTER 6

STABILITY AND UNFOLDINGS

In Chapter 2 we described contact transformations and unfoldings as devices for discovering the most general extension of the stationary phase portrait of a differential equation. In our definition of the term stationary phase portrait in Section 2.1, we were deliberately vague, and by it we meant only information regarding the location of the stationary states of a parameterized differential equation. Consequently, the stationary phase portrait does not include any direct information about the (local) stability of the stationary states: which states are stable, where states lose stability, what new states replace old ones, or when these new ones are themselves stable. Certainly, when two steady states meet, we know that an eigenvalue vanishes and that an exchange of stability might occur; in this way, some determination of stability can be made from the portrait in an *ad hoc* manner. But with the methods described in Chapter 2, there is no direct way to determine stability, and in order to add necessary stability information to the stationary phase portrait, we must include at least the location of Hopf bifurcation points (see Example 6 in Chapter 2).

In this chapter, we will present a technical refinement of the theory of Mather, which together with the older singularity theory of Thom (Levine, 1971), furnishes a computation of the most general extension of the stationary phase portrait of a parameterized differential equation to the stability phase portrait of the equation; this extension is accomplished in a manner very similar to that furnished by the theory in Chapter 2. The central purpose of Chapter 2 was not to find normal forms for the right sides of differential equations (the fold, the cusp, the swallowtail, etc.); these are convenient by-products of the theory. The central purpose is to determine how to add parameters to the right side of a differential equation so as to realize all possible stationary phase portraits arising from perturbations of the original differential equation; the astonishing result of the Mather-Thom theory is that, in general, *finitely* many parameters will suffice. Of course, the theory also tells us how to add these parameters, how to detect redundant parameters, and in what sense they are redundant. The central purpose of the present chapter is exactly the same as that of Chapter 2, with the stability phase portrait replacing the stationary phase portrait. Again, we are deliberately vague in our definition of the term stability phase portrait, meaning by it all information regarding both the location and the stability of the stationary states of a parameterized differential equation.

While the new computations will be very similar to those already presented, they will be much lengthier and the ideas behind them, especially those arising from Thom's theory, perhaps unfamiliar. Because both the exposition and the resulting computations are more difficult than those of Chapter 2, it may appear that the content of this chapter is deeper than that of Chapter 2. To correct this impression, we point out that in a very real mathematical sense, the theorems of Mather used in Chapter 2 perform an infinite amount of work, which, as in that chapter, may be interpreted to accomplish its central purpose. To accomplish the central purpose of the present chapter, we only need to perform a finite amount of additional work. This distinction can be appreciated by examining the original statements and proofs of Mather's Theorems given in Mather (1968).

The principal ideas of Thom's theory employed here are, first, the notion of an invariant subset of the space of $n \times n$ matrices, second, the notion of a smooth submanifold of Euclidean space, third, the notion of tangent space, and finally, the notion of transversality.

We need this complicated array of ideas for one purpose. We intend to distinguish stable from unstable stationary points; that is, we intend to partition the set of stationary points into the subset of stable points and the subset of unstable points. Inevitably, then, we must find the boundary between these two subsets, and we must find how this neutral set behaves under suitable transformation. As we will see, the notion of transversality, supported by the other notions, is ideally suited for this purpose.

The connection to physical application will come when we restrict attention to unfoldings satisfying a certain transversality condition, and, following Chapter 2, to the corresponding contact transformations that preserve that condition. These will be the <u>versal unfoldings of the first order</u>, and the corresponding contact transformations will be the <u>first-order contact transformations</u>. The crucial results of the transversality theory will be that a first-order contact transformation carries a first-order versal unfolding to another, that in some useful sense, it carries the neutral set of the first to that of the second, and that in the same sense it preserves the stability characteristics of the two sides of the neutral set. Thus we may reduce first-order versal unfoldings to normal forms, as we did in Chapter 2 with ordinary versal unfoldings. Of course, all this is to be done in the presence of control parameters; the ordinary versal unfoldings in Chapter 2 generally will not be versal unfoldings of the first order, but may become so on addition of new, independent, parameters.

6.1 Invariant Sets of Matrices

The first concept on the above list is that of an invariant subset of the space of $n \times n$ matrices. This concept does not require as much development as does transversality, and we may see right away how it arises naturally from the study of systems of autonomous differential equations.

To illustrate this fact, we consider the effect of an arbitrary coordinate transformation on a differential equation

$$\dot{x} = f(x) \tag{6.1}$$

with x an n-column vector as usual. Let ϕ be a transformation of coordinates, so that the vector x is represented in the new system by y. Thus, we have

$$x = \phi(y) \tag{6.2}$$

and equivalently

$$y = \phi^{-1}(x) \tag{6.3}$$

where ϕ^{-1} is the functional inverse of ϕ. Then

$$f(x) = \dot{x} = d\phi(y)\cdot\dot{y} \tag{6.4}$$

so that the differential equation (6.1) in the new coordinate system is given by

$$\dot{y} = [d\phi(y)]^{-1}\cdot f\big(\phi(y)\big) = g(y) \tag{6.5}$$

This leads to no problems: because ϕ is a coordinate transformation, we know that $[d\phi(y)]^{-1}$ exists. Of course, no information is lost in such a transformation; the entire phase portrait of (6.1) may be recovered from that of (6.5), and vice versa.

The above preservation of all information is in sharp contrast to the destruction of information that occurred in Example 6 of Chapter 2; there a contact transformation altered the stability of the trivial solution and eliminated the Hopf bifurcation point on it. We may conclude that not every contact transformation is of the form given by (6.5), although $[d\phi(y)]^{-1}\cdot f\big(\phi(y)\big)$ is a contact transformation of f(x). Clearly, we may hope to find a class of transformations that includes the class of all coordinate transformations and preserves the stability information of at least certain phase portraits, but is not as general as the class of all contact transformations.

Now suppose that x_0 is a stationary solution of (6.1); then of course $x_0 = \phi(y_0)$ and y_0 is a stationary solution of (6.5). The stability of x_0 is determined by the eigenvalues of the matrix $df(x_0)$ and that of y_0 by those of $dg(y_0)$. Because the stability of x_0 is the same as that of y_0, these two matrices ought to have the same eigenvalues. Let us check that they do; we begin by using (6.5) to obtain

$$dg(y_0) = d[d\phi(y_0)]^{-1}\cdot f(x_0) + [d\phi(y_0)]^{-1}\cdot df(x_0)\cdot d\phi(y_0) \tag{6.6}$$

Because $f(x_0) = 0$, we see from (6.6) that

$$dg(y_0) = [d\phi(y_0)]^{-1}\cdot df(x_0)\cdot d\phi(y_0) \tag{6.7}$$

and that, in terms of elementary matrix theory, the two matrices $dg(y_0)$ and $df(x_0)$ are conjugate. In fact the conjugating matrix is $d\phi(y_0)$. Then, we recall from elementary matrix theory that $dg(y_0)$ and $df(x_0)$ must have the same eigenvalues, and now we are satisfied that the stability of x_0 is the same as that of y_0 as we had expected.

This discussion suggests that it makes sense to single out for special attention the sets

$$M_n = \{\Gamma | \Gamma \text{ is a real } n \times n \text{ matrix}\} \tag{6.8}$$

and

$$GL_n = \{\Lambda | \Lambda \in M_n \text{ and } \Lambda \text{ is invertible}\}, \tag{6.9}$$

and to say that a subset $S \subset M_n$ is invariant if $\Lambda^{-1} \Gamma \Lambda \in S$ whenever $\Gamma \in S$ and $\Lambda \in GL_n$. That is, S contains all the conjugates of any of its members. The smallest invariant set containing a given matrix Γ is called the orbit of Γ, and we will denote it by

$$\text{Orb}(\Gamma) = \{\Lambda^{-1} \Gamma \Lambda | \Lambda \in GL_n\} \tag{6.10}$$

It consists of all the matrices having the same eigenvalue decomposition (or real Jordan form) as Γ. The natural subsets of M_n are invariant, as we illustrate in the following example; examples of sets $\text{Orb}(\Gamma)$ are most clearly understood after we discuss submanifolds in the next section (see Example 6).

Example 1. Some invariant subsets of M_2

Let

$$\Sigma = \{\Gamma \in M_2 | \det(\Gamma) = 0\} \tag{6.11}$$

$$T_w = \{\Gamma \in M_2 | \text{tr}(\Gamma) = 2w\} \tag{6.12}$$

and

$$H = \{\Gamma \in M_2 | \text{the eigenvalues of } \Gamma \text{ are pure imaginary}\} \tag{6.13}$$

Members of Σ have a zero eigenvalue and hence might represent differential equations whose steady solutions exhibit stability exchanges with other steady solutions; members of H have pure imaginary eigenvalues and might represent differential equations whose steady solutions exhibit stability exchanges with temporally periodic solutions. As a result, these are natural subsets of M_2 to study for stability preservation.

We have already recalled that conjugation preserves eigenvalues so that H is invariant. With the two formulas from linear algebra,

$$\mathrm{tr}(\Lambda_1\Lambda_2) = \mathrm{tr}(\Lambda_2\Lambda_1) \tag{6.14}$$

$$\det(\Lambda_1\Lambda_2) = \det(\Lambda_1)\,\det(\Lambda_2) \tag{6.15}$$

we see that

$$\begin{aligned} \mathrm{tr}[\Lambda^{-1}(\Gamma\Lambda)] &= \mathrm{tr}[(\Gamma\Lambda)\Lambda^{-1}] \\ &= \mathrm{tr}[\Gamma(\Lambda\ \Lambda^{-1})] \\ &= \mathrm{tr}[\Gamma] \end{aligned} \tag{6.16}$$

and that

$$\begin{aligned} \det[\Lambda^{-1}(\Gamma\Lambda)] &= \det[\Lambda^{-1}]\ \det[\Gamma]\ \det[\Lambda] \\ &= \det[\Lambda^{-1}\ \Lambda]\ \det[\Gamma] \\ &= \det[\Gamma] \end{aligned} \tag{6.17}$$

so that Σ and T_W are invariant. Consequently, we also have that $\Sigma \cap T_W$ is invariant. Upon writing

$$\Gamma = \begin{bmatrix} a & b \\ c & d \end{bmatrix} \tag{6.18}$$

we discover that M_2 is a 4-dimensional linear vector space and that T_0 is the 3-dimensional subspace given by setting a + d = 0. Instead of using the coordinates a, b, c, d to specify elements of M_2, we may introduce new coordinates by setting

$$\Gamma = \begin{bmatrix} w + y & x + z \\ x - z & w - y \end{bmatrix} \tag{6.19}$$

and we use these to identify M_2 with R^4 by identifying Γ with the column vector [x, y, z, w] via (for convenience, we no longer write it as a transpose)

$$\Gamma \leftrightarrow \begin{bmatrix} x \\ y \\ z \\ w \end{bmatrix} \tag{6.20}$$

Now Σ is identified with the 3-dimensional hypersurface given by

$$w^2 + z^2 = x^2 + y^2 \tag{6.21}$$

and T_0 is identified with the linear space R^3 given by

$$w = 0 \tag{6.22}$$

The constant w_0-section $w = w_0$ of M_2 is then identified via (6.12) with the translate T_{w_0} of R^3 in R^4

$$T_{w_0} \leftrightarrow \left\{ \begin{bmatrix} x \\ y \\ z \\ w_0 \end{bmatrix} \middle| \; x, y, z \in R \right\} \tag{6.23}$$

and the section $\Sigma \cap T_{w_0}$ of Σ consists of two cup-shaped surfaces (Fig. 6.1). As the value of w_0 decreases to zero, the top and bottom cups approach each other until, when $w_0 = 0$, they touch at the origin and their sides straighten (Fig. 6.2). Thus, $\Sigma \cap T_0$ is identified with a double cone. Finally, as the value of w_0 decreases from 0, we obtain the same sectional picture as that shown in Fig. 6.1. Clearly, each of these surfaces $\Sigma \cap T_{w_0}$ cuts the translate of R^3 containing the surface into three regions, two of which are convex. The points in the convex regions represent those matrices having complex eigenvalues, and the points in the remaining region represent those matrices having real eigenvalues (Fig. 6.1). In particular, the set H (6.13) is represented by the points within the double cone at the level $w = 0$ (Fig. 6.2).

Conjugation by a fixed matrix defines a map

$$\Gamma \to \Gamma' \tag{6.24}$$

that carries each level $w = w_0$ into itself because a matrix and any of its conjugates have the same trace. Pictorially, if Γ is represented by a point in Fig. 6.1, then so is Γ'.

For example, an easy calculation shows that conjugation of Γ by the rotation matrix

$$\mathrm{Rot}(\theta) = \begin{bmatrix} \cos\theta & \sin\theta \\ -\sin\theta & \cos\theta \end{bmatrix} \tag{6.25}$$

will be represented in R^4 by rotating the x and y coordinates through an angle 2θ and by leaving the z and w coordinates fixed. That is, we have

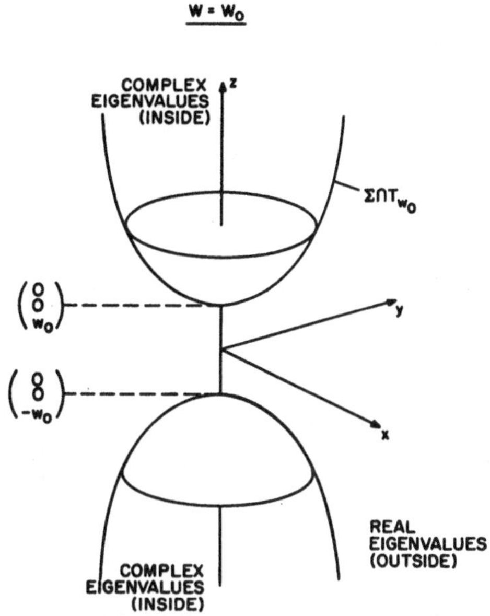

Fig. 6.1 Representation of the constant w_0-section of M_2 that is partitioned into three regions by the hypersurfaces $\Sigma \cap T_{w_0}$ and $\Sigma \cap T_{-w_0}$. Inside the cup-shaped regions are points representing matrices that have a conjugate pair of complex eigenvalues; outside are points representing matrices that have two real roots; and on the boundaries are points representing matrices that have two equal real roots.

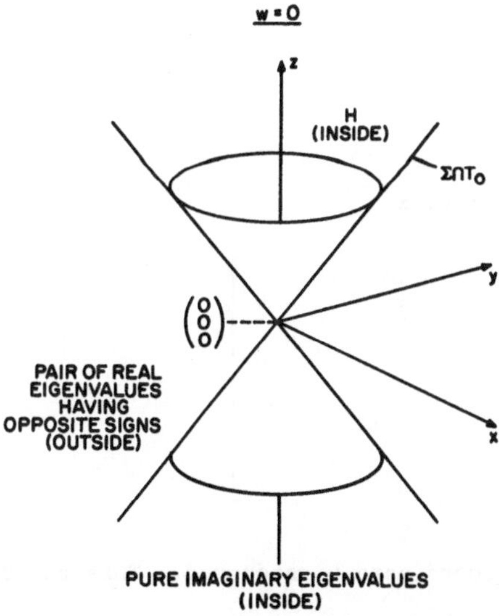

Fig. 6.2 Representation of the w = 0 section of M_2 that is partitioned into three regions by the hypersurface $\Sigma \cap T_0$. Inside the double cone are points of H representing matrices that have pure imaginary eigenvalues; points outside represent matrices that have two eigenvalues of equal magnitude but opposite signs; and points on the boundary represent matrices that have two zero eigenvalues.

$$\Gamma' = \left[\mathrm{Rot}(\theta)\right]^{-1}\cdot\Gamma\cdot\mathrm{Rot}(\theta) \tag{6.26}$$

and

$$\Gamma' = \left[\mathrm{Rot}(\theta)\right]^{-1}\begin{bmatrix} w+y & x+z \\ x-z & w-y \end{bmatrix}\mathrm{Rot}(\theta) \tag{6.27}$$

$$= \begin{bmatrix} w + y\cos(2\theta) - x\sin(2\theta) & y\sin(2\theta) + x\cos(2\theta) + z \\ y\sin(2\theta) + x\cos(2\theta) - z & w - y\cos(2\theta) + x\sin(2\theta) \end{bmatrix}$$

$$\leftrightarrow \begin{bmatrix} x\cos(2\theta) + y\sin(2\theta) \\ y\cos(2\theta) - x\sin(2\theta) \\ z \\ w \end{bmatrix} = \begin{bmatrix} \mathrm{Rot}(2\theta) & 0 \\ 0 & 1 \end{bmatrix}\begin{bmatrix} \begin{bmatrix} x \\ y \end{bmatrix} \\ \begin{bmatrix} z \\ w \end{bmatrix} \end{bmatrix}$$

In Fig. 6.1, this operation corresponds to a rotation about the z-axis. This rotation clearly carries the interior of either cup to itself, the exterior to itself, and the surface to itself. In particular, matrices having complex conjugate eigenvalues are carried to other such matrices; this correspondence also holds for matrices having real eigenvalues, and for ones having repeated eigenvalues.

However, the above pleasant result contrasts with the fact that another easy calculation shows that conjugation of Γ by the shear matrix

$$S_\gamma = \begin{bmatrix} 1 & \gamma \\ 0 & 1 \end{bmatrix} \tag{6.28}$$

is represented in R^4 by the matrix

$$U_\gamma = \begin{bmatrix} 1 - \frac{\gamma^2}{2} & \gamma & \frac{\gamma^2}{2} & 0 \\ -\gamma & 1 & \gamma & 0 \\ -\frac{\gamma^2}{2} & \gamma & 1 + \frac{\gamma^2}{2} & 0 \\ 0 & 0 & 0 & 1 \end{bmatrix} \tag{6.29}$$

that leaves only the last coordinate w untouched. This can be shown via (6.19) and (6.20) with the calculation

$$S_\gamma^{-1} \cdot \Gamma \cdot S_\gamma = \begin{bmatrix} w + y - \gamma(x - z) & x(1 - \gamma^2) + y(2\gamma) + z(1 + \gamma^2) \\ x - z & w - y + \gamma(x - z) \end{bmatrix} \tag{6.30}$$

$$\leftrightarrow \begin{bmatrix} (1 - \frac{\gamma^2}{2})\, x + \gamma y + \frac{\gamma^2}{2}\, z \\ y - \gamma x + \gamma z \\ -\frac{\gamma^2}{2}\, x + \gamma y + (1 + \frac{\gamma^2}{2}) z \\ w \end{bmatrix}$$

$$= U_\gamma \begin{bmatrix} x \\ y \\ z \\ w \end{bmatrix}$$

Once again we see algebraically from the preservation of trace and determinant by conjugation that the levels $w = w_0$ are preserved and that within each of these, the exterior, interior, and surfaces of the cups in Fig. 6.1 are each preserved. However, in the case of conjugation by the shear matrix S_γ, it is not as geometrically obvious as it was in the case of conjugation by the rotation matrix $Rot(\theta)$. #

6.2 Smooth Submanifolds of R^n

We see from Example 1 that even the most natural invariant subsets of M_2 will have fairly complicated structures. At most points, Σ is clearly a 3-dimensional hypersurface embedded in M_2. However, the trivial matrix $\begin{bmatrix} 0 & 0 \\ 0 & 0 \end{bmatrix}$ is a member of Σ, and at that point, Σ is obviously more complicated. Clearly this complication is related to the greater difficulties encountered in Chapter 2 with corank 2 singularities. Common sense suggests that we exclude points such as the trivial matrix from initial consideration and isolate the characterizing property of the other, simpler points. This property is embodied in the following definition.

Definition. A smooth k-submanifold of R^n is a subset that locally may be described smoothly and reversibly by k parameters whose values range over an open subset of R^k. Then we say that k is the dimension of the submanifold.

This definition simply generalizes from advanced calculus the notion of a surface in R^3. We will indicate a definition in better mathematical taste below.

Now follow three very simple examples, one of a set that is a smooth submanifold of R^3, another of a set that cannot be a submanifold, and finally one of a set that is a submanifold but cannot be a smooth submanifold.

Example 2. The sphere: a 2-submanifold of R^3

Let

$$S^2 = \{(x, y, z) \mid x^2 + y^2 + z^2 = 1\} \tag{6.31}$$

Let $(x_o, y_o, z_o) \in S^2$ be any point with $z_o \neq 1$. Then we may use polar coordinates to parameterize S^2 near (x_o, y_o, z_o). For suitable $0 < \theta_o < \pi$ and $0 < \phi_o < \pi$ with ϕ_o not an integral multiple of π, we have

$$\left.\begin{aligned} x_o &= \sin\phi_o \cos\theta_o \\ y_o &= \sin\phi_o \sin\theta_o \\ z_o &= \cos\phi_o \end{aligned}\right\} \tag{6.32}$$

Then the function defined by

$$\left.\begin{aligned} x(\theta, \phi) &= \sin\phi \cos\theta \\ y(\theta, \phi) &= \sin\phi \sin\theta \\ z(\theta, \phi) &= \cos\phi \end{aligned}\right\} \tag{6.33}$$

having as domain the open subset of R^2 given by

$$O = \{(\theta, \phi) \mid |\theta - \theta_o| < \pi, |\phi - \phi_o| < \min(\frac{\phi_o}{2}, \frac{\pi - \phi_o}{2})\} \tag{6.34}$$

is a parameterization for S^2. We have to check that x, y, z depend smoothly on θ, ϕ, and that θ, ϕ depend smoothly on x, y, z. The smoothness in the first case obviously follows from that of the functions sine and cosine; the smoothness in the second follows from the smoothness of the principal part function arccotangent, which has domain $(-\infty, \infty)$ and range $(0, \pi)$. In fact we have,

$$\left.\begin{aligned} \phi &= \operatorname{arccot}\left[z(x^2 + y^2)^{-1/2}\right] \\ \theta &= \operatorname{arccot}\left[xy^{-1}\right] \end{aligned}\right\} \tag{6.35}$$

To find a parameterization for S^2 near an arbitrary point (x_o, y_o, z_o), we may use polar coordinates with respect to axes which have been rotated so that (x_o, y_o, z_o) is the intersection of the new y-axis with the sphere S^2.

A parameterization near a point is not determined uniquely by that point; a useful parameterization near any point in the upper hemisphere of S^2 is also given by the function

$$\left.\begin{aligned} x(u, v) &= u \\ y(u, v) &= v \\ z(u, v) &= (1 - u^2 - v^2)^{1/2} \end{aligned}\right\} \qquad (6.36)$$

with domain

$$O' = \{(u, v) \mid u^2 + v^2 < 1\} \qquad (6.37)$$

#

Example 3. The double cone: a subset which is not a submanifold of R^3
Subsets of Euclidean space may fail to be submanifolds for a variety of reasons. Here we examine one such reason. Let

$$D = \{(x, y, z) \mid x^2 + y^2 = z^2\} \qquad (6.38)$$

Of course $(0, 0, 0) \in D$; if $(x_0, y_0, z_o) \neq (0, 0, 0)$, then we may parameterize D near (x_0, y_0, z_0) fairly easily. Suppose $z_0 > 0$; then we define a smooth function with domain

$$O = \{(u, v) \mid (u - x_0)^2 + (v - y_0)^2 < x_0{}^2 + y_0{}^2\} \qquad (6.39)$$

by setting

$$\left.\begin{aligned} x(u, v) &= u \\ y(u, v) &= v \\ z(u, v) &= (u^2 + v^2)^{1/2} \end{aligned}\right\} \qquad (6.40)$$

For $(x, y, z) \in D$ near (x_0, y_0, z_0), we may recover (u, v) from (x, y, z) by setting

$$\left.\begin{aligned} u(x, y, z) &= x \\ v(x, y, z) &= y \end{aligned}\right\} \qquad (6.41)$$

Thus $D - \{(0, 0, 0)\}$, the set D with the point (0, 0, 0) deleted, is a 2-submanifold of R^3.

However, D itself is not a 2-submanifold because there is no parameterization of D near (0, 0, 0). To see this fact, assume that we have a parameterization of D near (0, 0, 0) given by the function P_D

$$\left.\begin{aligned} x &= X(u, v) \\ y &= Y(u, v) \\ z &= Z(u, v) \end{aligned}\right\} \qquad (6.42)$$

with domain 0; that is, in (6.42),

$$(u, v) \in 0 \qquad (6.43)$$

Let (u_o, v_o) be the point of 0 carried to (0, 0, 0); that is,

$$0 = X(u_o, v_o) = Y(u_o, v_o) = Z(u_o, v_o) \qquad (6.44)$$

We may suppose that

$$0 = \{(u, v) \mid (u - u_o)^2 + (v - v_o)^2 < \varepsilon^2\} \qquad (6.45)$$

for some $\varepsilon > 0$. Then the function given by (6.42) carries $0 - \{(u_o, v_o)\}$ onto $N - \{(0, 0, 0)\}$ where N is a small neighborhood of (0, 0, 0) in D, so that $N - \{(0, 0, 0)\}$ meets both the upper and the lower cone (Fig. 6.3). Let

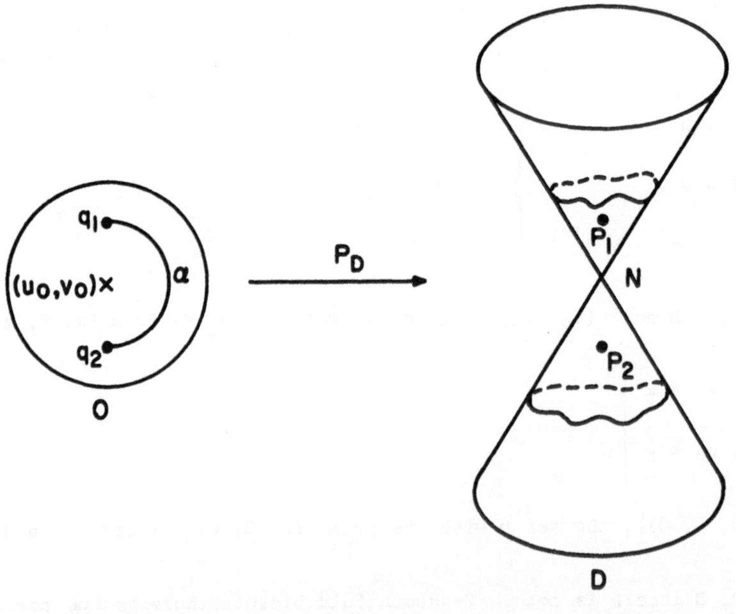

Fig. 6.3 The hypothesized parameterization P_D (6.42) that carries a curve α joining q_1 and q_2 in 0 (6.39) to a curve joining p_1 and p_2 in the subset $N - \{(0, 0, 0)\}$ of the double cone D. No curve is shown in N because, in order to join p_1 and p_2, it would have to pass through (0, 0, 0), a point which has been deleted.

$p_1 = (x_1, y_1, z_1)$ and $p_2 = (x_2, y_2, z_2)$ be points of N in the upper and lower cones respectively. Let $q_1 = (u_1, v_1)$ and $q_2 = (u_2, v_2)$ be the points in O carried to p_1 and p_2 respectively by P_D (6.42). Clearly, there is a curve α in $O - \{(u_o, v_o)\}$ joining q_1 to q_2. Then P_D (6.42) carries α to a curve β in $N - \{(0, 0, 0)\}$ joining p_1 to p_2. However, it is clear from Fig. 6.3 that <u>any</u> curve in D joining p_1 to p_2 must run through the point (0, 0, 0) and so the curve cannot lie entirely in $N - \{(0, 0, 0)\}$: the projection on the z-axis of such a curve would be a path joining $z_1 > 0$ to $z_2 < 0$, staying in the z-axis but skipping 0! Consequently there exists no parameterization P_D of D, and D is not a 2-submanifold of R^3. #

<u>Example 4. The cone: a subset which is not a smooth submanifold of R^3</u>

A more subtle reason for the failure of the smooth submanifold property is represented by the cone

$$C = \{(x, y, z) \mid x^2 + y^2 = z^2, \quad z \geq 0\} \tag{6.46}$$

As with the double cone in Example 3, it is easy to find a parameterization of C near $(x_o, y_o, z_o) \neq (0, 0, 0)$. But here the function with domain R^2 and given by

$$\left.\begin{aligned} x &= u \\ y &= v \\ z &= (u^2 + v^2)^{1/2} \end{aligned}\right\} \tag{6.47}$$

appears to be a parameterization. It fails to be a <u>smooth</u> parameterization because $z = (u^2 + v^2)^{1/2}$ is not <u>differentiable</u> at (0, 0).

Is it possible that there is some other function that smoothly parameterizes C near (0, 0, 0)? To see that the answer is <u>no</u>, suppose that we have a smooth parameterization P_C of C near (0, 0, 0),

$$\left.\begin{aligned} x &= X(u, v) \\ y &= Y(u, v) \\ z &= Z(u, v) \end{aligned}\right\} \tag{6.48}$$

with domain an open subset O of R^2; that is, in (6.48)

$$(u, v) \in O \tag{6.49}$$

and

$$0 = X(u_o, v_o) = Y(u_o, v_o) = Z(u_o, v_o) \tag{6.50}$$

Then any straight line segment β through (u_0, v_0) in O is carried by (6.48) into a smooth curve in R^3, lying in C and running through the vertex (0, 0, 0) (Fig. 6.4). Intuitively, such a curve is impossible because it must have a corner at (0, 0, 0) and so cannot be smooth. #

Digression. An inescapable concern is that definitions be clear and precise, and that alleged facts be tested against some standard of proof. The definitions and arguments above are heuristically, but not mathematically, legitimate. One path to mathematical legitimacy starts with the notion of a smooth map from an arbitrary subset of one Euclidean space to another. Let $A \subset R^n$ and $B \subset R^m$ be subsets; we say that a map

$$f: \quad A \to B \tag{6.51}$$

is smooth if there is a map

$$F: \quad R^n \to R^m \tag{6.52}$$

that is smooth in the usual sense of advanced calculus (i.e. infinitely differentiable), and for which

$$a \in A \text{ implies } f(a) = F(a) \tag{6.53}$$

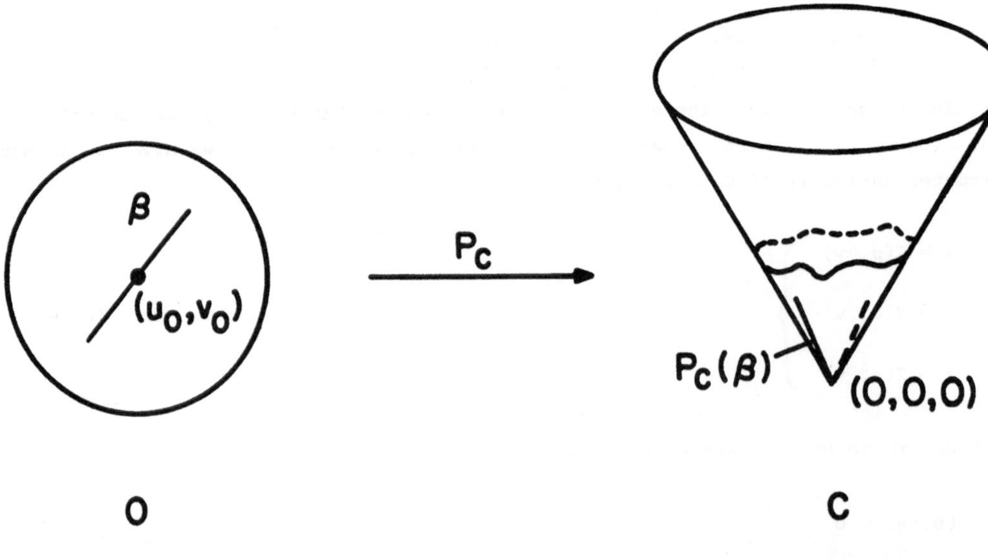

Fig. 6.4 The hypothesized parameterization P_C (6.48) that carries a curve β through (u_0, v_0) in O to a differentiable curve $P_C(\beta)$ through (0, 0, 0) on the cone C. Clearly, $P_C(\beta)$ must be nondifferentiable at the vertex (0, 0, 0) because the cone forms a corner at this point.

We then say that $f: A \to B$ is a diffeomorphism if there is also a smooth map $g: B \to A$ such that

$$\left.\begin{aligned} g\big(f(a)\big) &= a \quad \text{for all } a \in A \\ f\big(g(b)\big) &= b \quad \text{for all } b \in B \end{aligned}\right\} \tag{6.54}$$

We say that the subsets A and B are diffeomorphic in this case. Finally, we say that M is a smooth k-submanifold of R^n if, for each $x \in M$, there is an open subset U of R^n such that $x \in U \cap M$ and such that $U \cap M$ is diffeomorphic to some open subset V of R^k. Then there is a diffeomorphism $f: V \to U \cap M$, which is the parameterization of M near x in our heuristic definition. By continuing along this line of definition and argument, we could develop the theory of smooth manifolds efficiently and rigorously. That is not our purpose here; for a fully detailed exposition, see Milnor (1965).

Now we are in a position to say what we mean by an invariant submanifold of M_n. We may identify M_n with R^{n^2}, and then an invariant submanifold of M_n is simply an invariant subset of M_n that is also a smooth submanifold of R^{n^2}.

Example 5. Invariant submanifolds

Referring to Example 1, we see that Σ is not an invariant submanifold of R^4 (cf Figs. 6.2 and 6.3.) Let $0 \in \Sigma$ be the zero matrix; then $\Sigma - \{0\}$ is an invariant submanifold of R^4. T_w is an invariant submanifold of R^4 for every w, and H is an invariant submanifold of R^4. Of course these are three-dimensional sets so that

$$3 = \dim(\Sigma - \{0\}) = \dim(T_w) = \dim(H) \tag{6.55}$$

Another invariant set is cl(H) − H, which is the closure cl(H) of H with H deleted. This consists of all 2 × 2 matrices having both eigenvalues zero; a sample member is $\begin{bmatrix} 0 & 1 \\ 0 & 0 \end{bmatrix}$. It is easy to see that $cl(H) - H - \{0\}$ is a one-dimensional invariant submanifold of R^4. #

A much more interesting and difficult problem is to show that $Orb(\Gamma)$ defined in (6.10) is actually a submanifold of M_n.

Example 6. The orbit of a matrix

If Γ is an $n \times n$ matrix, then we have defined in (6.10) the subset

$$Orb(\Gamma) = \{\Lambda^{-1}\, \Gamma\, \Lambda \mid \Lambda \in GL_n\} \tag{6.56}$$

in M_n. It is the smallest invariant set containing Γ, and it is called the orbit of Γ. We wish to sketch in part the rather involved argument that it is a smooth submanifold of M_n. For simplicity, we assume that Γ may be diagonalized over the real numbers. Thus, there exists a real invertible matrix Λ_1 such that

$$\Gamma_1 = \Lambda_1^{-1} \Gamma \Lambda_1 = \begin{bmatrix} D_1 & & & \\ & D_2 & & \\ & & \ddots & \\ & & & D_r \end{bmatrix} \qquad (6.57)$$

where the square block D_k is the $n_k \times n_k$ scalar matrix with real diagonal entry λ_k. We have collected the diagonal entries so that $\lambda_k \neq \lambda_\ell$ for $k \neq \ell$. It is clear that

$$n_1 + \cdot \cdot \cdot + n_r = n \qquad (6.58)$$

Now, we notice that a matrix Γ_0 has the same eigenvalues λ_1 ..., λ_r, with each of them simple and with λ_k of multiplicity n_k, as Γ_1 if and only if there is an invertible matrix Λ_0 such that

$$\Lambda_0^{-1} \Gamma_0 \Lambda_0 = \Gamma_1 \qquad (6.59)$$

But then

$$\Gamma_0 = (\Lambda_1 \Lambda_0^{-1})^{-1} \Gamma (\Lambda_1 \Lambda_0^{-1}) \in \text{Orb}(\Gamma) \qquad (6.60)$$

and conversely $\Gamma \in \text{Orb}(\Gamma_0)$. Thus we have identified the elements of $\text{Orb}(\Gamma)$: They are exactly those matrices whose eigenvalues are the simple ones $\lambda_1, \ldots, \lambda_r$, in which the multiplicity of each λ_k is equal to n_k. However, such a matrix is completely determined by the list $V_1, \ldots, V_r$ of its eigenspaces, where V_k is the eigenspace of λ_k and has dimension n_k. These eigenspaces are independent and together span R^n. Conversely again, any such list determines one of the matrices in $\text{Orb}(\Gamma)$. Now we may let $(V_1^0, \ldots, V_r^0)$ be the list uniquely associated with Γ_0.

Our problem has become that of locally parameterizing near $(V_1^0, \ldots, V_r^0)$ the set of all such lists

$$O = \{(V_1, \ldots, V_r) \mid V_1, \ldots, V_r \text{ are independent vector subspaces of } R^n \text{ with } \dim V_k = n_k\} \qquad (6.61)$$

By solving this problem, we will solve our original problem of locally parameterizing $\text{Orb}(\Gamma)$ near Γ_0.

We begin by seeing how to locally parameterize the set

$$G_m = \{V \mid V \text{ is an m-dimensional vector subspace of } R^n\} \qquad (6.62)$$

because the device for parameterizing O is essentially the same. We wish to identify each m-dimensional vector subspace of R^n near to V_0 with a unique numerical label; that is, with a label which is essentially a list of numbers. Moreover, we wish to do so in such a way that <u>any</u> such list of numbers, whose entries lie within pre-arranged bounds, corresponds in turn to an m-dimensional vector subspace of R^n near V_0. Where are we to find such a list of numbers? The orthogonal complement W_0 of V_0 in R^n is an (n - m)-dimensional vector subspace of R^n; accordingly, after choosing a basis, we may identify the points of W_0 with unconstrained lists of n - m numbers, or with (n - m)-tuples.

We will make a construction using linear algebra, which assigns an m-dimensional vector subspace of R^n to each ordered m-tuple of points of W_0. Because we have identified points of W_0 with (n - m)-tuples of numbers, we are really assigning an m-dimensional vector subspace to each m-tuple of (n - m)-tuples of numbers; that is, to each unconstrained $m \times (n - m)$ matrix of numbers. This then is our goal, to assign an m-dimensional vector subspace to each $m \times (n - m)$ matrix in a one-one way; however, we may use m-tuples of points of W_0 instead of $m \times (n - m)$ matrices. Now we turn to our device for actually doing so.

We are given V_0 and W_0, its orthogonal complement; we choose a basis $e_1, \ldots, e_m$ for V_0. Now let any m points $x_k \in W_0$, k = 1, ..., m be given (see Fig. 6.5 for the case n = 3, m = 2). The m points

$$\varepsilon_1 = e_1 + x_1, \ldots, \varepsilon_m = e_m + x_m \tag{6.63}$$

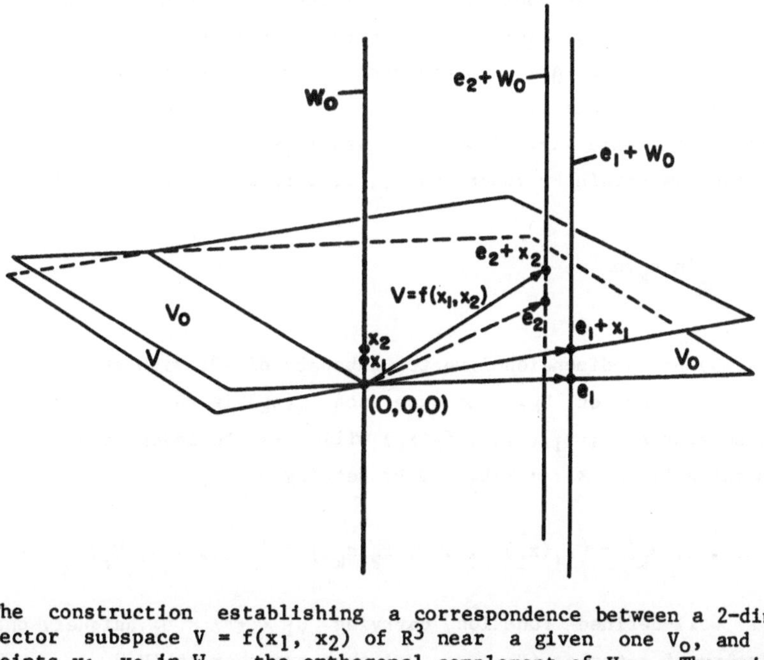

Fig. 6.5 The construction establishing a correspondence between a 2-dimensional vector subspace $V = f(x_1, x_2)$ of R^3 near a given one V_0, and pairs of points x_1, x_2 in W_0, the orthogonal complement of V_0. The points 0, e_1 and e_2 are independent in V_0, and the points x_1 and x_2 in W_0 are respectively the orthogonal projections on W_0 of the intersections $(e_1 + W_0) \cap V$ and $(e_2 + W_0) \cap V$.

of R^n are linearly independent and so span an m-dimensional vector subspace V of R^n. We assign V to $(x_1, \ldots, x_m)$ and we notice that this assignment is one-one; that if all x_k are near the origin of W_o, then V is near V_o; and finally that every V near V_o is assigned to some m-tuple $(x_1, \ldots, x_m)$ with all x_k near the origin of W_o. By writing

$$f(x_1, \ldots, x_m) = V \tag{6.64}$$

we have obtained a map

$$R^{m(n-m)} = \underbrace{W_o \times \cdots \times W_o}_{m \text{ factors}} \xrightarrow{f} G_m \tag{6.65}$$

which carries an open neighborhood U of the origin of $W_o \times \cdots \times W_o$ onto a neighborhood of V_o in G_m, in a one-one way. That is, f is a parameterization of G_m near W_o, and the x_k are the parameters; the points $(x_1, \ldots, x_m)$ are sometimes called Prüfer coordinates of V.

Referring to Fig. 6.5, we may repeat the construction in a more pictorial spirit: We let $0, e_1, \ldots, e_m$ be $(m + 1)$ independent points in V_o. We let W_o be the orthogonal linear space to V_o through 0, and we let $e_1 + W_o, \ldots, e_m + W_o$ be the spaces parallel to W_o, through $e_1, \ldots, e_m$ respectively. Choose points $x_1, \ldots, x_m \in W_o$ and construct for each k the rectangle with sides $(0, x_k)$ and $(0, e_k)$; write $e_k + x_k$ for the vertex opposite 0. Then the points $0, e_1 + x_1, \ldots, e_m + x_m$ are independent points, which must then span an m-dimensional linear space V.

In exactly the same way, if we are given $(V_1{}^o, \ldots, V_r{}^o) \in O$, where O is given by (6.61), then we obtain r functions $f_1, \ldots, f_r$ with open domains $U_1, \ldots, U_r$, with

$$0 \in U_k \subset R^{(n-n_k)n_k} \tag{6.66}$$

and with $f_k(z_k)$ an n_k-dimensional vector subspace of R^n, with $f_k(z_k)$ near $V_k{}^o$ for z_k a member of the open set U_k. Because each $f_k(z_k)$ is near the corresponding $V_k{}^o$, the vector subspaces $f_1(z_1), \ldots, f_r(z_r)$ will also be independent. Thus, we may define a mapping f: $U_1 \times \cdots \times U_r \to O$ by setting

$$f(z_1, \ldots, z_k) = \left(f_1(z_1), \ldots, f_k(z_k)\right) = (V_1, \ldots, V_r) \tag{6.67}$$

Then f is a well-defined function carrying $U_1 \times \cdots \times U_r$ uniquely onto a subset of O consisting of r-tuples $(V_1, \ldots, V_r)$ that are near to $(V_1{}^o, \ldots, V_r{}^o)$.

We still have not mapped a parameterizing domain into Orb(Γ); after all, Orb(Γ) is a set of matrices and not a set of r-tuples of vector spaces. We make the transition from O to Orb(Γ) by defining for each r-tuple $(V_1, \ldots, V_r) \varepsilon O$ a matrix $Mat(V_1, \ldots, V_r)$, which is completely determined by the requirement that

$$Mat(V_1, \ldots, V_r)\, x = \lambda_k\, x \quad \text{for all } x \,\varepsilon\, V_k, \quad k = 1, \ldots, r \tag{6.68}$$

In other words, if $e_1, \ldots, e_n$ is a basis for R^n with

$$e_{n_k+1}, \ldots, e_{n_{k+1}} \;\varepsilon\; V_k \tag{6.69}$$

and Λ is the n × n matrix whose columns are $e_1, \ldots, e_n$,

$$\Lambda = [e_1, \ldots, e_n] \tag{6.70}$$

then

$$\Lambda^{-1}\, Mat(V_1, \ldots, V_n)\, \Lambda = \begin{bmatrix} D_1 & & \\ & \ddots & \\ & & D_r \end{bmatrix} \tag{6.71}$$

or equivalently

$$Mat(V_1, \ldots, V_n) = \Lambda \begin{bmatrix} D_1 & & \\ & \ddots & \\ & & D_r \end{bmatrix} \Lambda^{-1} \tag{6.72}$$

Finally then,

$$g(z_1, \ldots, z_r) = Mat\big(f_1(z_1), \ldots, f_r(z_r)\big) \tag{6.73}$$

defines a parameterization of Orb(Γ) near Γ_o. We conclude from our parameterization, with the aid of (6.58) that

$$\begin{aligned} \dim\,[Orb(\Gamma)] &= n_1(n - n_1) + \cdots + n_r(n - n_r) \\ &= n^2 - n_1{}^2 - \cdots - n_r{}^2 \end{aligned} \tag{6.74}$$

For a matrix which is not diagonalizable over the real numbers, we may still find a decomposition (6.57), only now a block D_k is no longer a diagonal matrix. Instead D_k is now itself a diagonal block matrix,

$$D_k = \begin{bmatrix} \delta_k & & \\ & \ddots & \\ & & \delta_k \end{bmatrix} \tag{6.75}$$

with all diagonal block entries the same real Jordan matrix and with

$$\delta_k \neq \delta_\ell \quad \text{for} \quad k \neq \ell \tag{6.76}$$

Essentially the same argument as above will go through with the result that when the eigenvalues of Γ are all simple, the real ones appearing with multiplicities $n_1, \ldots, n_r$ respectively, and the complex ones with multiplicities $m_1, \ldots, m_s$ respectively, we have

$$\dim\left[\mathrm{Orb}(\Gamma)\right] = n^2 - n_1^{\ 2} - \cdot \cdot \cdot \cdot \cdot - n_r^{\ 2} - 2m_1^{\ 2} - \cdot \cdot \cdot \cdot \cdot - 2m_s^{\ 2} \tag{6.77}$$

This is the case in which only trivial Jordan forms appear, and it is all we shall need. The case in which non-trivial Jordan forms appear is considerably more complicated, and we by-pass it. #

That $\mathrm{Orb}(\Gamma)$ is a submanifold of M_n of known dimension is all we need later in this chapter, but a glance at particular small-dimensional examples of these objects will help clarify the underlying concepts. In the following we discuss four orbits in the set M_2 of 2×2 real matrices.

Example 7. Some orbits in M_2

We begin by considering $\mathrm{Orb}(\Gamma_1)$, where the matrix Γ_1 has eigenvalues 1 and -1. Then $\mathrm{Orb}(\Gamma_1)$ will consist of all 2×2 matrices A having these eigenvalues, and this may be expressed by

$$\mathrm{Orb}(\Gamma_1) = \{A \mid \mathrm{tr}(A) = 0, \ \det(A) = -1\} \tag{6.78}$$

That is, $\mathrm{Orb}(\Gamma_1)$ is a subset of T_0 defined in (6.12) in Example 1. Writing A in the form (6.19) as

$$A = \begin{bmatrix} w + y & x + z \\ x - z & w - y \end{bmatrix} \tag{6.79}$$

we see that $\mathrm{Orb}(\Gamma_1)$ may be identified with the subset O_1 of R^4 given by

$$O_1 = \{(x, y, z, w) \mid w = 0, \ z^2 = x^2 + y^2 - 1\} \tag{6.80}$$

Identifying R^3 with the subspace defined by $w = 0$, we see that O_1 lies entirely in R^3 and is a single-sheeted hyperboloid of rotation there (Fig. 6.6a).

Our second example is $\mathrm{Orb}(\Gamma_2)$, where Γ_2 is a 2×2 real matrix A having eigenvalues $1 \pm 2i$. Then we have

$$\mathrm{Orb}(\Gamma_2) = \{A \mid \mathrm{tr}(A) = 2, \det(A) = 5\} \tag{6.81}$$

and we see that $\mathrm{Orb}(\Gamma_2)$ may be identified with

$$O_2 = \{(x, y, z, w) \mid w = 1, z^2 = x^2 + y^2 + 4\} \tag{6.82}$$

Identifying R^3 with $w = 1$ this time, we see that O_2 is a two-sheeted hyperboloid of rotation (Fig. 6.6b)

Now let us consider the case of two real and equal simple eigenvalues. Then we may write

$$\Gamma_3 = \begin{bmatrix} 2 & 0 \\ 0 & 2 \end{bmatrix} \tag{6.83}$$

Because $\Lambda^{-1} \Gamma_3 \Lambda = \Gamma_3$ for all Λ, we conclude that $\mathrm{Orb}(\Gamma_3)$ consists of the single point Γ_3, and that it may be associated with the origin in R^3.

For our last example, we choose the nondiagonalizable matrix

$$\Gamma_4 = \begin{bmatrix} 2 & 1 \\ 0 & 2 \end{bmatrix} \tag{6.84}$$

Now the elements of $\mathrm{Orb}(\Gamma_4)$ are specified by

$$\mathrm{Orb}(\Gamma_4) = \{A \mid \mathrm{tr}(A) = 4, \det(A) = 4, A \neq \begin{bmatrix} 2 & 0 \\ 0 & 2 \end{bmatrix}\} \tag{6.85}$$

and may be identified with

$$O_4 = \{(x, y, z, w) \mid w = 2, z^2 = x^2 + y^2, z \neq 0\} \tag{6.86}$$

In this case, O_4 is precisely the double cone with the vertex at the origin removed (Fig. 6.6c). #

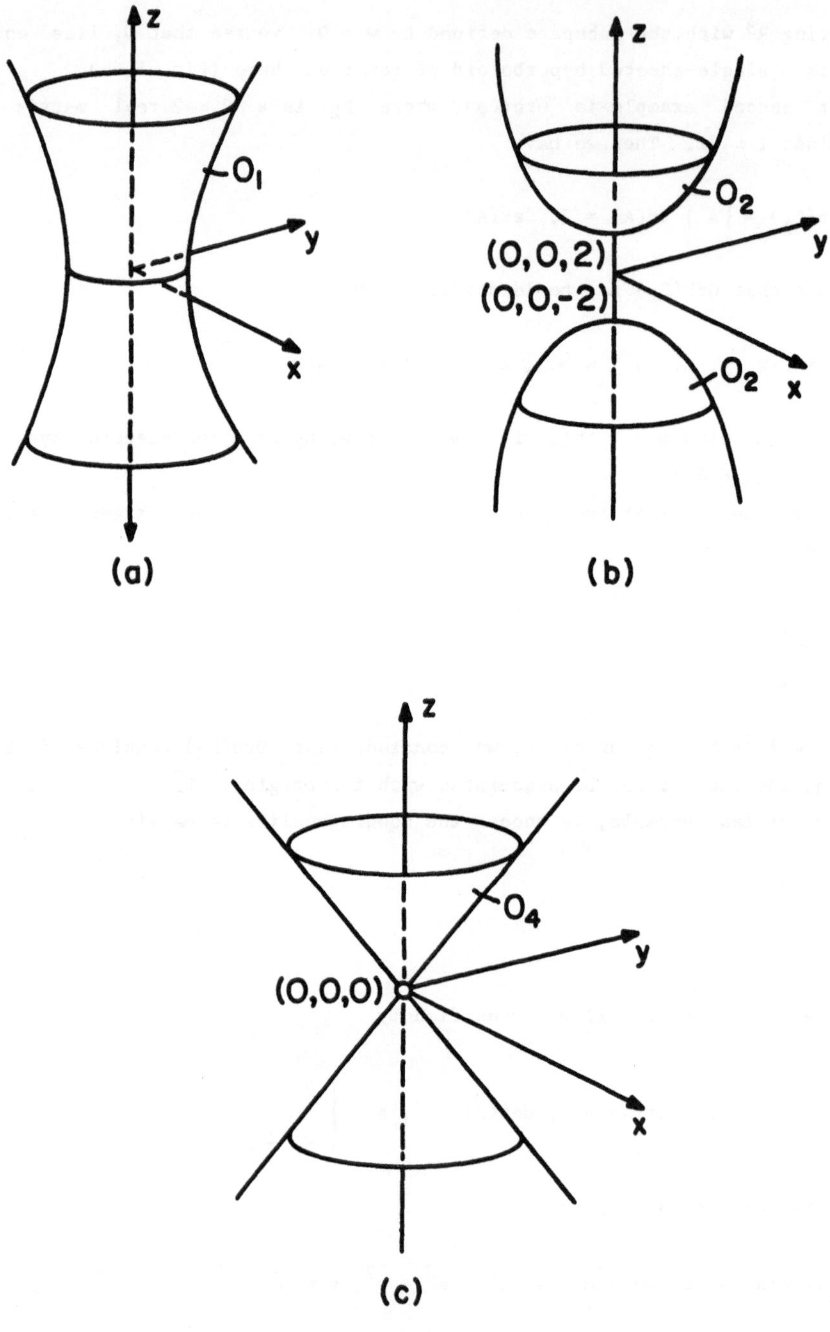

Fig. 6.6 Representation of three orbits of M_2: the hyperboloid of one sheet O_1 is associated with $Orb(\Gamma_1)$ of matrices having eigenvalues ± 1 (a); the hyperboloid of two sheets O_2, with $Orb(\Gamma_2)$ of matrices having eigenvalues $1 \pm 2i$ (b); the double cone with the vertex deleted O_4, with $Orb(\Gamma_4)$ of nondiagonalizable matrices having double eigenvalues 2 (c). Each of these lies in a 3-dimensional vector subspace given by $w = w_0$ for suitable w_0; here only that vector space is shown.

6.3 Transversality and Tangent Space

Our next goal is to introduce the notion of transversality of submanifolds and of maps. First we will discuss the notion informally. Unfortunately, such a discussion does not equip us with computational machinery adequate enough for using the concept in applications. Consequently, we will introduce the notion of tangent space, which leads to both a precise definition and to a computational detection of transversality.

Now we begin our informal discussion. We wish to use the word transversality to capture an essential property associated with curves and surfaces that cross. We do not simply use the word crossing, because we will encounter later submanifolds of R^n that are transversal but that do not "cross" because they do not have "sides". Nonetheless, the archetypical examples of transversality are given by curves and surfaces that do cross.

Example 8. Transversal curves and surfaces

Some of the statements we would like to make are

i) The x- and y-axes are transverse in the plane (Fig. 6.7a).

ii) The circle $S^1 = \{(x, y) \mid x^2 + y^2 = 1\}$ and the y-axis are transverse in the plane (Fig 6.7b).

iii) The sphere $S^2 = \{(x, y, z) \mid x^2 + y^2 + z^2 = 1\}$ and the z-axis are transverse in R^3 (Fig. 6.7c).

iv) The two spheres S^2 and $S^2 + (1, 0, 0)$ are transverse in R^3 (Fig. 6.8a).

The first three statements are intuitively clear because they involve sets that intersect at right angles (Fig. 6.7).

The fourth statement is illustrated in Fig. 6.8a, in which two spheres intersect, not necessarily at right angles, to form a circle. So far, the intuitive notion of crossing is conterminous with that of transversality. As illustrated in Fig. 6.8b, however, we would not wish to say that the tangent spheres S^2 and $S^2 + (2, 0, 0)$ are transverse in R^3. These two spheres only touch one another; neither enters the interior of the other, and so the surfaces do not cross. In fact, it appears by comparing Figs. 6.8a and 6.8b that tangency is the opposite of transversality.

Moreover, we will wish transversality to be a more inclusive condition than merely crossing; in some sense we will require that the two transverse surfaces span the space. In particular, we will wish to say that the x- and y-axes are not transverse in R^3. #

To correlate the various usages above, it is convenient to give a universal model of transversality.

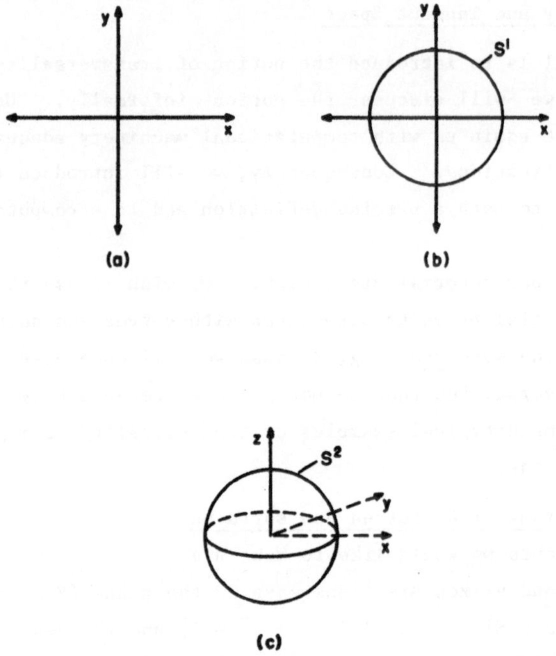

Fig. 6.7 Three examples of transverse curves and surfaces: the x- and y-axes in the plane (a); the circle S^1 and the y-axis in the plane (b); and the sphere S^2 and the z-axis in R^3 (c).

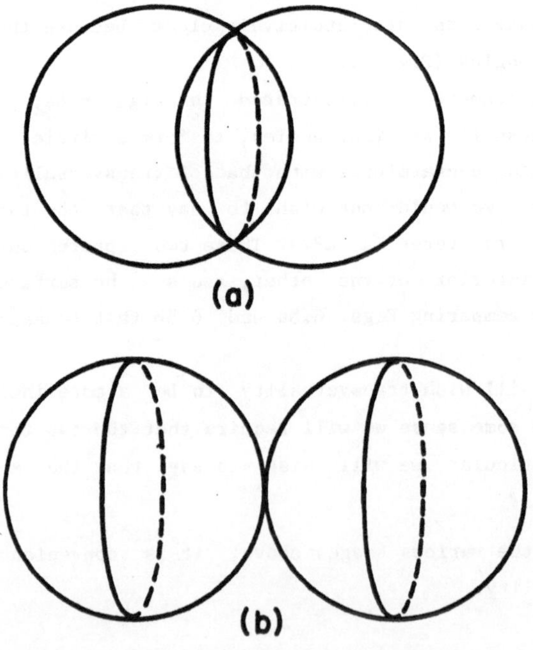

Fig. 6.8 Two unit spheres in R^3 intersecting transversally if their intersection forms a circle (a) but not transversally if they are tangent (b).

Universal Model of Transversality. Suppose V and W are two vector subspaces of R^n. Then V and W are transverse in R^n if and only if

$$R^n = V + W = \{v + w \mid v \in V, w \in W\} \qquad (6.87)$$

Now, heuristically, we may define two submanifolds P and Q, having dimensions p and q, respectively, in R^n to be transverse at a point α of their intersection

$$\alpha \in P \cap Q \qquad (6.88)$$

if there exist a p-dimensional vector subspace V and a q-dimensional vector subspace W of R^n such that i) V and W are transverse in R^n and ii) (0, V, W, R^n) models (α, P, Q, R^n) near α. Of course, the operative word is "models", but let us rely provisionally on its intuitive meaning; the following example will make clear this meaning.

Example 9. Transversality of two circles in the plane

Let S^1 be the unit circle in the plane, and let $S^1 + (n, 0)$ be the result of translating S^1 horizontally by n units (Fig. 6.9). Then the point $\alpha = (5/2, \sqrt{3}/2)$ is in the intersection of $S^1 + (2, 0)$ and $S^1 + (3, 0)$. Let T_2 be the line tangent to $S^1 + (2, 0)$ at α and let T_3 be the line tangent to $S_1 + (3, 0)$ at α. Let W be the

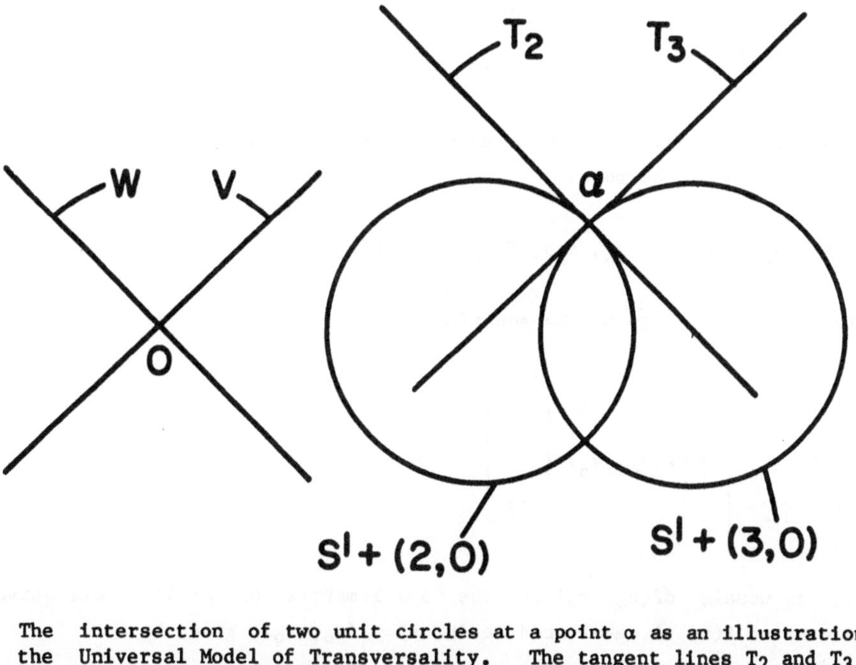

Fig. 6.9 The intersection of two unit circles at a point α as an illustration of the Universal Model of Transversality. The tangent lines T_2 and T_3 are parallel to the lines W and V that together span the plane R^2; moreover, the lines T_2 and T_3 are linear approximations to the circles and are therefore first-order models for the circles. Consequently, the two circles are (provisionally) transverse at the point α.

line through the origin parallel to T_2 and let V be the line through the origin parallel to T_3. Then V and W are vector subspaces of R^2; clearly $V + W = R^2$, so that V and W are transverse in R^2.

On the other hand, we know that T_2 is a first-order (linear) approximation to $S^1 + (2, 0)$ near α, and T_3 to $S^1 + (3, 0)$, so that the quadruple $(0, T_2, T_3, R^2)$ is a first-order approximation to $(0, S^1 + (2, 0), S^1 + (3, 0), R^2)$ near α. And because the quadruple $(0, W, V, R^2)$ is visibly congruent to $(0, T_2, T_3, R^2)$ in Fig. 6.9, it is reasonable to say that $(0, W, V, R^2)$ is some kind of a first-order model for $(0, S^1 + (2, 0), S^1 + (3, 0), R^2)$ in the vicinity of α. This is the intended sense for the word "model", and thus our provisional definition for transversality results in the declaration that $S^1 + (2, 0)$ and $S^1 + (3, 0)$ are transverse at $\alpha = (5/2, \sqrt{3}/2)$ in R^2. #

Our notion of transversality is still not precise enough to allow us to make computations, however. As Example 9 above indicates, the concept we need in order to obtain the necessary precision is that of the <u>tangent space to a manifold</u>. We recall from advanced calculus how to obtain the equation of the plane tangent to a parameterized surface.

Suppose a surface S is given parametrically by

$$\begin{bmatrix} x \\ y \\ z \end{bmatrix} = \begin{bmatrix} X(u, v) \\ Y(u, v) \\ Z(u, v) \end{bmatrix} = f(u, v) \tag{6.89}$$

In Fig. 6.10 we show the simple case $X(u, v) = u$ and $Y(u, v) = v$. Then the plane τ_o tangent to S at the point

$$(x_o, y_o, z_o) = \big(X(u_o, v_o), Y(u_o, v_o), Z(u_o, v_o)\big) \tag{6.90}$$

is given parametrically by the equation

$$\begin{bmatrix} x - x_o \\ y - y_o \\ z - z_o \end{bmatrix} = df(u_o, v_o)\cdot\begin{bmatrix} \xi_1 \\ \xi_2 \end{bmatrix} \tag{6.91}$$

where, as usual, $df(u_o, v_o)$ is the 3×2 matrix of partial derivatives of the components of $f(u, v)$, evaluated at $(u, v) = (u_o, v_o)$, given by

$$df(u_o, v_o) = \begin{bmatrix} \partial_u X(u_o, v_o) & \partial_v X(u_o, v_o) \\ \partial_u Y(u_o, v_o) & \partial_v Y(u_o, v_o) \\ \partial_u Z(u_o, v_o) & \partial_v Z(u_o, v_o) \end{bmatrix} \tag{6.92}$$

For the case illustrated in Fig. 6.10, we have

$$df(u_o, v_o) = \begin{bmatrix} 1 & 0 \\ 0 & 1 \\ \partial_u Z(u_o, v_o) & \partial_v Z(u_o, v_o) \end{bmatrix} \tag{6.93}$$

In this case, the vector

$$N_o = [- \partial_u Z(u_o, v_o), - \partial_v Z(u_o, v_o), 1] \tag{6.94}$$

is perpendicular to both columns of $df(u_o, v_o)$ in (6.93). Thus the tangent plane to the surface S through the point $(u_o, v_o, Z(u_o, v_o))$ may be pictured (Fig. 6.10) as the plane through that point and perpendicular to N_o.

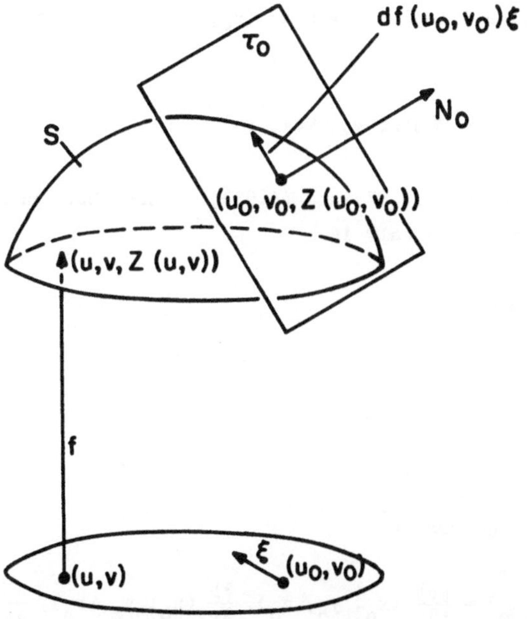

Fig. 6.10 A surface S parameterized by $f(u, v) = (u, v, Z(u, v))$. Here τ_o is the plane tangent to S through the point $(u_o, v_o, Z(u_o, v_o))$. The vector ξ is in R^2; the corresponding vector $df(u_o, v_o)\cdot\xi$ is in R^3 and tangent to S at $(u_o, v_o, Z(u_o, v_o))$. The vector N_o is normal to the surface S at $(u_o, v_o, Z(u_o, v_o))$.

We note that the condition, that the surface S be reversibly described by (u, v) near (x_0, y_0, z_0), is computationally expressed by the requirement that

$$\text{rank}\left[df(u_0, v_0)\right] = 2 \tag{6.95}$$

In fact, (6.95) implies that some 2 × 2 submatrix of $df(u_0, v_0)$ is invertible; say it is the upper one. Then from the Implicit Function Theorem it follows that for (x, y) near (x_0, y_0), the equation

$$\left.\begin{aligned} X(u, v) &= x \\ Y(u, v) &= y \end{aligned}\right\} \tag{6.96}$$

has a unique solution

$$\left.\begin{aligned} u &= U(x, y) \\ v &= V(x, y) \end{aligned}\right\} \tag{6.97}$$

Thus, the point (x, y, z), actually (x, y) alone, on the surface S determines (u, v).

Having reversed the parameterization, we may proceed one step further and express the surface S implicitly near (x_0, y_0, z_0). First we define a function F by setting

$$F(x, y, z) = z - Z\big(U(x, y), V(x, y)\big) \tag{6.98}$$

Then, combining (6.89), (6.97) and (6.98), we see that near (x_0, y_0, z_0), the point (x, y, z) is in S if and only if

$$F(x, y, z) = 0 \tag{6.99}$$

Finally, because

$$\frac{\partial F}{\partial z} = 1 \tag{6.100}$$

it is clear that the vector

$$dF(x_0, y_0, z_0) = \left[\frac{\partial F}{\partial x}(x_0, y_0, z_0), \frac{\partial F}{\partial y}(x_0, y_0, z_0), \frac{\partial F}{\partial z}(x_o, y_o, z_o)\right] \neq 0 \tag{6.101}$$

Moreover, because

$$F\big(f(u, v)\big) = 0 \tag{6.102}$$

we see by applying the chain rule to (6.102) that

$$dF(x_o, y_o, z_o)\cdot df(u_o, v_o) = 0 \tag{6.103}$$

This equation implies that any vector of the form (cf. (6.91))

$$\zeta = df(u_o, v_o)\cdot\xi \tag{6.104}$$

is perpendicular to $dF(x_0, y_0, z_0)$; that is, $dF(x_0, y_0, z_0)$ is perpendicular to the surface S at (x_0, y_0, z_0). This is the familiar fact that grad(f) is normal to $F = 0$.

Somewhat easier to handle than the tangent plane τ_0 is the tangent space $T_{p_0}S$ to the surface S at the given point $p_0 = (x_0, y_0, z_0)$; this is the plane through the origin and parallel to the tangent plane τ_0. It does not appear tangent, but it has the advantage of being a vector space, and the tangent plane may be recovered from it by parallel translation from the origin to the given point p_0. Here the tangent space is the set of all 3-element column vectors given by

$$\begin{bmatrix} x \\ y \\ z \end{bmatrix} = df(u_o, v_o)\cdot\begin{bmatrix} u \\ v \end{bmatrix} \tag{6.105}$$

where the 2-element column vector ranges over all R^2. Using (6.103) we see that

$$T_{p_o}S = \{ \begin{bmatrix} x \\ y \\ z \end{bmatrix} \mid dF(p_o)\cdot\begin{bmatrix} x \\ y \\ z \end{bmatrix} = 0\} \tag{6.106}$$

Because the tangent plane at a point of a surface is a first-order approximation to that surface near the point, it is reasonable to define two surfaces in R^3 to be transverse at a point of their intersection if and only if their tangent spaces at that point are transverse in the sense of the above Universal Model of Transversality.

To see that this definition is the correct one, consider two surfaces P and Q in R^3 that are transverse at every point of intersection (Fig. 6.11a). Suppose $p_0 = (x_0, y_0, z_0)$ is such a point; we express the surface P near p_0 implicitly; near p_0, the point (x, y, z) is in P if and only if

$$F(x, y, z) = 0 \tag{6.107}$$

We express the surface Q near p_0 parametrically,

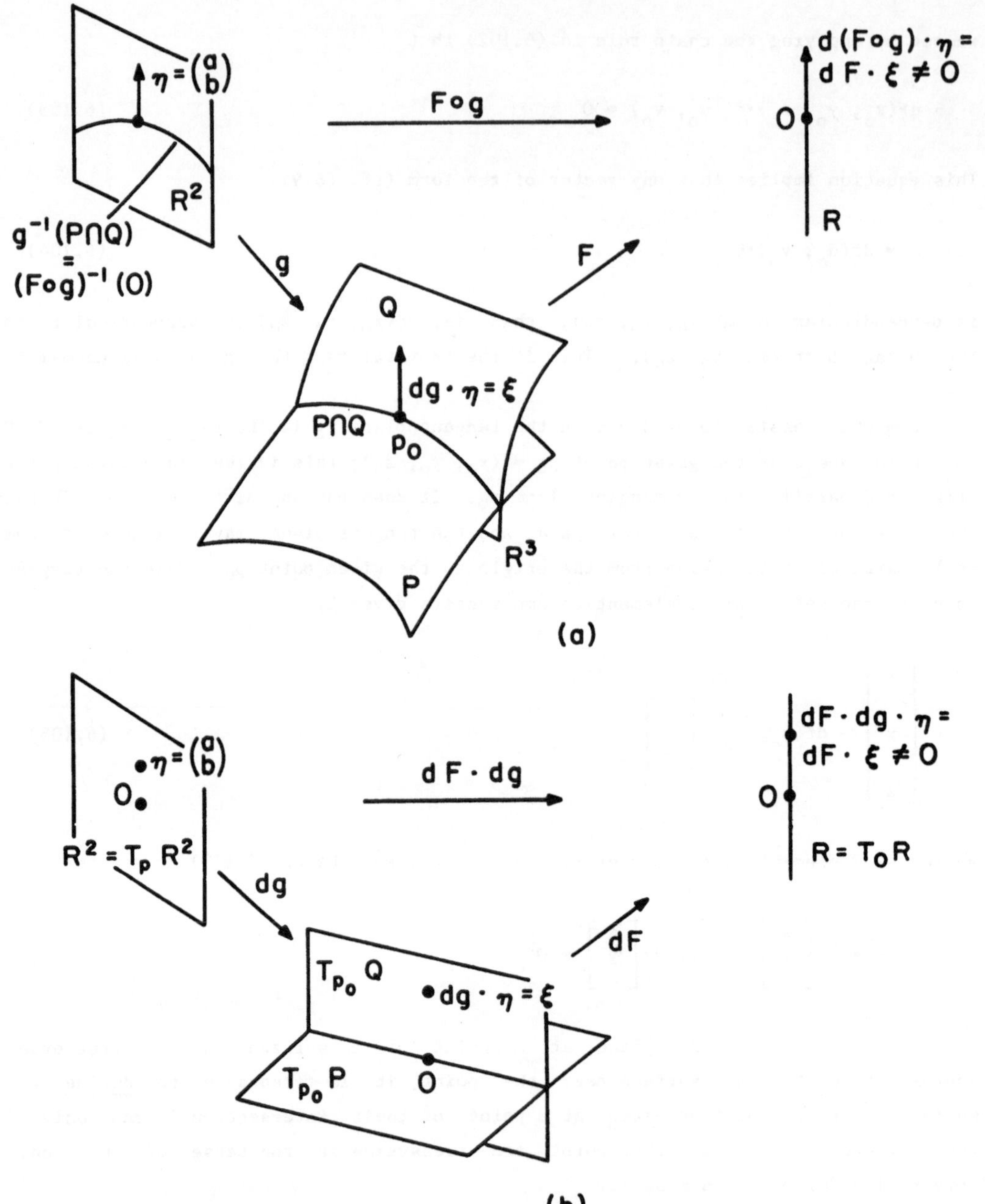

Fig. 6.11 Two transverse surfaces P and Q meet in a smooth curve $P \cap Q$. In (b), $T_{p_0}P + T_{p_0}Q = R^3$. Because $dF:R^3 \to R$ is non-trivial, but is trivial on $T_{p_0}P$, there exists $\xi \in T_{p_0}Q$ as shown such that $dF\cdot\xi \neq 0$. But $g:R^2 \to Q$ is a parameterization, as shown in (a), so that $T_{p_0}Q = dg\cdot R^2$ and there is some η such that $\xi = dg\cdot\eta$. Then $0 \neq dF\cdot\xi = dF\cdot dg\cdot\eta$. Now $F\circ g = 0$ defines the smooth curve $g^{-1}(P \cap Q)$ in (a), so that $P \cap Q$ is also a smooth curve.

$$\begin{bmatrix} x \\ y \\ z \end{bmatrix} = g(u, v) \qquad (6.108)$$

and $(x_0, y_0, z_0) = g(u_0, v_0)$. Then $g(u, v) \in P \cap Q$ if and only if (Fig. 6.11a)

$$F\big(g(u, v)\big) = 0 \qquad (6.109)$$

We would like to solve this equation for either u or v in terms of the other variable. To do so, we use the Implicit Function Theorem aided by the transversality condition,

$$T_{p_0} P + T_{p_0} Q = R^3 \qquad (6.110)$$

It follows from (6.110) that there must be an element $\xi \in T_{p_0}Q$ such that $\xi \notin T_{p_0}P$ (Fig. 6.11b). Thus from (6.105) we have

$$\xi = dg(u_0, v_0)\cdot\begin{bmatrix} a \\ b \end{bmatrix} \qquad (6.111)$$

and

$$dF(p_0)\cdot\xi \neq 0 \qquad (6.112)$$

Consequently

$$dF(x_0, y_0, z_0)\cdot dg(u_0, v_0)\cdot\begin{bmatrix} a \\ b \end{bmatrix} \neq 0 \qquad (6.113)$$

and (6.113) must continue to hold with either $a = 1$ and $b = 0$ or $a = 0$ and $b = 1$. Suppose that it is the second case that holds. Then we have

$$\partial_v F\big(g(u_0, v_0)\big) = dF(x_0, y_0, z_0)\cdot dg(u_0, v_0)\cdot\begin{bmatrix} 0 \\ 1 \end{bmatrix} \neq 0 \qquad (6.114)$$

and we may solve (6.109) for v in terms of u near (u_0, v_0); say $v = V(u)$. But then from (6.108) we see that

$$\begin{bmatrix} x \\ y \\ z \end{bmatrix} = g(u, V(u)) \tag{6.115}$$

is a parameterization for the intersection $P \cap Q$ near p_0. Consequently, the intersection of two transverse surfaces is a smooth 1-submanifold, or a nonsingular curve, in R^3.

These definitions work in general, with exactly the same effect: If $x = f(u)$ is a parameterization of the k-submanifold P of R^n, then we define the <u>tangent space</u> $T_{x_0}P$ of P at $x_0 = f(u_0)$ to be given by

$$T_{x_0} P = \{df(u_0)\cdot u \mid u \in R^k\} \tag{6.116}$$

If a k-submanifold P and an ℓ-submanifold Q meet at $x_0 = f(u_0) = g(v_0)$, where f parameterizes P near x_0, and g parameterizes Q near x_0, then we say that <u>P and Q are transverse at p_0</u> if

$$T_{x_0} P + T_{x_0} Q = R^n \tag{6.117}$$

If P and Q are transverse, that is, transverse at every intersection point, then the Implicit Function Theorem tells us that $P \cap Q$ is a smooth $(k + \ell - n)$-submanifold of R^n.

<u>Example 10. The tangent space at the fold on a cusp surface</u>

Let

$$P = \{(x, \alpha, \beta) \mid x^3 - \alpha x - \beta = 0\} \tag{6.118}$$

The parameterization by (x, α)

$$f(x, \alpha) = \begin{bmatrix} x \\ \alpha \\ \beta \end{bmatrix} = \begin{bmatrix} x \\ \alpha \\ x^3 - \alpha x \end{bmatrix} \tag{6.119}$$

shows that P is a manifold. Then

$$df(x, \alpha) = \begin{bmatrix} 1 & 0 \\ 0 & 1 \\ 3x^2 - \alpha & -x \end{bmatrix} \tag{6.120}$$

and for $p_o = (x_o, \alpha_o, \beta_o) \in P$, the tangent space at p_o is given parametrically by

$$T_{p_o} P = \left\{ \begin{bmatrix} u \\ v \\ (3x_o^2 - \alpha_o)u - x_o v \end{bmatrix} \middle| \; u, v \in R \right\} \tag{6.121}$$

Along the fold curve we expect $T_{p_o}P$ to be perpendicular to the (α, β)-plane, which has basis $\{(0, 1, 0), (0, 0, 1)\}$. But the fold curve is given by the further requirement that $3x^2 - \alpha = 0$. Then for p_o in the fold curve, we see from (6.121) that the vector (1, 0, 0) is in T_{p_o}, and the tangent space is indeed perpendicular to the (α, β) plane. The tangent plane X, which is parallel to $T_{p_o}P$, is depicted in Fig. 6.12. #

Example 11. The tangent space of Orb(Γ).

Let

$$\text{Orb}(\Gamma) = \{\Lambda^{-1} \Gamma \Lambda \mid \Lambda \in GL_n\} \tag{6.122}$$

in which GL_n, the set of invertible matrices, is specified by (6.9). We have seen that Orb(Γ) is a smooth submanifold of M_n of dimension given by (6.77)

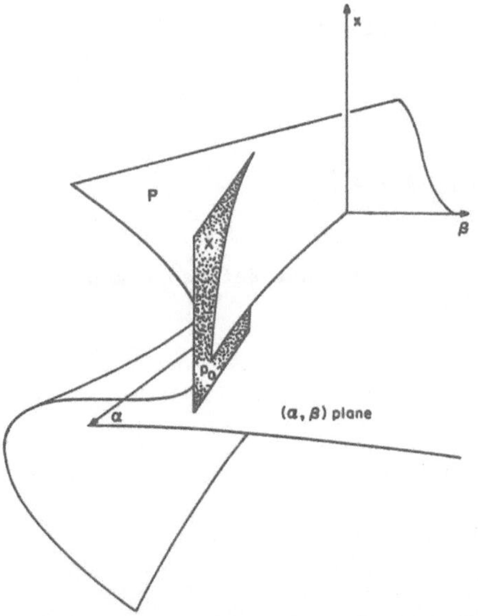

Fig. 6.12 The plane X tangent to the cusp surface P at the point $p_o = (x_o, \alpha_o, \beta_o)$. The point p_o is on the fold curve. The tangent plane X is perpendicular to the (α, β)-plane and is parallel to the tangent space $T_{p_o}P$. For visibility, the cusp surface is cut off at $\alpha = \alpha_o$.

$$\dim\left[\mathrm{Orb}(\Gamma)\right] = n^2 - n_1{}^2 - \cdot\cdot\cdot\cdot - n_r{}^2 - 2m_1{}^2 - \cdot\cdot\cdot\cdot - 2m_s{}^2 \tag{6.123}$$

in the case where all the eigenvalues of Γ are simple. Here $n_1, \ldots, n_r$ are the multiplicities of the real eigenvalues and $m_1, \ldots, m_s$ those of the complex conjugate pairs.

We wish to compute the tangent space $T_{\Gamma_0}\mathrm{Orb}(\Gamma)$ of $\mathrm{Orb}(\Gamma)$ at the point Γ_0 of $\mathrm{Orb}(\Gamma)$. For simplicity, we assume that $m_1 = m_2 = \cdots = m_s = 0$. Because we already know the dimension of $\mathrm{Orb}(\Gamma)$ is that of $T_{\Gamma_0}\mathrm{Orb}(\Gamma)$, we only need find a vector subspace of $T_{\Gamma_0}\mathrm{Orb}(\Gamma)$ having that same dimension: But this vector subspace will be exactly $T_{\Gamma_0}\mathrm{Orb}(\Gamma)$ because a proper vector subspace of another must have a properly smaller dimension.

Let $A \in M_n$ be an arbitrary $n \times n$ matrix. Then we define a curve $\Lambda(t)$ in GL_n passing through the identity I by setting

$$\Lambda(t) = I + t\,A \tag{6.124}$$

Then we define a path Γ_t in $\mathrm{Orb}(\Gamma)$ passing through Γ_0 by setting

$$\Gamma_t = \Lambda(t)^{-1}\,\Gamma_0\,\Lambda(t) \tag{6.125}$$

which we may write as

$$\Gamma_t = (I - t\,A + \cdot\cdot\cdot)\,\Gamma_0(I + t\,A) \tag{6.126}$$

Then the matrix

$$L_{\Gamma_0}(A) = \frac{d}{dt}\bigg|_0 \Gamma_t = \Gamma_0\,A - A\,\Gamma_0 \tag{6.127}$$

is a member of $T_{\Gamma_0}\mathrm{Orb}(\Gamma)$. Thus we have defined a map

$$L_{\Gamma_0} : M_n \to M_n \tag{6.128}$$

with the property that

$$L_{\Gamma_0}(M_n) \subset T_{\Gamma_0}\mathrm{Orb}(\Gamma) \tag{6.129}$$

On the other hand, L_{Γ_0} is clearly a linear map so that $L_{\Gamma_0}(M_n)$ is a vector subspace of $T_{\Gamma_0}\mathrm{Orb}(\Gamma)$. Finally, from linear algebra we know that

$$\dim\left[L_{\Gamma_0}(M_n)\right] = n^2 - \dim\left\{A \mid L_{\Gamma_0}(A) = 0\right\} \tag{6.130}$$

To calculate

$$\dim \{A \mid L_{\Gamma_o}(A) = 0\} = \dim \{A \mid \Gamma_o A - A \Gamma_o = 0\} \tag{6.131}$$

we notice that conjugation by $\Lambda \in GL_n$,

$$A \to \Lambda^{-1} A \Lambda = B \tag{6.132}$$

carries the vector space $\{A \mid \Gamma_o A - A \Gamma_o = 0\}$ linearly and isomorphically onto the vector space $\{B \mid \Lambda^{-1} \Gamma_o \Lambda B - B \Lambda^{-1} \Gamma_o \Lambda = 0\}$. However, for a suitable choice Λ_2 of Λ we may write

$$\Gamma_1 = \Lambda_2^{-1} \Gamma_o \Lambda_2 = \begin{bmatrix} D_1 & & \\ & \ddots & \\ & & D_r \end{bmatrix} \tag{6.133}$$

where D_k is the $n_k \times n_k$ diagonal matrix corresponding to multiplication by λ_k, with $\lambda_k \neq \lambda_\ell$ for $k \neq \ell$. We may write a matrix B in block form corresponding to (6.133)

$$B = \begin{bmatrix} B_{11} & B_{12} & \cdots & B_{1r} \\ \vdots & & & \vdots \\ B_{r1} & B_{r2} & \cdots & B_{rr} \end{bmatrix} \tag{6.134}$$

Then an easy calculation shows that

$$\Gamma_1 B - B \Gamma_1 = 0 \tag{6.135}$$

if and only if

$$B_{ij} = 0 \qquad \text{for } i \neq j \tag{6.136}$$

However, the entries of the diagonal blocks B_{ii} are unconstrained. Thus we find that

$$\dim \{B \mid \Gamma_1 B - B \Gamma_1 = 0\} = n_1^2 + \cdots + n_r^2 \tag{6.137}$$

and consequently via (6.132) that

$$\dim \{A \mid \Gamma_o A - A \Gamma_o = 0\} = n_1^2 + \cdots + n_r^2 \tag{6.138}$$

so that (6.130) now gives us

$$\dim\left[L_{\Gamma_o}(M_n)\right] = n^2 - n_1^2 - \cdots - n_r^2 \tag{6.139}$$

Thus $L_{\Gamma_o}(M_n)$ is a vector subspace of $T_{\Gamma_o}\mathrm{Orb}(\Gamma)$ of the same dimension, so that we must have

$$L_{\Gamma_o}(M_n) = T_{\Gamma_o}\mathrm{Orb}(\Gamma) \tag{6.140}$$

and we have computed $T_{\Gamma_o}\mathrm{Orb}(\Gamma)$ to be composed of matrices of the form (6.127). #

We turn now to a further refinement of the idea of transversality. Suppose P is a smooth p-submanifold of R^m and that Q is a smooth q-submanifold of R^n. Let ϕ: $P \to R^n$ be a smooth map. Then

$$P_\phi = \{(x, \phi(x)) \mid x \in P\} \subset R^m \times R^n = R^{m+n} \tag{6.141}$$

is one smooth p-submanifold of R^{m+n}; in fact, the map $x \to (x, f(x))$ is a diffeomorphism. Another smooth submanifold is the smooth $(m + q)$-submanifold $R^m \times Q$ of R^{m+n}; that is

$$R^m \times Q \subset R^{m+n} \tag{6.142}$$

We will say that the map ϕ: $P \to R^n$ is <u>transverse</u> to Q if P_ϕ and Q are transverse in $R^m \times Q$ as smooth submanifolds. Then $P_\phi \cap (R^m \times Q)$ will be a smooth submanifold of R^{n+m} with dimension $p + (m + q) - (n + m) = p + q - n$. Notice that the map $x \to (x, \phi(x))$ carries $\phi^{-1}(Q)$ diffeomorphically onto the intersection $P_\phi \cap (R^m \times Q)$, so that $\phi^{-1}(Q)$ is a submanifold of P of dimension $p + q - n$.

<u>Example 12. The spaces associated with transversality of a map on the cusp surface</u>

In the definition above, let $P \subset R^3$ be the cusp surface we have considered before,

$$P = \{(x, \alpha, \beta) \mid x^3 - \alpha x - \beta = 0\} \tag{6.143}$$

let $R^n = R$, and let $Q = \{0\} \subset R$. For the map ϕ: $P \to R$ we take the map defined by setting

$$\phi(x, \alpha, \beta) = 3x^2 - \alpha \tag{6.144}$$

Later, in Example 13, we will see that ϕ is transverse to $\{0\} \subset R$; here we only wish to identify the spaces appearing in the definition of the transversality of ϕ. The smooth submanifold P_ϕ is given by

$$P_\phi = \{(x, \alpha, \beta, 3x^2 - \alpha) \mid x^3 - \alpha x - \beta = 0\} \tag{6.145}$$

and P_ϕ is transverse to $R^3 \times \{0\}$; that is not obvious, but the fact that

$$P_\phi \cap (R^3 \times \{0\}) = \{(x, \alpha, \beta, 0) \mid x^3 - \alpha x - \beta = 0,\ 3x^2 - \alpha = 0\} \tag{6.146}$$

is indeed a smooth 1-submanifold can be seen from the parameterization

$$x \to (x, 3x^2, -2x^3, 0) \tag{6.147}$$

Dropping the terminal 0 gives us a parameterization for the smooth curve $\phi^{-1}(0)$ of fold points in P. #

In Example 12, we have omitted verification of the transversality of P_ϕ and $R^3 \times \{0\}$. We could carry out this verification with the machinery presented so far, but we can do so much more easily by using the differential of a smooth map $\phi\colon P \to R^n$. First we note that any smooth map $\phi\colon P \to R^n$ is the restriction to $P \subset R^m$ of a smooth map $\Phi\colon R^m \to R^n$. For Φ we may calculate the matrix $d\Phi(x)$ of partial derivatives at a point $x \in P$. Then, $d\Phi(x)$ defines a linear map for which we use the same symbol,

$$d\Phi(x)\colon\ R^m \to R^n \tag{6.148}$$

Finally, we define $d\phi(x)\colon\ T_xP \to R^n$ by setting

$$d\phi(x) = d\Phi(x) \mid T_xP \tag{6.149}$$

That is, $d\phi(x)$ is the restriction of $d\Phi(x)$ to T_xP.

Of course, $\phi\colon P \to R^n$ has many extensions Φ, and we must check that the definition of $d\phi(x)$ depends only on the map ϕ and not on the extension Φ. To see this independence (Fig. 6.13), let $\xi \in R^m$ and let $\alpha(t)$ be any path such that $\alpha(0) = x$ and such that the velocity vector

$$\left.\frac{d\,\alpha(t)}{dt}\right|_{t=0} = \xi \tag{6.150}$$

Then the Chain Rule implies that

$$d\Phi(x)\cdot\xi = \left.\frac{d\,\Phi\big(\alpha(t)\big)}{dt}\right|_{t=0} \tag{6.151}$$

For $\xi \in T_xP$, we may choose the path $\alpha(t)$ to lie entirely in P (to show this analytically, use a parameterization of P). But then

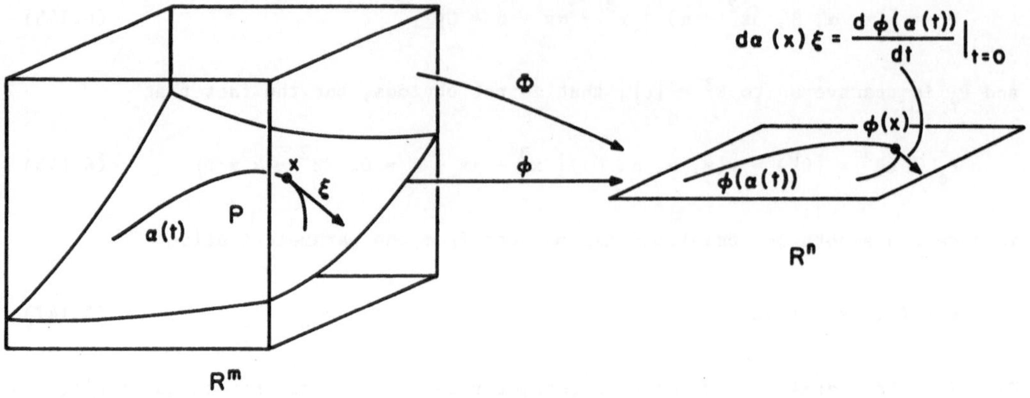

Fig. 6.13 The differential of a smooth map from a smooth submanifold P of R^m to R^n. The smooth map Φ restricted to the submanifold P of R^m defines a smooth map ϕ: $P \to R^n$. The vector ξ is the tangent velocity vector to the curve $\alpha(t)$ in P at t=0. The point x is $\alpha(0)$. Then $d\phi(x)\cdot\xi$ is the velocity vector tangent to the curve $\phi(\alpha(x))$ at t=0.

$$d\phi(x)\cdot\xi = d\Phi(x)\cdot\xi = \frac{d\ \Phi(\alpha(t))}{dt}\Bigg|_{t=0} = \frac{d\ \phi(\alpha(t))}{dt}\Bigg|_{t=0} \tag{6.152}$$

and the right-most quantity visibly depends only on ϕ (Fig. 6.13).

For our improved, and computational, definition of transversality, we say that ϕ is <u>transverse</u> <u>to</u> <u>Q</u> at $x \in \phi^{-1}(Q)$ if the composition of linear maps

$$T_xP \xrightarrow{d\phi(x)} R^n \xrightarrow{\text{quot}} R^n/T_{\phi(x)}Q \tag{6.153}$$

carries the vector space T_xP <u>onto</u> $R^n/T_{\phi(x)}Q$; here

$$R^n \xrightarrow{\text{quot}} R^n/T_{\phi(x)}Q \tag{6.154}$$

is the quotient map of vector spaces (see Section 2.6). And finally we say that ϕ is <u>transverse</u> <u>to</u> <u>Q</u> if ϕ is transverse to Q at every $x \in \phi^{-1}(Q)$.

Example 13. Computational verification of transversality of a map on the cusp surface

In the previous example, we did not check that the map ϕ (6.144) on the cusp surface is actually transverse. Here we do so by using the differential. Because $Q = \{0\}$, the tangent space T_0Q is the zero-dimensional vector subspace of R, and the quotient R/T_0Q is naturally identified with R. Thus we have to check that the map

$$T_{(x,\alpha,\beta)}P \xrightarrow{d\phi(x,\alpha,\beta)} R \tag{6.155}$$

is onto for every (x, α, β) such that $\phi(x, \alpha, \beta) = 0$. Because $\phi(x, \alpha, \beta) = 3x^2 - \alpha$, we have that

$$d\phi(x, \alpha, \beta) = (6x, -1, 0) \tag{6.156}$$

Applying $d\phi$ to a typical tangent vector (6.121) in $T_{p_0}P$, for p_0 in the fold curve, produces

$$d\phi(p_0)\cdot\begin{bmatrix} u \\ v \\ -x_0 v \end{bmatrix} = 6\, x_0\, u - v \tag{6.157}$$

However, the left side of (6.157) may be made to equal any number, so

$$d\phi(p_0): \; T_{p_0}P \to R \tag{6.158}$$

is onto for any p_0 in the fold curve, and we conclude that ϕ: $P \to R$ is transverse to $\{0\}$. #

Thus far, our examples of transversality have been fairly superficial. In subsequent sections of this chapter, we will be particularly interested in the transversality of maps

$$P \xrightarrow{f} R^n \tag{6.159}$$

to submanifolds of R^n, where the submanifold P of R^{m+k} is itself defined by a transversality condition. In the following example we see how this situation arises in a natural way.

Example 14. Transversality of maps associated with the hyperbolic umbilic. The hyperbolic umbilic is the unfolding H defined by (Example 19, Chapter 2, and Table 2.2)

$$H(x, y, \alpha, \varepsilon, \gamma, \delta) = \begin{bmatrix} xy + \alpha + \varepsilon x \\ x^2 - y^2 + \gamma + \delta y \end{bmatrix} \tag{6.160}$$

with dynamical variables x, y and control parameters $\alpha, \varepsilon, \gamma, \delta$. Its stationary points are the solutions of $H = 0$. It is easy to check that the map $H: R^6 \to R^2$ is transverse to zero in R^2. Consequently, the set $S_H = H^{-1}(0)$ of stationary points is a smooth 4-submanifold of R^6.

An easier way to see that S_H is a smooth 4-submanifold is to use the parameterization

$$f(x, y, \varepsilon, \delta) = \begin{bmatrix} x \\ y \\ -xy - \varepsilon x \\ \varepsilon \\ -x^2 + y^2 - \delta y \\ \delta \end{bmatrix} \tag{6.161}$$

Now it is easy to check that for $p \in S_H$, the tangent space T_pS_H is given by

$$T_pS_H = \{\xi \mid dH(p)\cdot\xi = 0\} \tag{6.162}$$

and via (6.161) is also given by

$$T_pS_H = \begin{bmatrix} 1 & 0 & 0 & 0 \\ 0 & 1 & 0 & 0 \\ -y-\varepsilon & -x & -x & 0 \\ 0 & 0 & 1 & 0 \\ -2x & 2y-\delta & 0 & -y \\ 0 & 0 & 0 & 1 \end{bmatrix} \cdot R^4 \tag{6.163}$$

where $p = (x, y, -xy - \varepsilon x, \varepsilon, -x^2 + y^2 - \delta y, \delta)$. In fact, a little algebra shows that for $p \in S_H$, we have

$$dH(p)\cdot\xi = 0 \tag{6.164}$$

if and only if

$$\xi = df(x, y, \varepsilon, \delta)\cdot\eta \tag{6.165}$$

for some $\eta \in R^4$, where $df(x, y, \varepsilon, \delta)$ is the matrix in (6.163). Thus the null space of dH(p) is 4-dimensional, and the domain of dH(p) is 6-dimensional. Recall from linear algebra that for a linear map such as dH(p),

$$\dim(\text{null space}) + \dim(\text{image}) = \dim(\text{domain}) \tag{6.166}$$

But then we see that dH(p) is onto R^2, and transversality of H to $\{0\} \subset R^2$ follows.

A more interesting transversal map is provided by the differential of H with respect to the dynamical variables,

$$d_{xy}H(p) = \begin{bmatrix} y + \varepsilon & x \\ 2x & \delta - 2y \end{bmatrix} \tag{6.167}$$

This differential defines a mapping from the smooth 4-manifold S_H to the space M_2 of 2×2 matrices; recall from Section 6.1 that we may identify M_2 with R^4. Thus, we have a fairly natural map

$$d_{xy}H: \quad S_H \to M_2 = R^4 \tag{6.168}$$

and we seek an interesting submanifold $Q \subset M_2$ to which $d_{xy}H$ is transverse. Here "interesting" means invariant, and we try

$$Q = \{M \mid \mathrm{tr}(M) = 0\} \tag{6.169}$$

first. To check transversality of $d_{xy}H/S_H$, we have to calculate the differential of $d_{xy}H$ itself, and then restrict it to the tangent space of S_H. We begin by aligning the four coordinates of $d_{xy}H$, as prescribed by an identification of M_2 with R^4,

$$d_{xy}H \leftrightarrow \begin{bmatrix} y + \varepsilon \\ x \\ 2x \\ \delta - 2y \end{bmatrix} \tag{6.170}$$

Then we see that

$$d(d_{xy}H) \leftrightarrow \begin{array}{c} \begin{matrix} x & y & \alpha & \varepsilon & \gamma & \delta \end{matrix} \\ \begin{bmatrix} 0 & 1 & 0 & 1 & 0 & 0 \\ 1 & 0 & 0 & 0 & 0 & 0 \\ 2 & 0 & 0 & 0 & 0 & 0 \\ 0 & -2 & 0 & 0 & 0 & 1 \end{bmatrix} \end{array} \qquad (6.171)$$

We will apply this matrix only to T_pS_H where from (6.161)

$$p = (x,\ y,\ -xy - \varepsilon x,\ \varepsilon,\ -x^2 + y^2 - \delta y,\ \delta) \in S_H \qquad (6.172)$$

According to equation (6.163), we have $\xi \in T_pS_H$ if and only if ξ is a linear combination of the four columns of the matrix appearing in that equation. Thus, the vector space $d(d_{xy}H)\cdot T_pS_H$ will be spanned by the four vectors resulting when the matrix in (6.171) is applied to the four columns from (6.163). These four nontrivial vectors are

$$\eta_1 = \begin{bmatrix} 0 \\ 1 \\ 2 \\ 0 \end{bmatrix},\ \eta_2 = \begin{bmatrix} 1 \\ 0 \\ 0 \\ -2 \end{bmatrix},\ \eta_3 = \begin{bmatrix} 1 \\ 0 \\ 0 \\ 0 \end{bmatrix},\ \eta_4 = \begin{bmatrix} 0 \\ 0 \\ 0 \\ 1 \end{bmatrix} \qquad (6.173)$$

(It is only a coincidental property of this particular example that these four columns are exactly the nontrivial columns of (6.171)). But $\eta_1 - \eta_4$ do not span R^4 and so $d_{xy}H$ is not transverse to $\{0\} \subset R^4$.

However, we may hope that $\eta_1 - \eta_4$ do span the quotient space $R^4/T_{d_{xy}H(p)}T_o$, where T_o is the space of matrices with trace 0. Then via (6.153), we will conclude that $d_{xy}H$ is transverse to T_o.

We note that for any vector subspace V of R^n and $p \in V$ we have

$$T_pV = V \qquad (6.174)$$

Thus, because T_o is a subvector space of $R^4 = M_2$, we may write

$$T_{d_{xy}H(p)}T_o = T_o \qquad (6.175)$$

and a basis for T_o is given by

$$\zeta_1 = \begin{bmatrix} 1 \\ 0 \\ 0 \\ -1 \end{bmatrix}, \quad \zeta_2 = \begin{bmatrix} 0 \\ 1 \\ 0 \\ 0 \end{bmatrix}, \quad \zeta_3 = \begin{bmatrix} 0 \\ 0 \\ 1 \\ 0 \end{bmatrix} \tag{6.176}$$

Consequently, the quotient space $R^4/T_{d_{xy}H(p)}T_o$ is spanned by the single vector represented by [1, 0, 0, 0]. But this representative is exactly η_3 in (6.173); thus we conclude that for $\mathrm{tr}\big(d_{xy}H(p)\big) = 0$ and $H(p) = 0$, we have that the composition

$$T_pS_H \xrightarrow{d(d_{xy}H)} R^4 \xrightarrow{\text{quot}} R^4/T_{d_{xy}H(p)}T_o \tag{6.177}$$

is onto. Thus we conclude that the map

$$d_{xy}H\colon \; S_H \to R^4 \tag{6.178}$$

is transverse to the submanifold T_o of R^4 consisting of matrices having trace 0. It follows that the set

$$D_H = \{p \in S_H \mid \mathrm{tr}\big(d_{xy}H(p)\big) = 0\} \tag{6.179}$$

is a smooth 3-submanifold of the smooth 4-submanifold S_H of R^6.

We interpret this result dynamically: For the parameterized differential equation

$$\left.\begin{aligned} \dot{x} &= xy + \alpha + \varepsilon x \\ \dot{y} &= x^2 - y^2 + \gamma + \delta x \end{aligned}\right\} = H(x, y, \alpha, \varepsilon, \gamma, \delta) \tag{6.180}$$

S_H is the 4-dimensional set of stationary points and $D_H \subset S_H$ is the 3-dimensional subset of stationary points at which the two eigenvalues of $d_{xy}H$ are either pure imaginary or real and opposite (cf Fig. 6.2); thus D_H contains the Hopf bifurcation points as an open subset.

As before, we may recover this result without the use of transversality. Using (6.161) and (6.167) we note that the parameterization (6.161) carries $\{(x, y, \varepsilon, \delta) \mid y = \varepsilon + \delta\}$ onto D_H, and thus it parameterizes D_H. #

If it is so easy, as in the last paragraph of the preceeding example, to see that D_H is a smooth 3-submanifold, then why use transversality? The answer to this question is that we are not interested merely in the fact that D_H is a smooth 3-submanifold of R^6. We wish to be able to control the behavior of D_H when suitable contact transformations are applied to the unfolding H. We wish to be able to select a class of contact transformations with the property that, for any transform H' of H by one of these, the set $D_{H'}$, which consists of points in $S_{H'}$ at which $d_{xy}H'$ has trace zero, is again a non-empty smooth submanifold of dimension 3. In particular, we wish to avoid the situation of Example 6 of Chapter 2, in which the set of points at which the corresponding differential has pure imaginary eigenvalues is destroyed by an insufficiently restricted contact transformation. To use transversality to achieve this goal is the concern of the next section.

6.4 Versal Unfoldings and Contact Transformations of the First Order

The concept of transversality developed in the last section is the one we need to select a family of contact transformations together with a companion family of versal unfoldings for which the simple but drastic pathology of Example 6 of Chapter 2, cannot occur. The difficulty in Example 6 of Chapter 2 arises from the great generality of the concept of parameterized contact transformations. As indicated earlier in this chapter, these transformations are generalizations of ordinary parameterized coordinate transformations. Thus, we seek a subclass of contact transformations that is smaller than the entire class but still contains the coordinate transformations. Such a subclass is given by the class of contact transformations of order k.

Recall how a contact transformation at $(0, 0) \in R^n \times R^p$ is given (see Section 2.2). First it consists of an invertible $n \times n$ matrix $M(x, \lambda)$ depending smoothly on $x \in R^n$ and $\lambda \in R^p$ near $(0, 0)$. Second, it consists of a function $\phi: R^n \times R^p \to R^n$ such that i) $\phi(0, 0) = 0$, ii) $\phi(x, \lambda)$ depends smoothly on $x \in R^n$ and $\lambda \in R^p$ near $(0, 0)$, and iii) $d_x\phi(0, 0)$ is invertible. Finally, it consists of a smooth function $\beta: R^p \to R^p$ such that i) $\beta(0) = 0$, ii) $\beta(\lambda)$ depends smoothly on λ near 0, and iii) $d\beta(0)$ is invertible. That is, $\phi(x, \lambda)$ is a parameterized change of coordinates, and $\beta(\lambda)$ is a change of coordinates.

The contact transformation determined by $[M, \phi, \beta]$ then transforms an unfolding

$$U: \quad R^n \times R^p \to R^n \tag{6.181}$$

into a new unfolding

$$U': \quad R^n \times R^p \to R^n \tag{6.182}$$

defined by

$$U'(x, \lambda) = M(x, \lambda) \cdot U(\phi(x, \lambda), \beta(\lambda)) \tag{6.183}$$

Now our interest in unfoldings arises from the parameterized differential equations that they define. And as we have seen in Section 6.1, the coordinate transformation

$$\left.\begin{aligned} y &= \phi(x, \lambda) \\ \mu &= \beta(\lambda) \end{aligned}\right\} \tag{6.184}$$

applied to the differential equation

$$\dot{y} = U(y, \mu) \tag{6.185}$$

produces the differential equation

$$\dot{x} = \left[d_x\phi(x, \lambda)\right]^{-1}\cdot U\left(\phi(x, \lambda), \beta(\lambda)\right) \tag{6.186}$$

Thus, the effect of the contact transformation determined by (M, ϕ, β) on a differential equation is the same as that of a coordinate transformation if and only if

$$M(x, \lambda) = \left[d_x\phi(x, \lambda)\right]^{-1} \tag{6.187}$$

Because we are interested only in the effects of transformation on differential equations, we might as well identify a coordinate transformation with the contact transformation it determines. Then it makes sense to compare contact transformations with coordinate transformations. In particular, we will say that the contact transformation determined by (M, ϕ, β) is a <u>coordinate transformation up to the kth order at the origin</u> if the Taylor expansions of $\left[d_x\phi(x, \lambda)\right]^{-1}$ and $M(x, \lambda)$ about the origin agree up to terms of the kth order; more briefly we will say that (M, ϕ, β) is a <u>kth-order contact transformation</u>. Thus, the contact transformations of Chapter 2 are ordinary transformations. As we saw in Example 6 of Chapter 2, they may not preserve the stability data of an unfolding. Our objectives in the remainder of this chapter are, first, to see that first-order contact transformations do preserve the stability data of suitable unfoldings, and, second, to develop the appropriate unfoldings following the theory of Mather.

These unfoldings, companions to first-order contact transformations, will be <u>versal unfoldings of the first order</u>. To define these, we begin by noting that if the unfolding

$$U:\ R^n \times R^p \to R^n \tag{6.188}$$

is versal, then the map (6.188) is transverse to $\{0\}$ in R^n near the origin of $R^n \times R^p$. We may restate the Transversality Condition (2.53) of Chapter 2 as the requirement that for any smooth n-vector function $f(x)$ defined near 0, there exist constants $a_1, \ldots, a_p$ and a smooth n-vector function $G(x)$ such that

$$f(x) = d_xU(x, 0)\cdot G(x) + H(x)\cdot f(x) + \sum_{j=1}^{p} a_j \left.\frac{\partial U(x, \lambda)}{\partial\lambda_j}\right|_{\lambda=0} \tag{6.189}$$

where, as usual, d_xU denotes the matrix of partial derivatives of U with respect to the first n variables $x_1, \ldots, x_n$. In comparing (6.189) with (2.53), we notice that the sum in (6.189) corresponds to $N(x)\cdot\gamma$ of (2.53). By taking f(x) in (6.189) to be successively the constant functions $f(x) = e_i$, where $e_1, \ldots, e_n$ is the standard basis for R^n, we find constants a_{ij} for $i = 1, \ldots, n$ and $j = 1, \ldots, p$, and n-vector functions $G_i(x)$ and matrix functions $H_i(x)$ for $i = 1, \ldots, n$ such that

$$e_i = d_xU(x, 0)\cdot G_i(x) + H_i(x)\cdot f(x) + \sum_{j=1}^{p} a_{ij} \frac{\partial U(x, 0)}{\partial \lambda_j} \quad \text{for } i = 1, \ldots, n \tag{6.190}$$

We set x = 0 in (6.190) and conclude that

$$e_i = dU(0, 0)\cdot\begin{bmatrix} \xi_i \\ \eta_i \end{bmatrix} \tag{6.191}$$

where

$$dU(0, 0) = \left(d_xU(0, 0), d_\lambda U(0, 0)\right) \tag{6.192}$$

f(0) = 0, and

$$\left.\begin{aligned} \eta_i &= \begin{bmatrix} a_{i1} \\ \vdots \\ a_{ip} \end{bmatrix} \\ \xi_i &= G_i(0) \end{aligned}\right\} \tag{6.193}$$

Thus the map

$$dU(0, 0): R^n \times R^p \to R^n \tag{6.194}$$

is onto, and we conclude that $U: R^n \times R^p \to R^n$ is transverse to $\{0\}$ in R^n near the origin of $R^n \times R^p$.

From the fact that $U: R^n \times R^p \to R^n$ is transverse to $\{0\}$ in R^n near the origin of $R^n \times R^p$, it follows that the stationary set

$$S_U = \{(x, \lambda) \mid U(x, \lambda) = 0\} \tag{6.195}$$

is a smooth submanifold of $R^n \times R^p$ near the origin of $R^n \times R^p$. Already

$$d_xU: R^n \times R^p \to M_n \tag{6.196}$$

is a smooth map. By restricting $d_x U$ to S_U, we obtain a smooth map

$$d_x U: \quad S_U \to M_n \tag{6.197}$$

Now we may make our definition: The versal unfolding U is <u>versal of the first order</u> if it satisfies, in addition to (6.189), the additional transversality condition that the map (6.197) is transverse to $\mathrm{Orb}\left(d_x U(0, 0)\right)$ in M_n.

We note that although the Transversality Condition (2.53) of Chapter 2 is so called because it implies the transversality of a great many maps (such as U), it does <u>not</u> imply the transversality of the map (6.197).

At the end of Example 14, we asked a question: Why use transversality? We now have enough machinery to produce the answer, but before doing so we will add a parameter to an ordinary versal unfolding to obtain an example of a versal unfolding of the first order.

<u>Example 15. An extended hyperbolic umbilic</u>

Once again we consider the unfolding given by

$$H(x, y, \alpha, \varepsilon, \gamma, \delta) = \begin{bmatrix} xy + \alpha + \varepsilon x \\ x^2 - y^2 + \gamma + \delta y \end{bmatrix} \tag{6.198}$$

We have already calculated $d_{xy}H$ in (6.167) and we see that

$$d_{xy}H(0) = 0 \tag{6.199}$$

Then

$$\mathrm{Orb}\left(d_{xy}H(0)\right) = \mathrm{Orb}(0) = \{0\} \subset R^4 \tag{6.200}$$

We have already seen in Chapter 2 that H is a versal unfolding of $\begin{bmatrix} xy \\ x^2-y^2 \end{bmatrix}$. Then H will be versal of the first order according to our definition if and only if

$$S_H \overset{d_{xy}H}{\to} M_2 = R^4 \tag{6.201}$$

is transverse to $\mathrm{Orb}\left(d_{xy}H(0)\right) = \{0\}$ in $M_2 = R^4$. In turn, the map $d_{xy}H$ is transverse to $\{0\}$ if and only if the map

$$d(d_{xy}H)(0): \quad T_0 S_H \to R^4 \tag{6.202}$$

is onto. Unfortunately, we have already seen in Example 14 that the map $d(d_{xy}H)(0)$ has a 3-dimensional image, even when extended from $T_0 S_H$ to all of $R^2 \times R^4$; thus it cannot be onto.

Now, we add a parameter μ to H to produce a new unfolding

$$H_e(x, y, \alpha, \varepsilon, \gamma, \delta, \mu) = \begin{bmatrix} xy + \alpha + \varepsilon x + \mu y \\ x^2 - y^2 + \gamma + \delta y \end{bmatrix} \tag{6.203}$$

Identifying M_2 with R^4 as in (6.170) in Example 14, we see that we may write

$$d_{xy}H_e \leftrightarrow \begin{bmatrix} y + \varepsilon \\ x + \mu \\ 2x \\ \delta - 2y \end{bmatrix} \tag{6.204}$$

Then we have

$$d(d_{xy}H_e) \leftrightarrow \begin{array}{c} \begin{matrix} x & y & \alpha & \varepsilon & \gamma & \delta & \mu \end{matrix} \\ \begin{bmatrix} 0 & 1 & 0 & 1 & 0 & 0 & 0 \\ 1 & 0 & 0 & 0 & 0 & 0 & 1 \\ 2 & 0 & 0 & 0 & 0 & 0 & 0 \\ 0 & -2 & 0 & 0 & 0 & 1 & 0 \end{bmatrix} \end{array} \tag{6.205}$$

Thus

$$d(d_{xy}H_e): \; R^2 \times R^5 \to R^4 \tag{6.206}$$

is onto because $d(d_{xy}H_e)$ has rank 4.

But is the restriction of $d(d_{xy}H_e)$ to $T_0S_{H_e}$,

$$d(d_{xy}H_e): \; T_0S_{H_e} \to R^4 \tag{6.207}$$

also onto? In analogy with (6.161) we observe from (6.203) that S_{H_e} may be parameterized by the function f_e defined by setting

$$f_e(x, y, \varepsilon, \delta, \mu) = \begin{bmatrix} x \\ y \\ -xy - \varepsilon x - \mu y \\ \varepsilon \\ -x^2 + y^2 - \delta y \\ \delta \\ \mu \end{bmatrix} \tag{6.208}$$

so that $T_0 S_{H_e}$ is spanned by the columns of the matrix

$$df_e(0) = \begin{bmatrix} 1 & 0 & 0 & 0 & 0 \\ 0 & 1 & 0 & 0 & 0 \\ 0 & 0 & 0 & 0 & 0 \\ 0 & 0 & 1 & 0 & 0 \\ 0 & 0 & 0 & 0 & 0 \\ 0 & 0 & 0 & 1 & 0 \\ 0 & 0 & 0 & 0 & 1 \end{bmatrix} \tag{6.209}$$

Now it is easy to see from (6.205) and (6.209) that the composition or matrix product is

$$d\left(d_{xy}H_e(0)\right)\cdot df_e(0) = \begin{bmatrix} 0 & 1 & 1 & 0 & 0 \\ 1 & 0 & 0 & 0 & 1 \\ 2 & 0 & 0 & 0 & 0 \\ 0 & -2 & 0 & 1 & 0 \end{bmatrix} \tag{6.210}$$

As before, it follows that the map of (6.207) is onto.

Adding a parameter to a versal unfolding produces a versal unfolding; thus H_e is versal. On the other hand, because the map of (6.207) is onto, the map is transverse to $\{0\} = \mathrm{Orb}\left(d_{xy}H_e(0)\right)$ in R^4. Thus H_e is versal of the first order according to our definition. #

We see from this example that a first-order versal unfolding $U:R^n \times R^p \to R^n$ about the origin of $R^n \times R^p$ comes equipped with a pair of natural smooth submanifolds S_U and D_U^o of $R^n \times R^p$ that are related by

$$S_U \supset D_U^o \tag{6.211}$$

The submanifold S_U of $R^n \times R^p$ is defined by setting

$$S_U = \{(x, \mu) \mid U(x, \mu) = 0\} \tag{6.212}$$

and the submanifold D_U^o of $R^n \times R^p$ is defined by setting

$$D_U^o = \{(x, \mu) \mid U(x, \mu) = 0 \text{ and } d_x U(x,\mu) \in \mathrm{Orb}(d_x U(0, 0))\} \tag{6.213}$$

That these are submanifolds of $R^n \times R^p$ near the origin follows from the transversality conditions in the definition of first-order versality; their dimensions are given by

$$\dim(S_U) = p \tag{6.214}$$

and

$$\dim(D_U^o) = n + p - n^2 + \dim\left[\mathrm{Orb}(d_x U(0, 0))\right] \tag{6.215}$$

An ordinary versal unfolding $V:R^n \times R^p \to R^n$ has only one natural submanifold S_V of $R^n \times R^p$ associated with it. As we saw in Chapter 2, this set S_V has the indispensable property that if V' is obtained from V by applying an ordinary contact transformation to V, then the transformation carries S_V to $S_{V'}$ in a very strong sense. In the same way, we wish the pair (S_U, D_U^o) to have the property that if U' is obtained from U by applying a first-order contact transformation, then the transformation carries (S_U, D_U^o) to $(S_{U'}, D_{U'}^o)$ in some reasonably strong sense. Because a first-order contact transformation is an ordinary transformation, we already have a strong sense in which S_U is carried to $S_{U'}$. Thus it is D_U^o that we must examine.

The first problem that may arise is that, like the transformation of Example 6 in Chapter 2, the first-order contact transformation may "destroy" the set D_U^o. We regard D_U^o as "destroyed" if, for one example, $D_{U'}^o$ has different dimension from D_U^o or if, for another example, $d_x U'(0, 0)$ is not in $\mathrm{Orb}(d_x U(0, 0))$. The following proposition then shows that at least D_U^o is not destroyed in any of the senses that we may attempt to make formal.

Proposition 6.1 (Preservation Lemma). Let $U:R^n \times R^p \to R^n$ be a smooth versal unfolding near (0, 0), and let P be a smooth invariant submanifold of M_n containing $d_x U(0, 0)$. Suppose that the map

$$S_U \to M_n \tag{6.216}$$

defined by

$$(x, \mu) \to d_x U(x, \mu) \tag{6.217}$$

is transverse to P near (0, 0). Let (M, ϕ, β) be a first-order contact transformation, and let $U': R^n \times R^p \to R^n$ be obtained by applying the contact transformation (M, ϕ, β) to U. Then we have the following two results:

1) $d_x U'(0, 0) \in \mathrm{Orb}\big(d_x U(0, 0)\big)$ (6.218)

2) Near the origin, the map

$$S_{U'} \to M_n \tag{6.219}$$

defined by

$$(x, \mu) \to d_x U'(x, \mu) \tag{6.220}$$

is transverse to P in M_n.

We notice first that this proposition does not apply only to first-order versal unfoldings. Because the submanifold P of R^n need not be $\mathrm{Orb}\big(d_x U(0, 0)\big)$, the proposition applies, for example, to the hyperbolic umbilic of Example 14. To apply the proposition in that case, we may use $P = T_0$, the set of matrices having trace zero.

Second, we notice that in the case $P = \mathrm{Orb}\big(d_x U(0, 0)\big)$, the proposition implies that if U is a first-order versal unfolding, then so is U'. That is, a first-order contact transformation carries first-order versal unfoldings to first-order versal unfoldings.

Finally, we notice, again in the case that $P = \mathrm{Orb}\big(d_x U(0, 0)\big)$, that D_U^o is "destroyed" in neither of the two senses we suggested. The first conclusion of the proposition tells us that $d_x U(0, 0)$ is conjugate to $d_x U'(0, 0)$; thus

$$\mathrm{Orb}\big(d_x U(0, 0)\big) = \mathrm{Orb}\big(d_x U'(0, 0)\big) \tag{6.221}$$

Then the second conclusion, together with the calculation (6.215) of the dimension of D_U^o and with (6.221), tells us that

$$\dim(D_U^o) = \dim(D_{U'}^o) \tag{6.222}$$

Now we examine two first-order versal unfoldings.

Example 16. First-order contact transformations of the extended hyperbolic umbilic.

We begin with a first-order contact transformation of the extended hyperbolic umbilic (6.203) discussed in Example 15,

$$H_e(x, y, \alpha, \varepsilon, \gamma, \delta, \mu) = \begin{bmatrix} xy + \alpha + \varepsilon x + \mu y \\ x^2 - y^2 + \gamma + \delta y \end{bmatrix} \tag{6.223}$$

The transformation will be the coordinate transformation defined by (M, ϕ, β) with

$$\left.\begin{aligned} &M(x, y, \alpha, \varepsilon, \gamma, \delta, \mu) = 1 \\ &\begin{bmatrix} x' \\ y' \end{bmatrix} = \phi(x, y, \alpha, \varepsilon, \gamma, \delta, \mu) = \begin{bmatrix} x + \mu \\ y \end{bmatrix} \\ &(\alpha', \varepsilon', \gamma', \delta', \nu) = \beta(\alpha, \varepsilon, \gamma, \delta, \mu) = (\alpha - \varepsilon\mu, \varepsilon, \gamma + \mu^2, \delta, -2\mu) \end{aligned}\right\} \tag{6.224}$$

We check by substituting (6.224) into (6.223) that the transformed unfolding H_e' is given by

$$H_e'(x', y', \alpha', \varepsilon', \gamma', \delta', \nu) = \begin{bmatrix} x'y' + \alpha' + \varepsilon'x' \\ x'^2 - y'^2 + \gamma' + \delta'y' + \nu x' \end{bmatrix} \tag{6.225}$$

Proposition 6.1 tells us that transformation of a first-order versal unfolding by a first-order contact transformation yields a first-order versal unfolding. Consequently, (6.225) gives us an alternative to (6.223) for extending the hyperbolic umbilic to a first-order versal unfolding. #

Our next example is an application of Proposition 6.1 to a case in which the unfolding is not first-order versal.

Example 17. First-order contact transformation of the hyperbolic umbilic.

For this example, we consider the ordinary hyperbolic umbilic,

$$H(x, y, \alpha, \varepsilon, \gamma, \delta) = \begin{bmatrix} xy + \alpha + \varepsilon x \\ x^2 - y^2 + \gamma + \delta y \end{bmatrix} \tag{6.226}$$

We apply to H the first-order contact transformation defined by

$$\left.\begin{aligned} M(x, y, \alpha, \varepsilon, \gamma, \delta) &= \begin{bmatrix} 1+\varepsilon^3 & x^2 \\ 0 & 1+\delta^2\gamma^3 \end{bmatrix} \\ \phi(x, y, \alpha, \varepsilon, \gamma, \delta) &= \begin{bmatrix} x \\ y \end{bmatrix} \\ \beta(\alpha, \varepsilon, \gamma, \delta) &= (\alpha, \varepsilon, \gamma, \delta) \end{aligned}\right\} \tag{6.227}$$

We obtain a new unfolding defined by (6.228)

$$H'(x, y, \alpha, \varepsilon, \gamma, \delta) = \begin{bmatrix} xy + x^4 - x^2y^2 + \alpha + \alpha\varepsilon^3 + (\varepsilon+\varepsilon^4)x + \varepsilon^3xy + \gamma x^2 + \delta x^2y \\ x^2 - y^2 + \gamma + \delta^2\gamma^4 + (\delta+\delta^3\gamma^3)y + \delta^2\gamma^3x^2 - \delta^2\gamma^3y^2 \end{bmatrix}$$

As usual we set

$$S_{H'} = \{(x, y, \alpha, \varepsilon, \gamma, \delta) \mid H'(x, y, \alpha, \varepsilon, \gamma, \delta) = 0\} \tag{6.229}$$

and Proposition 6.1 assures us that near the origin $S_{H'}$ is a smooth submanifold of $R^2 \times R^4$ of dimension 4. We may check this prediction by observing that the Implicit Function Theorem allows us to solve

$$H'(x, y, \alpha, \varepsilon, \gamma, \delta) = 0 \tag{6.230}$$

near the origin for α and γ in terms of the remaining four variables. If the solution is given by

$$\begin{aligned} \alpha &= \hat{\alpha}(x, y, \varepsilon, \delta) \\ \gamma &= \hat{\gamma}(x, y, \varepsilon, \delta) \end{aligned} \tag{6.231}$$

then we see that a parameterization of $S_{H'}$ near the origin is given by

$$f(x, y, \varepsilon, \delta) = \begin{bmatrix} x \\ y \\ \hat{\alpha}(x, y, \varepsilon, \delta) \\ \varepsilon \\ \hat{\gamma}(x, y, \varepsilon, \delta) \\ \delta \end{bmatrix} \tag{6.232}$$

Proposition 6.1 goes further and assures us that, near the origin, the subset D of $S_{H'}$ given by

$$D = \{(x, y, \alpha, \varepsilon, \gamma, \delta) \mid H'(x, y, \alpha, \varepsilon, \gamma, \delta) = 0 \quad \text{and} \quad tr(d_{xy}H'(x, y, \alpha, \varepsilon, \gamma, \delta))\} = 0 \tag{6.233}$$

is a non-empty smooth submanifold of dimension 3. Again, by writing down $d_{xy}H'(x, y, \alpha, \varepsilon, \gamma, \delta)$ and inspecting the two equations in the definition (6.233) we see that we may solve them near the origin for α, γ, y in terms of the remaining three variables,

$$\left.\begin{aligned} \alpha &= \tilde{\alpha}(x, \varepsilon, \delta) \\ \gamma &= \tilde{\gamma}(x, \varepsilon, \delta) \\ y &= \tilde{y}(x, \varepsilon, \delta) \end{aligned}\right\} \tag{6.234}$$

Then we see that a parameterization of $S_{H'}$ near the origin is given by

$$g(x, \varepsilon, \delta) = \begin{bmatrix} x \\ \tilde{y}(x, \varepsilon, \delta) \\ \tilde{\alpha}(x, \varepsilon, \delta) \\ \varepsilon \\ \tilde{\gamma}(x, \varepsilon, \delta) \\ \delta \end{bmatrix} \tag{6.235}$$

#

Thus we have seen in the above example that the Implicit Function Theorem may be used to show that $S_{H'}$ and D are smooth submanifolds of $R^2 \times R^4$. But, then, why use Proposition 6.1? Proposition 6.1 is useful because it provides exactly the blanket assurance that, for every unfolding $U(x, \mu)$ satisfying its hypotheses, it will be possible to argue as above that, near the origin, both S_U and

$$D_U(P) = \{(x, \mu) \mid U(x, \mu) = 0 \quad \text{and} \quad d_x U(x, \mu) \in P\} \tag{6.236}$$

are smooth submanifolds of $R^n \times R^p$. That is, having checked the hypotheses, we do not need to find on an ad hoc basis the variables (such as α and γ, or α, γ and y), for which to solve in order to parameterize S_U or D.

Finally, we have to see in what sense a first-order contact transformation (M, ϕ, β) preserves the smooth submanifold associated with an unfolding. We begin by stating a proposition.

Proposition 6.2 (Preservation of Natural Manifolds). Let

$$U: \quad R^n \times R^p \to R^n \tag{6.237}$$

be a versal unfolding. Let

$$P \subset M_n \tag{6.238}$$

be an invariant smooth submanifold of M_n containing the matrix $d_xU(0, 0)$. Let

$$S_U = \{(x, \mu) \mid U(x, \mu) = 0\} \tag{6.239}$$

be the stationary set of U. Suppose that the restricted map

$$d_xU: \quad S_U \to M_n \tag{6.240}$$

is transverse to P in M_n near the origin. Let (M, ϕ, β) be a first-order contact transformation, and let the unfolding U' be obtained by applying (M, ϕ, β) to U. Let

$$S_{U'} = \{(x, \mu) \mid U'(x, \mu) = 0\} \tag{6.241}$$

be the stationary set of U', and define the secondary sets

$$D_U(P) = \{(x, \mu) \mid (x, \mu) \in S_U \quad \text{and} \quad d_xU(x, \mu) \in P\} \tag{6.242}$$

$$D_{U'}(P) = \{(x, \mu) \mid (x, \mu) \in S_{U'} \quad \text{and} \quad d_xU'(x, \mu) \in P\} \tag{6.243}$$

Then, in addition to the conclusions of Proposition 6.1, we have that

$$(\phi, \beta): \quad R^n \times R^p \to R^n \times R^p \tag{6.244}$$

is a diffeomorphism near the origin,

$$(\phi, \beta)S_{U'} = S_U \quad \text{near the origin} \tag{6.245}$$

and

$$(\phi, \beta)D_{U'}(P) \text{ agrees with } D_U(P) \text{ to the first order near the origin} \tag{6.246}$$

Conclusion (6.244) is merely a restatement of the fact that (ϕ, β) is a coordinate transformation. Conclusion (6.245) is obvious from the fact that

(M, ϕ, β) already must be an ordinary contact transformation. The essential content of the proposition lies in conclusion (6.246), which, by definition, states that $D_U(P)$ and $(\phi, \beta)D_{U'}(P)$ are smooth submanifolds of $R^n \times R^p$, both containing the origin and having the same tangent plane there (Figure 6.14). The agreement of two submanifolds up to a given order at a point p_o is a classical concept paralleling the agreement of two functions up to a given order at a point in both their domains. Agreement to the null order means that the surfaces share at least the point p_o; equality to the first order means that they are at least tangent at p_o; equality to the second order means that they osculate at p_o; and so on. Here we are interested only in first-order agreement. We note that to obtain higher-order contact, we use higher-order contact transformations. Also, we note that for a contact transformation defined by a coordinate transformation, we obtain identity of $(\phi, \beta)D_{U'}(P)$ and $D_U(P)$ near the origin.

We see that the sense (6.246), in which a first-order contact transformation carries the secondary set $D_{U'}(P)$ into the secondary set $D_U(P)$, is considerably weaker than the sense (6.245). As we will see in the remaining sections, this weaker sense is still strong enough to be useful.

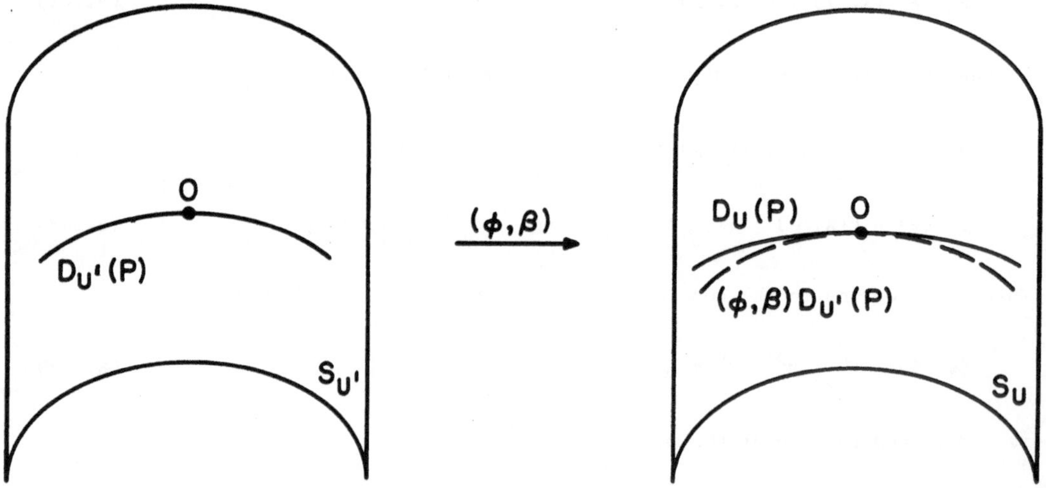

Fig. 6.14 The effect of the first-order contact transformation (M, ϕ, β). The coordinate transformation (ϕ, β) carries the stationary set $S_{U'}$ of the unfolding U' precisely onto that of U, near the origin. However, it carries the secondary set $D_{U'}(P)$ only onto one which approximates $D_U(P)$ to the first order near the origin.

6.5 Stability and First-Order Versal Unfoldings and Contact Transformations

As suggested by the examples at the end of the preceeding section, any first-order contact transformation $t_0 = (M, \phi, \beta)$ may be decomposed into two transformations of two pure kinds: We may first apply the contact transformation

$$t_1 = (d\phi^{-1}, \phi, \beta) \tag{6.247}$$

induced by the coordinate transformation (ϕ, β), and then we may apply the contact transformation

$$t_2 = (M \cdot d\phi, id, id) \tag{6.248}$$

where id is the identity map.

For convenience we express this observation formally.

Proposition 6.3 (Factorization Lemma). Any first-order contact transformation $t_0 = (M, \phi, \beta)$ may be uniquely expressed as the composition

$$t_0 = t_2 t_1 \tag{6.249}$$

of two first-order contact transformations, one, t_1, induced by a coordinate transformation, and the other, t_2, defined by

$$t_2 = (M', id, id) \tag{6.250}$$

with $M' = 1$ to the first order.

For another point of view, we recall that in Chapter 2 we obtained some useful information, notably the Lyapunov-Schmidt splitting decomposition, about ordinary contact transformations. We would like to transfer this information to first-order transformations, and a reasonably efficient way to do so is to observe that an ordinary transformation may be turned into a first-order transformation by multiplying it by an invertible $n \times n$ matrix

$$\Gamma = \Gamma_0 + \Gamma_1(x) + \Gamma_2(\mu) \tag{6.251}$$

where Γ_0 is constant, $\Gamma_1(x)$ is linear in x, and $\Gamma_2(\mu)$ is linear in μ. Specifically, if

$$t = (M, \phi, \beta) \tag{6.252}$$

is an ordinary contact transformation, then

$$t' = (\Gamma \cdot M, \phi, \beta) \tag{6.253}$$

is a first-order contact transformation, where

$$\Gamma = \text{const} + \text{lin}(x) + \text{lin}(\mu) \tag{6.254}$$

is the first-order Maclaurin expansion of $d_x\phi(x, \mu)\cdot M(x, \mu)^{-1}$.

We apply the above simple observation to obtain a first-order version of the parameterized Lyapunov-Schmidt Splitting Lemma.

Proposition 6.4 (First-Order Lyapunov-Schmidt Splitting Lemma). Suppose that $F: R^n \times R^p \to R^n$ is an unfolding, and suppose that

$$\text{rank}\left[d_x F(0, 0)\right] = r \tag{6.255}$$

Then for any $s \le r$, there is an invertible $n \times n$ matrix Γ, depending linearly on x and μ as in (6.254), an unfolding $V: R^{n-s} \times R^p \to R^{n-s}$, and a first-order contact transformation carrying F to

$$F'(x, \mu) = \Gamma \cdot \begin{bmatrix} x_1 \\ \vdots \\ x_s \\ V(x_{s+1}, \ldots, x_n, \mu_1, \ldots, \mu_p) \end{bmatrix} \tag{6.256}$$

As with the original Lyapunov-Schmidt Lemma, we give a proof because the proof provides the means for applying the lemma.

Proof: The ordinary Lyapunov-Schmidt Splitting Procedure produces an ordinary contact transformation (M, ϕ, β) and an unfolding $V: R^{n-s} \times R^p \to R^{n-s}$ such that

$$\begin{bmatrix} x_1 \\ \vdots \\ x_s \\ V(x_{s+1}, \ldots, x_n, \mu_1, \ldots, \mu_p) \end{bmatrix} = M(x, \mu)\cdot F\left(\phi(x, \mu), \beta(\mu)\right) \tag{6.257}$$

Let Γ be defined by the first-order Maclaurin expansion (6.254). Then $(\Gamma\cdot M, \phi, \beta)$ is a first-order contact transformation. If F' is the unfolding to which $(\Gamma\cdot M, \phi, \beta)$ carries F, then we have (6.256) and the proposition is proved.

An occasionally useful observation is contained in the following proposition.

Proposition 6.5 (Extension Lemma). The unfolding

$$F(u, v, \mu) = \begin{bmatrix} u \\ G(v, \mu) \end{bmatrix} \tag{6.258}$$

is first-order versal if and only if $G(v, \mu)$ is first-order versal.

Although there is not much opportunity to apply Proposition 6.5, because normally the "twist" Γ in (6.256) is non-trivial, the following proposition is very useful.

Proposition 6.6 (Expansion Lemma). Let

$$U: \ R^n \times R^p \to R^n \tag{6.259}$$

be a first-order versal unfolding. Let

$$P \subset M_n \tag{6.260}$$

be a smooth invariant submanifold of M_n containing $d_xU(0, 0)$ and let S_U be the stationary set (6.239) of U. Then the restriction

$$d_xU: \ S_U \to M_n \tag{6.261}$$

is transverse to P in M_n.

Another simple proposition that will be very useful is the following one.

Proposition 6.7 (Change of Coordinates Lemma). If the unfolding U' is obtained from the unfolding U by applying a coordinate transformation, then U' is first-order versal if and only if U is first-order versal.

Now we present an example that will illustrate both the relative simplicity of ordinary versality preserved by our extended point of view and the additional complexity that point of view brings to the problem.

Example 18. The modified Lorenz system unfolded further

In Section 3.2, we derived a physically interpretable versal unfolding of the Lorenz system. This unfolding was given by (3.63) - (3.65), which can be written as

$$\left.\begin{aligned} \dot{x}_1 &= -\sigma x_1 + \sigma x_2 + \sigma \mu \\ \dot{x}_2 &= -x_1x_3 + (\nu + 1)\, x_1 - x_2 \\ \dot{x}_3 &= x_1x_2 - bx_3 + (2\mu/3)x_1 \end{aligned}\right\} = L(x, \mu, \nu) \tag{6.262}$$

Here $\sigma > 0$ and $b > 0$ are held fixed while μ and ν are the control parameters d_0 and

d_1. We seek a first-order versal unfolding which extends (6.262) about the stationary point $x = 0$, $\mu = 0$, $\nu = 0$. To do so we find a Lyapunov-Schmidt splitting of (6.262) with

$$\left.\begin{aligned} u_1 &= x_1 \\ u_2 &= x_3 \\ v &= x_2 \end{aligned}\right\} \tag{6.263}$$

Then the stationary set S_L of (6.262) is described by

$$\left.\begin{aligned} x_1 &= x_2 + \mu \\ x_3 &= (x_2 + \mu)(3x_2 + 2\mu)(3b)^{-1} \\ 0 &= \left(x_2 + \frac{8\mu}{9}\right)^3 - \left(\frac{\mu^2}{27} + b\nu\right)\left(x_2 + \frac{8\mu}{9}\right) - \left(\frac{2\mu^3}{729} + \frac{b\nu\mu}{9} + b\mu\right) \end{aligned}\right\} \tag{6.264}$$

That is,

$$S_L = \{(x, \mu, \nu) \mid (6.264) \text{ holds}\} \tag{6.265}$$

To parameterize S_L, it is convenient to observe that the Implicit Function Theorem implies that

$$\left.\begin{aligned} \gamma &= \frac{\mu^2}{27} + b\nu \\ \delta &= \frac{2\mu^3}{729} + \frac{b\nu\mu}{9} + b\mu \end{aligned}\right\} \tag{6.266}$$

may be solved uniquely near the origin for μ and ν in terms of γ and δ; that is,

$$\left.\begin{aligned} \mu &= \hat{\mu}(\gamma, \delta) \\ \nu &= \hat{\nu}(\gamma, \delta) \end{aligned}\right\} \tag{6.267}$$

Then a function f parameterizing S_L near the origin in terms of the variables γ and

$$z = x_2 + 8\mu/9 \tag{6.268}$$

is given by

$$\left.\begin{aligned}
x_1 &= z + \tfrac{1}{9}\hat{\mu}(\gamma,\ z^3 - \gamma z)\\
x_2 &= z - \tfrac{8}{9}\hat{\mu}(\gamma,\ z^3 - \gamma z)\\
x_3 &= \left[z + \tfrac{1}{9}\hat{\mu}(\gamma,\ z^3 - \gamma z)\right]\left[3z - \tfrac{2}{3}\hat{\mu}(\gamma,\ z^3 - \gamma z)\right](3b)^{-1}\\
\mu &= \hat{\mu}(\gamma,\ z^3 - \gamma z)\\
\nu &= \hat{\nu}(\gamma,\ z^3 - \gamma z)
\end{aligned}\right\} = f(z,\ \gamma) \tag{6.269}$$

Having parameterized S_L, we seek the additional parameters necessary to make the map

$$S_L \to M_3 \tag{6.270}$$

given by

$$(x,\ \mu,\ \nu) \to d_x L(x,\ \mu,\ \nu) \tag{6.271}$$

transverse to $\mathrm{Orb}\big(d_x L(0,\ 0,\ 0)\big)$. For simplicity, we assume that

$$b \neq \sigma + 1 \tag{6.272}$$

We calculate

$$d_x L = \begin{bmatrix} -\sigma & \sigma & 0 \\ \nu + 1 - x_3 & -1 & -x_1 \\ \frac{2\mu}{3} + x_2 & x_1 & -b \end{bmatrix} \tag{6.273}$$

so that

$$d_x L(0,\ 0,\ 0) = \begin{bmatrix} -\sigma & \sigma & 0 \\ 1 & -1 & 0 \\ 0 & 0 & -b \end{bmatrix} \tag{6.274}$$

Hence the eigenvalues of $d_x L(0,\ 0,\ 0)$ are 0, $-\sigma - 1$, and $-b$, all distinct by (6.272). Consequently, we discover via (6.77) that

$$\dim\left[\mathrm{Orb}\big(d_x L(0,\ 0,\ 0)\big)\right] = 3^2 - 1^2 - 1^2 - 1^2 = 6 \tag{6.275}$$

For our transversality calculation, we will need the tangent space $T_A \mathrm{Orb}\big(d_x L(0, 0, 0)\big)$ in which

$$A = d_x L(0, 0, 0) \tag{6.276}$$

We recall from Example 11, that this tangent space is the image of the linear map

$$L_A: M_3 \to M_3 \tag{6.277}$$

given by (6.127) as

$$L_A(X) = A X - X A \tag{6.278}$$

The vector space M_3 has for a basis the nine 3 × 3 matrices

$$\varepsilon_1 = \begin{bmatrix} 1 & 0 & 0 \\ 0 & 0 & 0 \\ 0 & 0 & 0 \end{bmatrix}, \varepsilon_2 = \begin{bmatrix} 0 & 1 & 0 \\ 0 & 0 & 0 \\ 0 & 0 & 0 \end{bmatrix}, \ldots, \varepsilon_9 = \begin{bmatrix} 0 & 0 & 0 \\ 0 & 0 & 0 \\ 0 & 0 & 1 \end{bmatrix} \tag{6.279}$$

Accordingly, the vector space $L_A(M_3) = T_A \mathrm{Orb}\big(d_x L(0, 0, 0)\big)$ will be spanned by the nine matrices

$$\left.\begin{aligned} L_A(\varepsilon_1) &= A \begin{bmatrix} 1 & 0 & 0 \\ 0 & 0 & 0 \\ 0 & 0 & 0 \end{bmatrix} - \begin{bmatrix} 1 & 0 & 0 \\ 0 & 0 & 0 \\ 0 & 0 & 0 \end{bmatrix} A \\ &\ \ \vdots \\ L_A(\varepsilon_9) &= A \begin{bmatrix} 0 & 0 & 0 \\ 0 & 0 & 0 \\ 0 & 0 & 1 \end{bmatrix} - \begin{bmatrix} 0 & 0 & 0 \\ 0 & 0 & 0 \\ 0 & 0 & 1 \end{bmatrix} A \end{aligned}\right\} \tag{6.280}$$

But with (6.275) we may find a basis for $T_A \mathrm{Orb}\big(d_x L(0, 0, 0)\big)$ by selecting a subset of six linearly independent matrices. Thus, we discover that a basis for $T_A \mathrm{Orb}\big(d_x L(0, 0, 0)\big)$ is given by

$$L_A(\varepsilon_1) = \begin{bmatrix} 0 & -\sigma & 0 \\ 1 & 0 & 0 \\ 0 & 0 & 0 \end{bmatrix} \tag{6.281}$$

$$L_A(\varepsilon_2) = \begin{bmatrix} -1 & 1-\sigma & 0 \\ 0 & 1 & 0 \\ 0 & 0 & 0 \end{bmatrix} \tag{6.282}$$

$$L_A(\varepsilon_3) = \begin{bmatrix} 0 & 0 & b-\sigma \\ 0 & 0 & 1 \\ 0 & 0 & 0 \end{bmatrix} \tag{6.283}$$

$$L_A(\varepsilon_6) = \begin{bmatrix} 0 & 0 & \sigma \\ 0 & 0 & b-1 \\ 0 & 0 & 0 \end{bmatrix} \tag{6.284}$$

$$L_A(\varepsilon_7) = \begin{bmatrix} 0 & 0 & 0 \\ 0 & 0 & 0 \\ \sigma-b & -\sigma & 0 \end{bmatrix} \tag{6.285}$$

$$L_A(\varepsilon_8) = \begin{bmatrix} 0 & 0 & 0 \\ 0 & 0 & 0 \\ -1 & 1-b & 0 \end{bmatrix} \tag{6.286}$$

Then the three-dimensional quotient space

$$Q_A = M_3/T_A\mathrm{Orb}\big(d_xL(0,\ 0,\ 0)\big) \tag{6.287}$$

has for a basis the three elements represented by

$$\alpha_1 = \begin{bmatrix} 0 & 0 & 0 \\ 0 & 0 & 0 \\ 0 & 0 & 1 \end{bmatrix},\ \alpha_2 = \begin{bmatrix} 0 & 0 & 0 \\ 0 & 1 & 0 \\ 0 & 0 & 0 \end{bmatrix},\ \alpha_3 = \begin{bmatrix} 0 & 1 & 0 \\ 0 & 0 & 0 \\ 0 & 0 & 0 \end{bmatrix} \tag{6.288}$$

so that (6.281)-(6.286) and (6.288) form an alternate basis for M_3.

To determine transversality, we must calculate the differential at the origin of the map $S_L \xrightarrow{d_xL} M_3$ given by (6.273). Because the map $R^2 \xrightarrow{f} S_L$ is a diffeomorphism near the origin, it suffices to calculate the differential of the composition

$$R^2 \xrightarrow{f} S_L \xrightarrow{d_x L} M_3 \tag{6.289}$$

which we write as F,

$$F = d_x L \circ f \tag{6.290}$$

so that from (6.273) and (6.290) we have

$$F(\gamma, z) = \begin{bmatrix} -\sigma & \sigma & 0 \\ \hat{\nu}+1 - f_3(\gamma, z) & -1 & -f_1(\gamma, z) \\ \frac{2\hat{\mu}}{3} + f_2(\gamma, z) & f_1(\gamma, z) & -b \end{bmatrix} \tag{6.291}$$

in which f_1, f_2 and f_3 denote the first, second and third components of f (6.269). For our purposes, it suffices to calculate the two images

$$\begin{aligned} \xi_1 &= dF(0, 0) \begin{bmatrix} 1 \\ 0 \end{bmatrix} = \frac{\partial F}{\partial \gamma}(0, 0) \\ \xi_2 &= dF(0, 0) \begin{bmatrix} 0 \\ 1 \end{bmatrix} = \frac{\partial F}{\partial z}(0, 0) \end{aligned} \tag{6.292}$$

We note that these are two 3 × 3 matrices. To find ξ_1 and ξ_2, we must find first the differentials

$$\left.\begin{aligned} df_k(0, 0) \begin{bmatrix} 1 \\ 0 \end{bmatrix} &= \partial_\gamma f_k(0, 0) \\ df_k(0, 0) \begin{bmatrix} 0 \\ 1 \end{bmatrix} &= \partial_z f_k(0, 0) \end{aligned}\right\} \quad \text{for } k = 1, 2, 3 \tag{6.293}$$

To determine these, we must evaluate at the origin the derivatives of $\hat{\mu}(\gamma, \delta)$ and $\hat{\nu}(\gamma, \delta)$ with respect to γ and δ. Using (6.266), we see that

$$\begin{bmatrix} \frac{\partial \gamma}{\partial \mu}(0, 0) & \frac{\partial \gamma}{\partial \nu}(0, 0) \\ \frac{\partial \delta}{\partial \mu}(0, 0) & \frac{\partial \delta}{\partial \nu}(0, 0) \end{bmatrix} = \begin{bmatrix} 0 & b \\ b & 0 \end{bmatrix} \tag{6.294}$$

but we seek the inverse of this matrix; that is,

$$\begin{bmatrix} \frac{\partial\hat{\mu}}{\partial\gamma}(0, 0) & \frac{\partial\hat{\mu}}{\partial\delta}(0, 0) \\ \frac{\partial\hat{\nu}}{\partial\gamma}(0, 0) & \frac{\partial\hat{\nu}}{\partial\delta}(0, 0) \end{bmatrix} = \begin{bmatrix} 0 & \frac{1}{b} \\ \frac{1}{b} & 0 \end{bmatrix} \tag{6.295}$$

Consequently, from (6.269), (6.295) and the chain rule we have

$$\left.\begin{aligned} \frac{\partial f_1}{\partial\gamma}(0, 0) = 0 \quad &, \quad \frac{\partial f_1}{\partial z}(0, 0) = 1 \\ \frac{\partial f_2}{\partial\gamma}(0, 0) = 0 \quad &, \quad \frac{\partial f_2}{\partial z}(0, 0) = 1 \\ \frac{\partial f_3}{\partial\gamma}(0, 0) = 0 \quad &, \quad \frac{\partial f_3}{\partial z}(0, 0) = 0 \end{aligned}\right\} \tag{6.296}$$

so that with the aid of (6.291)-(6.292) we calculate

$$\xi_1 = \begin{bmatrix} 0 & 0 & 0 \\ \frac{1}{b} & 0 & 0 \\ 0 & 0 & 0 \end{bmatrix} \tag{6.297}$$

$$\xi_2 = \begin{bmatrix} 0 & 0 & 0 \\ 0 & 0 & -1 \\ 1 & 1 & 0 \end{bmatrix} \tag{6.298}$$

But from (6.283)-(6.286) we discover that ξ_2 is in $T_A\mathrm{Orb}(d_xL(0, 0, 0))$ and so represents zero in our quotient space $Q_A = M_3/T_A\mathrm{Orb}(d_xL(0, 0, 0))$.

At this point in our calculations, we see that in order to achieve first-order versality we must find two parameters and two associated basis vectors which together with ξ_1 form an alternate basis for Q_A. One of the simplest ways to do so is to regard b as a control parameter and to regard σ as another. Then extending the column vector (γ, z) to (γ, z, b, σ), we see that for

$$q_o = (0, 0, b_o, \sigma_o) \tag{6.299}$$

$$dF(q_o)\begin{bmatrix} 0 \\ 0 \\ 1 \\ 0 \end{bmatrix} = \frac{\partial F}{\partial b}(q_o) = \begin{bmatrix} 0 & 0 & 0 \\ 0 & 0 & 0 \\ 0 & 0 & -1 \end{bmatrix} = \eta_1 \tag{6.300}$$

$$dF(q_o)\begin{bmatrix}0\\0\\0\\1\end{bmatrix} = \frac{\partial F}{\partial \sigma}(q_o) = \begin{bmatrix}-1 & 1 & 0\\ 0 & 0 & 0\\ 0 & 0 & 0\end{bmatrix} = \eta_2 \tag{6.301}$$

It is easy to see that

$$\left.\begin{aligned} &-\frac{1}{\sigma}L_A(\varepsilon_1) + \frac{b}{\sigma}\xi_1 = \alpha_3 \\ &L_A(\varepsilon_2) - L_A(\varepsilon_1) - (\eta_2 - b\xi_1) = \alpha_2 \\ &-\eta_1 = \alpha_1 \end{aligned}\right\} \tag{6.302}$$

so that ξ_1, η_1 and η_2 also form a basis for Q_A because $L_A(\varepsilon_1)$ and $L_A(\varepsilon_2)$ represent zero there. Consequently, the extended map of (6.291)

$$F(\gamma, z, b, \sigma) = \begin{bmatrix} -\sigma & \sigma & 0 \\ \hat{\nu}+1 - f_3 & -1 & -f_1 \\ \frac{2\hat{\mu}}{3} + f_2 & f_1 & -b \end{bmatrix} \tag{6.303}$$

is transverse to $\mathrm{Orb}(d_xL(0, 0, 0))$ in M_3 near q_o. We remark only that we have used the fact that

$$\frac{\partial\hat{\mu}}{\partial b}(q_o) = \frac{\partial\hat{\nu}}{\partial b}(q_o) = 0 \tag{6.304}$$

which is demonstrated easily via an argument similar to that used to obtain (6.295).

Finally, it is easy to write the unfolding that leads to (6.303), and it is given by

$$L^e(x, \mu, \nu, b, \sigma) = \begin{cases} -\sigma x_1 + \sigma x_2 + \sigma\mu \\ -x_1x_3 + (\nu+1)x_1 - x_2 \\ x_1x_2 - bx_3 + (2\mu/3)x_1 \end{cases} \tag{6.305}$$

Thus we have found a first-order versal unfolding which extends the ordinary versal unfolding (6.262). And, very pleasantly, we needed to adjoin only parameters that were originally given but previously viewed as unimportant qualitatively. #

We recall that originally, in pursuing the notion of higher-order versality, we intended to speak about stability and its preservation under transformations. Thus two particular types of submanifolds of M_n concern us: those where matrices have a single vanishing eigenvalue and those where matrices have a single complex conjugate pair of pure imaginary eigenvalues. Thus, we are interested primarily in transversality to submanifolds of M_n of one of these types. The fortunate circumstance is that first-order versality implies transversality to these two types of submanifolds of M_n as the following rather trivial proposition and its first corollary show.

Proposition 6.8 (Contagion of Transversality). If

$$f: M \to N \tag{6.306}$$

is transverse to $P \subset N$ near $x_0 \in M$, and Q is a submanifold of N containing P, then f is transverse to $Q \subset N$ near $x_0 \in M$.

The following corollary is the Expansion Lemma, Proposition 6.6.

Corollary 6.8.1. If $f: R^n \times R^p \to R^n$ is first-order versal, S_F is the stationary set of F, and Q is an invariant submanifold of M_n containing $d_xF(0, 0)$, then the restriction

$$d_xF: S_F \to M_n \tag{6.307}$$

is transverse to Q in M_n.

To be more specific we define two invariant smooth submanifolds, the transition sets, of M_n which together determine all elementary losses of stability. We set

$$N_n^r = \{A \in M_n \mid \text{one eigenvalue is zero with multiplicity one; all other eigenvalues have negative real part}\} \tag{6.308}$$

$$N_n^i = \{A \in M_n \mid \text{one pair of eigenvalues is pure imaginary with multiplicity one; all others have negative real part}\} \tag{6.309}$$

Notice that for $n = 2$ in Example 1, N_n^r is a subset of Σ (6.11) and $N_n^i = H$ (6.13). For the study of stability, these are the two most important invariant submanifolds of M_n. They each have dimension one less than that of M_n; that is

$$\dim(N_n^r) = \dim(N_n^i) = n^2 - 1 \tag{6.310}$$

Consequently, each divides M_n locally into two pieces. In one, all the eigenvalues have negative real part, and in the other, at least one eigenvalue has positive real part.

Furthermore, because of (6.310), transversality to either N_n^i or N_n^r implies "crossing" in the sense that we used for our primitive notion of transversality in Section 6.3. In particular, suppose that we have a map f from a smooth submanifold

S of R^m to M_n

$$f: S \to M_n \tag{6.311}$$

with

$$f(x_o) \in N_n^s \quad , \quad s = i \text{ or } r \tag{6.312}$$

which is transverse to N_n^s in M_n near x_o. Then we already know that near x_o

$$D = f^{-1}(N_n^s) \tag{6.313}$$

will be a submanifold of S of dimension one less than S. However, the transversality of f to N_n^s in M_n also implies that near x_o, D divides S into two pieces, with all the eigenvalues of f(x) having negative real part for x in one piece, and at least one eigenvalue of f(x) having a positive real part for x in the other piece (see Fig. 6.15).

Applying this reasoning to a first-order versal unfolding

$$F: R^n \times R^p \to R^n \tag{6.314}$$

with $S = S_F$ and

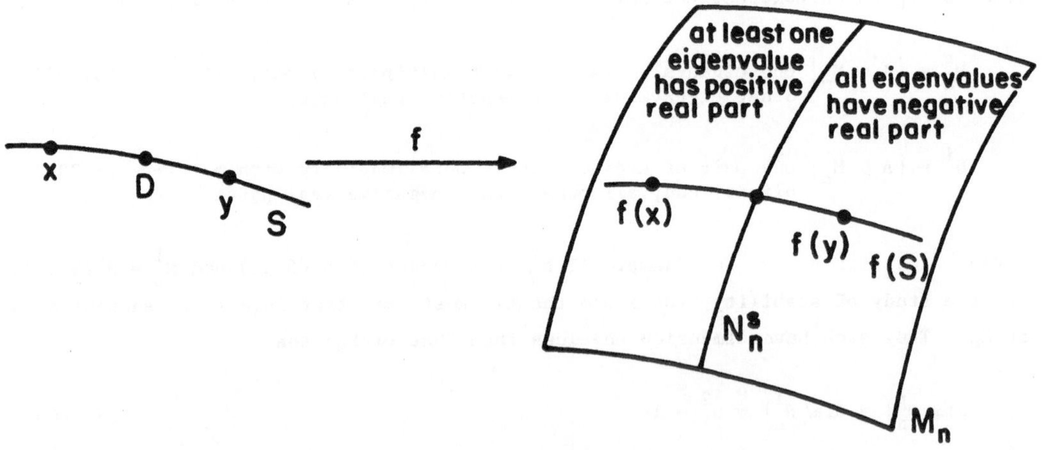

Fig. 6.15 The map $f{:}S \to M_n$ is transverse to n in M_n, and N_n^s divides M_n locally into two pieces as shown. Then $D = f^{-1}(N_n^s)$ divides S locally into two pieces, which f carries to the corresponding pieces of M_n .

$$f(x, \mu) = d_x F(x, \mu) \tag{6.315}$$

for $(x, \mu) \in S_F$, we obtain the following corollary to Proposition 6.8.

Corollary 6.8.2. Suppose that F is a first-order versal unfolding with $d_x F(0, 0) \in N_n^s$, $s = i$ or r. Then near the origin

$$D_F = \{(x, \mu) \in S_F \mid d_x F(x, \mu) \in N_n^s\} \tag{6.316}$$

is a smooth submanifold of dimension one less than S_F, and D_F divides S_F near the origin into two pieces, such that (x, μ) is a stable stationary point for (x, μ) in one piece and (x, μ) is an unstable stationary point for (x, μ) in the other piece.

For terminological convenience, we will refer to S_F as the stationary set of F and to D_F as the neutral set of F. Together these are part of the stability phase portrait of F.

Now we apply this information to our first-order versal unfolding of the modified Lorenz system.

Example 19. The stability phase portrait of a first-order versal unfolding of the Lorenz system.

In Example 18, we found a first-order versal unfolding $L^e(x, \mu, \nu, b, \sigma)$ (6.305) of the Lorenz system for (x, μ, ν, b, σ) near

$$p_o = (0, 0, 0, b_o, \sigma_o) \tag{6.317}$$

with $b_o > 0$, $\sigma_o > 0$ and $b_o \neq \sigma_o + 1$. We have

$$d_x L^e(p_o) \in N_3^r \tag{6.318}$$

Notice that N_n^r is always disjoint from the closure of N_n^i. Thus (6.318) implies that for p near p_o, we will always have $d_x L(p) \notin N_3^i$, and thus we may ignore N_3^i in the rest of this example.

Because L^e is first-order versal, Corollary 6.8.1 tells us that

$$d_x L^e\colon S_{L^e} \to M_3 \tag{6.319}$$

is transverse to N_3^r in M_3. In fact, Corollary 6.8.2 tells us that

$$D_{L^e} = \{p = (x, \mu, \nu, b, \sigma) \mid p \in S_{L^e} \text{ and } d_x L^e(p) \in N_3^r\} \tag{6.320}$$

divides S_{L^e} near p_o into two pieces, one consisting of stable stationary points and the other other consisting of unstable points.

Equation (6.318) implies that for stationary points $p \in S_{L^e}$ near p_o, $d_x L^e(p)$ will have eigenvalues near those of $d_x L(p_o)$. Thus two eigenvalues will be negative, and the neutral set D_{L^e} will coincide with the set

$$D_{L^e} = \{p \mid p \in S_{L^e} \text{ and } \det(d_x L^e(p)) = 0\} \tag{6.321}$$

Referring to (6.290) and (6.303), we see that this set may be described as the set where $\det(d_x L^e) = \det(F) = 0$, and that is given by

$$b = g(\gamma, z) \tag{6.322}$$

where $g(\gamma, z)$ is a suitable function, so that D_{L^e} is indeed a 3-submanifold of the 4-submanifold S_{L^e} of R^7.

More specifically, we may find from (6.264) a first-order Lyapunov-Schmidt splitting of (6.303) so that S_{L^e} is the set on which

$$\left.\begin{aligned} x_1 &= x_2 + \mu \\ x_3 &= (x_2 + \mu)(3x_2 + 2\mu)(3b)^{-1} \\ 0 &= \left(x_2 + \frac{8\mu}{9}\right)^3 - \left(\frac{\mu^2}{27} + b\nu\right)\left(x_2 + \frac{8\mu}{9}\right) - \left(\frac{2\mu^3}{729} + \frac{b\nu\mu}{9} + b\mu\right) \end{aligned}\right\} \tag{6.323}$$

For each fixed value of (b, σ) near (b_o, σ_o), (6.323) describes a cusp surface with natural parameters (6.266)

$$\left.\begin{aligned} \gamma &= \frac{\mu^2}{27} + b\nu \\ \delta &= \frac{2\mu^3}{729} + \frac{b\nu\mu}{9} + b\mu \end{aligned}\right\} \tag{6.324}$$

or "unnatural" parameters, μ and ν, and with the singular point at the origin. Then S_{L^e} is the 4-dimensional submanifold of R^7 swept out by these cusp surfaces as the values of (b, σ) vary near (b_o, σ_o). More formally, and parallel to Example 18, equations (6.324) may be solved for μ and ν in terms of γ, δ, and b, to produce

$$\left.\begin{aligned} \mu &= \hat{\mu}(\gamma, \delta, b) \\ \nu &= \hat{\nu}(\gamma, \delta, b) \end{aligned}\right\} \tag{6.325}$$

and then S_{L^e} may be parameterized near the origin in terms of the four variables $z = x_2 + 8\mu/9$, γ, b, and σ by the function f

$$f(z,\gamma,b,\sigma)=\begin{bmatrix}x_1\\x_2\\x_3\\\mu\\\nu\\b\\\sigma\end{bmatrix}=\begin{bmatrix}z+\frac{1}{9}\hat{\mu}(\gamma,z^3-\gamma z,b)\\ z-\frac{8}{9}\hat{\mu}(\gamma,z^3-\gamma z,b)\\ \left[z+\frac{1}{9}\hat{\mu}\right]\left[3z-\frac{2}{3}\hat{\mu}\right](3b)^{-1}\\ \hat{\mu}(\gamma,z^3-\gamma z,b)\\ \hat{\nu}(\gamma,z^3-\gamma z,b)\\ b\\ \sigma\end{bmatrix} \tag{6.326}$$

Each of the cusp surfaces in S_{Le} contains its own fold curve; the subset Δ_{Le} of S_{Le} swept out by the fold curves by varying the values of b and σ near b_0 and σ_0 is described by adjoining to (6.323) the single equation

$$3(x_2+8\mu/9)^2=\frac{\mu^2}{27}+b\nu \tag{6.327}$$

We do not display a parameterization of Δ_{Le} near the origin, but it is clear that a parameterization of Δ_{Le} as a 3-submanifold of S_{Le} may be obtained from (6.326). However, we may still make a geometric observation about Δ_{Le} by beginning with the fact that it is swept out by the fold curves of cusp surfaces. This observation is that for $\tilde{p}=(\tilde{x},\tilde{\mu},\tilde{\nu},\tilde{b},\tilde{\sigma})$ in Δ_{Le} there exist two points $p'=(x',\mu',\nu',b',\sigma')$ and $p''=(x'',\mu',\nu',b',\sigma')$ in S_{Le} arbitrarily near to $\tilde{p}$ with different dynamical variables $x'\neq x''$ but with the same control variables (μ',ν',b',σ'). That is, near $\tilde{p}\in\Delta_{Le}$, the equation

$$L^e(x,\mu,\nu,b,\sigma)=0 \tag{6.328}$$

defining S_{Le} <u>cannot</u> be solved uniquely for x in terms of the control variables. On the other hand, the conditions

$$L^e(\tilde{p})=0 \quad\text{and}\quad \det\left(d_xL^e(\tilde{p})\right)\neq 0 \tag{6.329}$$

form the hypothesis in the Implicit Function Theorem guaranteeing that near $\tilde{p}$, equation (6.328) <u>can</u> be solved uniquely for x in terms of the control variables. Thus we must have $\det(d_xL^e)=0$ on Δ_{Le}. Finally, we have already argued that (6.321) holds, so that $\Delta_{Le}\subset D_{Le}$. However, because both are smooth 3-submanifolds of R^7, they must coincide. Thus D_{Le} is the set swept out by the fold curves of the cusp surfaces.

Finally, it is a routine matter to determine the stable and unstable sides of D_{L^e} in S_{L^e}; we have only to calculate the eigenvalues of $d_x L^e(p)$ at two points $p = p_1$ and $p = p_2$, one on each side of $D_{L^e} = \Delta_{L^e}$. The result is most easily expressed in terms of the individual cusp surfaces: The middle pleat consists of unstable points and the outer two of stable points (Fig. 6.16). #

Now that we have completed these calculations it is advisable to review what they tell us. We have not only computed the stationary set S_{L^e} and the neutral set D_{L^e}; we have computed these in such a way that we know that they will be preserved by any first-order contact transformation of L^e. Merely to compute the stationary and neutral sets for an individual unfolding is a finite procedure that may be carried out routinely in specific cases. However, to compute these sets in such a way that they are preserved by first-order contact transformations amounts to computing the sets for an infinite family of related unfoldings. As we will see in the next section, the freedom we thus acquire, to pass from one unfolding to another without recomputing the stationary and neutral sets, is precisely the freedom we need to arrive at an unfolding containing enough physically interpretable parameters.

It is still reasonable to seek simpler calculations to generate usable stability information that is invariant under first-order transformations. These are available because first-order transformations preserve transversality to invariant submanifolds of M_n, and the crucial manifolds for determining the

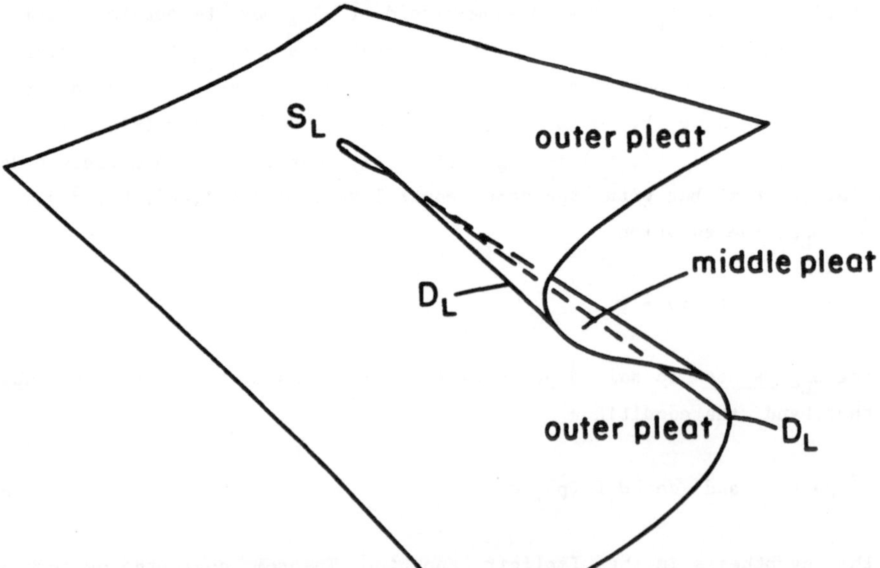

Fig. 6.16 Schematic representation of S_L. The submanifold D_L of S_L divides S_L into the set of stable points, coinciding with the outer pleats and the set of unstable points, coinciding with the middle pleat.

stability and neutral sets are N_n^r and N_n^i rather than Orb(A). In the following example, we will examine such a calculation for the original modified Lorenz unfolding L. However, this simpler method will not suffice to support the technique of first-order alterations in the next section. Presentation of that technique is the central objective of this chapter.

Example 20. The stability phase portrait of the original unfolding of the modified Lorenz system.

The original modified Lorenz unfolding is given by (3.63)-(3.69) or (6.262),

$$L(x, \mu, \nu) = \begin{cases} -\sigma x_1 + \sigma x_2 + \sigma \mu \\ -x_1 x_3 + (\nu + 1)\, x_1 - x_2 \\ x_1 x_2 - b x_3 + (2\mu/3)\, x_1 \end{cases} \tag{6.330}$$

in which μ and ν are the only unfolding parameters. The stationary set S_L is parameterized by (6.269) and we wish to check that the map

$$d_x L: \quad S_L \to M_3 \tag{6.331}$$

is transverse to N_3^r near the origin.

As in Example 18, we compose the map $d_x L$ in (6.331) with the parameterizing diffeomorphism f in (6.269) to produce

$$F = d_x L \circ f \tag{6.332}$$

Then $d_x L$ is transverse to N_3^r in M_3 if and only if F is transverse to N_3^r.

To determine whether $d_x L$ is transverse, we must begin by calculating the tangent space $T_A N_3^r$, where (cf. (6.274))

$$A = d_x L(0, 0, 0) = \begin{bmatrix} -\sigma & \sigma & 0 \\ 1 & -1 & 0 \\ 0 & 0 & -b \end{bmatrix} \tag{6.333}$$

To do so, we follow and use an unpublished method of Milnor that is analogous to the Lyapunov-Schmidt procedure. First we observe that interchanging the second and third basis vectors conjugates N_3^r into itself and A into

$$A' = \begin{bmatrix} -\sigma & 0 & \sigma \\ 0 & -b & 0 \\ 1 & 0 & -1 \end{bmatrix} \tag{6.334}$$

Thus any matrix B' near A' will have the form

$$B' = \begin{bmatrix} X & y \\ z & u \end{bmatrix} \tag{6.335}$$

with X a 2 × 2 invertible matrix, y a 2-column, z a 2-row and u a scalar. Now, B' will have the same rank as

$$\begin{bmatrix} X^{-1} & 0 \\ - z X^{-1} & 1 \end{bmatrix} \begin{bmatrix} X & y \\ z & u \end{bmatrix} = \begin{bmatrix} 1 & X^{-1}y \\ 0 & - z X^{-1}y + u \end{bmatrix} \tag{6.336}$$

and this rank will be 2 if and only if the lower right corner vanishes. Thus, for B' near A', we have that

$$B' \in N_3^r \quad \text{if and only if rank}(B') = 2 \tag{6.337}$$

or via (6.336) that

$$B' \in N_3^r \quad \text{if and only if} \quad u - z X^{-1}y = 0 \tag{6.338}$$

Now, the function defined for B' near A' by setting

$$G(B') = u - z X^{-1}y \tag{6.339}$$

is transverse to $\{0\}$ in R near A'. Thus, near A', N_3^r is given implicitly by

$$N_3^r = \{B' \mid G(B') = 0\} \tag{6.340}$$

and therefore

$$T_{A'}N_3^r = \left\{ \begin{bmatrix} X_1 & y_1 \\ z_1 & u_1 \end{bmatrix} \;\middle|\; dG(A') \begin{bmatrix} X_1 & y_1 \\ z_1 & u_1 \end{bmatrix} = 0 \right\} \tag{6.341}$$

Calculating dG(A') with the aid of (6.334) and (6.339), we see that

$$dG(A')\begin{bmatrix} X_1 & y_1 \\ z_1 & u_1 \end{bmatrix} \tag{6.342}$$

$$= u_1 - z_1 X^{-1}y + z X^{-1} X_1X^{-1}y - z X^{-1}y_1 \qquad \text{at} \begin{bmatrix} X & y \\ z & u \end{bmatrix} = A'$$

$$= u_1 - z_1 \begin{bmatrix} -\sigma & 0 \\ 0 & -b \end{bmatrix}^{-1} \begin{bmatrix} \sigma \\ 0 \end{bmatrix}$$

$$+ (1, 0) \begin{bmatrix} -\sigma & 0 \\ 0 & -b \end{bmatrix}^{-1} X_1 \begin{bmatrix} -\sigma & 0 \\ 0 & -b \end{bmatrix}^{-1} \begin{bmatrix} \sigma \\ 0 \end{bmatrix}$$

$$- (1, 0) \begin{bmatrix} -\sigma & 0 \\ 0 & -b \end{bmatrix}^{-1} y_1$$

$$= u_1 + z_1 \begin{bmatrix} 1 \\ 0 \end{bmatrix} + (1, 0) \begin{bmatrix} \sigma^{-1} & 0 \\ 0 & b^{-1} \end{bmatrix} X_1 \begin{bmatrix} 1 \\ 0 \end{bmatrix} + (\sigma^{-1}, 0)\, y_1$$

Because $T_{A'}N_3^r$ is 8-dimensional, we must find 8 linearly independent matrices satisfying

$$dG(A')\begin{bmatrix} X_1 & y_1 \\ z_1 & u_1 \end{bmatrix} = 0 \tag{6.343}$$

and our strategy is clear: specify X_1, y_1, z_1 in eight ways and let (6.343) with (6.342) determine u_1. With this approach, we obtain as a basis for $T_{A'}N_3^r$ the eight matrices

$$\zeta_1' = \begin{bmatrix} 1 & 0 & 0 \\ 0 & 0 & 0 \\ 0 & 0 & -\sigma^{-1} \end{bmatrix}, \zeta_2' = \begin{bmatrix} 0 & 1 & 0 \\ 0 & 0 & 0 \\ 0 & 0 & 0 \end{bmatrix}, \zeta_3' = \begin{bmatrix} 0 & 0 & 1 \\ 0 & 0 & 0 \\ 0 & 0 & -\sigma^{-1} \end{bmatrix} \tag{6.344}$$

$$\zeta_4' = \begin{bmatrix} 0 & 0 & 0 \\ 1 & 0 & 0 \\ 0 & 0 & 0 \end{bmatrix}, \zeta_5' = \begin{bmatrix} 0 & 0 & 0 \\ 0 & 1 & 0 \\ 0 & 0 & 0 \end{bmatrix}, \zeta_6' = \begin{bmatrix} 0 & 0 & 0 \\ 0 & 0 & 1 \\ 0 & 0 & 0 \end{bmatrix}$$

$$\zeta_7' = \begin{bmatrix} 0 & 0 & 0 \\ 0 & 0 & 0 \\ 1 & 0 & -1 \end{bmatrix}, \zeta_8' = \begin{bmatrix} 0 & 0 & 0 \\ 0 & 0 & 0 \\ 0 & 1 & 0 \end{bmatrix}$$

By re-permuting the second and third basis vectors in (6.344), we obtain a basis for $T_A N_3^r$

$$\zeta_1 = \begin{bmatrix} 1 & 0 & 0 \\ 0 & -\sigma^{-1} & 0 \\ 0 & 0 & 0 \end{bmatrix}, \zeta_2 = \begin{bmatrix} 0 & 0 & 1 \\ 0 & 0 & 0 \\ 0 & 0 & 0 \end{bmatrix}, \zeta_3 = \begin{bmatrix} 0 & 1 & 0 \\ 0 & -\sigma^{-1} & 0 \\ 0 & 0 & 0 \end{bmatrix} \tag{6.345}$$

$$\zeta_4 = \begin{bmatrix} 0 & 0 & 0 \\ 0 & 0 & 0 \\ 1 & 0 & 0 \end{bmatrix}, \zeta_5 = \begin{bmatrix} 0 & 0 & 0 \\ 0 & 0 & 0 \\ 0 & 0 & 1 \end{bmatrix}, \zeta_6 = \begin{bmatrix} 0 & 0 & 0 \\ 0 & 0 & 0 \\ 0 & 1 & 0 \end{bmatrix}$$

$$\zeta_7 = \begin{bmatrix} 0 & 0 & 0 \\ 1 & -1 & 0 \\ 0 & 0 & 0 \end{bmatrix}, \zeta_8 = \begin{bmatrix} 0 & 0 & 0 \\ 0 & 0 & 1 \\ 0 & 0 & 0 \end{bmatrix}$$

Finally, since $T_A N_3^r$ is 8-dimensional, the quotient space $M_3/T_A N_3^r$ is 1-dimensional, and then

$$\xi_1 = dF(0, 0) \begin{bmatrix} 1 \\ 0 \end{bmatrix} = \frac{\partial F}{\partial \gamma}(0, 0) \tag{6.346}$$

represents a non-zero element if and only if $dF(0, 0) \begin{bmatrix} 1 \\ 0 \end{bmatrix}$ adjoined to (6.345) produces a basis for M_3. Now, we saw in (6.297) that

$$dF(0,\ 0)\begin{bmatrix} 1 \\ 0 \end{bmatrix} = \begin{bmatrix} 0 & 0 & 0 \\ \frac{1}{b} & 0 & 0 \\ 0 & 0 & 0 \end{bmatrix} \tag{6.347}$$

A glance at (6.345) reveals that this matrix indeed is not in the span of (6.345), so that the adjunction of this matrix to (6.345) produces a basis for M_3. Then it follows that

$$dF(0,\ 0):\quad R^2 \to M_3/T_A N_3^r \tag{6.348}$$

is onto, and from that fact it follows that

$$F:\quad R^2 \to M_3 \tag{6.349}$$

is transverse near the origin to N_3^r in M_3. And finally via (6.332) it follows that

$$d_x L:\quad S_L \to M_3 \tag{6.350}$$

is transverse near the origin to N_3^r .

From this transversality we conclude that

$$D_L = \{(x,\ \mu,\ \nu) \mid (x,\ \mu,\ \nu) \in S_L \text{ and } d_x L(x,\ \mu,\ \nu) \in N_3^r\} \tag{6.351}$$

is a smooth submanifold of S_L of dimension one. We may simplify the argument that led up to $\Delta_{Le} \subset D_{Le}$ to show that the fold curve $\Delta_L \subset D_L$. Because both Δ_L and D_L are curves, we conclude that $\Delta_L = D_L$. That is, D_L <u>is</u> the fold curve near the origin. And by calculating $d_x L(x,\ \mu,\ \nu)$ for two choices of $(x,\ \mu,\ \nu) \in S_L$, one from each side of D_L (Fig. 6.16), we see that we may conclude that the middle pleat of S_L consists of unstable points and that the two outer pleats consist of stable points.

We conclude that for the unfolding L given by (6.330), the differential map

$$d_x L:\quad S_L \to M_3 \tag{6.352}$$

is transverse to N_3^r . Consequently, a first-order contact transformation T will actually transform the stability information of L into that of TL in the sense of (6.246). #

As we have already mentioned, an unfolding $F: R^n \times R^p \to R^n$ satisfying the condition that the differential map

$$dF:\quad S_F \to M_n \tag{6.353}$$

is transverse to N_n^s for s = r or i, already has the property that first-order contact transformations will preserve its stationary and neutral sets. But in general such an unfolding will not be first-order versal, and it will be necessary to add parameters to make it so. It is then reasonable to ask what advantage is to be gained from adding parameters to an unfolding until it is first-order versal rather than only adding parameters until the differential map from its stationary set to M_n is transverse to the neutral set $N_n = N_n^r \cup N_n^i$.

To determine this advantage we must first decide just what we consider to be an advantage. This decision is complicated by the fact that our two principal goals are in conflict. One is to minimize computations, and the other is to maximize extracted information. Depending on the trade-off we are willing to accept between these two goals, this section and the next offer us three procedures to choose from.

The first choice involves the least complication, and yet still informs us about the location of stability in the stationary set. This is the choice to unfold our system until the differential map from the stationary set to M_n is transverse to $N_n^r \cup N_n^i$. As we have seen, this choice allows us the freedom to perform first-order contact transformations without losing the stability information. The second choice involves greater complication, but it informs us not only about the location of stability but also about the distribution of other kinds of qualitative behavior; for instance, it distinguishes the set of sinks from that of nodes, and so on. This is the choice to unfold our system until we have a first-order versal unfolding. This choice enables us to keep track, during first-order contact transformations, of the sets describing different kinds of qualitative behavior because the boundaries between these are defined by the invariant submanifolds of M_n and the differential from the stationary set of M_n is transverse to all of these submanifolds.

Both of the first two choices suffer from the possibility that the unfolding we find may not be general enough; a different unfolding may exhibit a radically different distribution either of stability or of some other qualitative characteristic. The third choice does not suffer from this possibility; each unfolding it generates already contains all the information about the stability phase portrait of any unfolding of our original system. This choice results from a mild variant of Mather's Theory, to be presented in the following section. As a by-product, this variant of Mather's Theory will justify a first-order version of elementary alterations; we have already found in Chapters 3 to 5 that the ordinary version of these is crucial to the process of finding physically interpretable versal unfoldings.

Finally, as we have mentioned, there is a trade-off. The first two choices lead to unfoldings with a small number of parameters; but this number of parameters may not suffice for the unfolding to be optimal in the sense that it contains all the information about the stability phase portrait that any other comparable unfolding contains. The third choice leads to unfoldings having a large number of control parameters that is of the order of the cube of the number of dynamical variables. Unfortunately, such a number of parameters is too large, on both intuitive and practical grounds, to be physically interpretable. Because our applications consist of dynamical systems obtained by truncating Fourier series, we expect that the number of physically interpretable parameters has the same order of magnitude as n. For example, n of the parameters could be the Fourier coefficients of an imposed heating-rate function; see Chapter 4 for specific details. From the mathematical point of view, we note that generically the codimension of Orb(A) in M_n is n; therefore we need a number of parameters of the order of the number n of dynamical variables to achieve transversality to Orb(A) and so first-order versality. Thus, we do expect a large number of control parameters, but we expect it to be of the order of n and not that of n^3. Accordingly, we need a procedure intermediate between the second and third ones. Ideally, this procedure would produce unfoldings that have relatively few control parameters, as does the second choice, and yet have sufficient generality, as does the third. We sketch such a procedure at the end of the following section.

6.6 First-Order Mather Theory

In this section we will describe the Mather Theory of first-order versal unfoldings and first-order contact transformations. It is partly parallel to that for ordinary versal unfoldings and contact transformations in Chapter 2. One of the central ideas we have already introduced is the first-order version of a versal unfolding. But another, the idea of a first-order versal unfolding, we have introduced only in part, in order to generate examples. We have not yet defined either the first-order version of pulling-back, or the notion on which that depends, the first-order version of a contact map.

A first-order contact map satisfies the same conditions as a first-order contact transformation except for the requirement that the control parameters transform invertibly. More specifically, a first-order contact map consists of an invertible n × n matrix M(x, λ) depending smoothly on (x, λ), a parameterized coordinate transformation y = y(x, λ),

$$y: \quad R^n \times R^p \to R^n \tag{6.354}$$

satisfying y(0, 0) = 0, and smooth functions μ = μ(λ),

$$\mu: \quad R^p \to R^q \tag{6.355}$$

satisfying μ(0) = 0. These parallel the conditions imposed in Chapter 2 for

$$T = \left[M(x, \lambda); \quad y(x, \lambda); \quad \mu(\lambda)\right] \tag{6.356}$$

to be a contact map. In addition, for T to be a first-order contact map, we impose the condition that $M(x, \lambda)$ agrees with $\left(d_x y(x, \lambda)\right)^{-1}$ to the first order; that is

$$M(x, \lambda) = \left(d_x y(x, \lambda)\right)^{-1} + O(2) \tag{6.357}$$

The effect of T on an unfolding

$$V: \quad R^n \times R^q \to R^n \tag{6.358}$$

is defined in exactly the same way as it was in Chapter 2; that is,

$$T\,V(x, \lambda) = M(x, \lambda)\cdot V\left(y(x, \lambda), \mu(\lambda)\right) \tag{6.359}$$

and again we say that V <u>pulls back</u> to T V, or that T V is a <u>pull-back</u> of V.

Now we notice that our definition of first-order versality in the beginning of Section 6.4 is not parallel to our definition in Section 2.3 of an ordinary versal unfolding. Our first objective in this section is to give a definition that is more closely parallel to the one in Chapter 2. The new definition will be more algebraic in nature, apparently lacking the geometric content of the one in Section 6.4. However, we will show that the geometric content actually follows from the algebraic definition, so that the latter is the stronger definition.

As in Chapter 2, we begin our algebraic definition with an algebraic version of a transversality condition.

<u>The First-Order Transversality Condition.</u> The $n \times p$ matrix $N(y)$ depending smoothly on y satisfies the First-Order Transversality Condition with respect to the n-vector $f(y)$ depending smoothly on y if and only if the following two conditions hold:

(i) Every n-column vector $Y(y)$ depending smoothly on y may be written as

$$Y(y) = d_y f(y)\cdot G(y) + H(y)\cdot f(y) + N(y)\cdot\Lambda \tag{6.360}$$

near the origin, where $H(y)$ is an $n \times n$ matrix depending smoothly on y, $G(y)$ is an n-column depending smoothly on y, and Λ is a constant p-column.

(ii) The matrix $H(y)$ and the column $G(y)$ are related by the condition that

$$H(y) + dG(y) = O(2) \tag{6.361}$$

That is, we require that the sum $G(y) + dH(y)$ vanishes to the first order at the origin.

The First-Order Transversality Condition is what we need to re-define algebraically the notion of first-order versality. We will say that the unfolding $U(y, \mu)$ is <u>first-order Versal</u> if and only if $d_\mu U(y, 0)$ satisfies the First-Order

Transversality Condition with respect to $U(y, 0)$. We note that we capitalize the word "versal" to distinguish this, algebraic, definition from the geometric one given in Section 6.4.

How are the algebraic and the geometric definitions related? Part of the answer to this question is given by the following proposition.

Proposition 6.9 (De-Capitalization Lemma). If the unfolding $U(y, \mu)$ is first-order Versal, then it is first-order versal.

As an illustrative exercise in the mechanics of these concepts, we prove Proposition 6.9 here.

Proof. We begin by observing that any constant $n \times n$ matrix Γ may be written as

$$\Gamma = dY(0) \tag{6.362}$$

for a suitable n-column $Y(y)$ depending smoothly on y and satisfying

$$Y(0) = 0 \tag{6.363}$$

Because $U(y, \mu)$ is first-order Versal, there exist an $n \times n$ matrix $H(y)$ depending smoothly on y, an n-column $G(y)$ depending smoothly on y, and a constant p-column Λ such that

$$Y(y) = d_y U(y, 0)\cdot G(y) + H(y)\cdot U(y,0) + d_\mu U(y, 0)\cdot\Lambda \tag{6.364}$$

Evaluating (6.364) at $y = 0$, we see that

$$0 = d_y U(0, 0)\cdot G(0) + d_\mu U(0, 0)\cdot\Lambda \tag{6.365}$$

in which we have used the fact that $U(0, 0) = 0$.

In order to show that $U(y, \mu)$ is first-order versal, we must show that $d_y U: S_U \to M_n$ is transverse to $\mathrm{Orb}\big(d_y U(0, 0)\big)$ in M_n. We begin by claiming that the vector

$$\xi_o = \begin{bmatrix} G(0) \\ \Lambda \end{bmatrix} \tag{6.366}$$

is in $T_0 S_U$. To verify this claim, we recall from Section 6.3 that because the stationary set S_U is given implicitly near the origin by the equation

$$S_U = \{(y, \mu) \mid U(y, \mu) = 0\} \tag{6.367}$$

its tangent space $T_0 S_U$ at the origin is given by

$$T_oS_U = \{\xi \mid dU(0, 0)\cdot\xi = 0\} \tag{6.368}$$

But from (6.366), we find that

$$dU(0, 0)\cdot\xi_o = d_yU(0, 0)\cdot G(0) + d_\mu U(0, 0)\cdot\Lambda \tag{6.369}$$

so that

$$dU(0, 0)\cdot\xi_o = 0 \tag{6.370}$$

and $\xi_o \in T_oS_U$. Thus our claim that

$$\xi_o = \begin{bmatrix} G(0) \\ \Lambda \end{bmatrix} \in T_oS_U \tag{6.371}$$

is verified.

To continue with our argument, we take the differential with respect to y of Y in (6.364), and evaluate the result at y = 0 to obtain

$$\begin{aligned} \Gamma = {} & d_yU(0, 0)\cdot d_yG(0) + d_y{}^2U(0, 0)\cdot G(0) + H(0)\cdot d_yU(0, 0) \\ & + d_\mu\big(d_yU(0, 0)\big)\cdot\Lambda \end{aligned} \tag{6.372}$$

Condition (ii), given by (6.361) in the definition of the First-Order Transversality Condition, implies that

$$H(0) = - d_yG(0) \tag{6.373}$$

On the other hand, the differential

$$d(d_yU)(0, 0)\colon\ T_oS_U \to M_n \tag{6.374}$$

of the map

$$d_yU\colon\ S_U \to M_n \tag{6.375}$$

may be evaluated at $\xi_o \in T_oS_U$, to obtain

$$d(d_yU)(0, 0)\begin{bmatrix} G(0) \\ \Lambda \end{bmatrix} = d_y{}^2U(0, 0)\cdot G(0) + d_\mu\big(d_yU(0, 0)\big)\cdot\Lambda \tag{6.376}$$

Combining (6.372), (6.373) and (6.376) we arrive at the following equation

$$\Gamma = d_yU(0, 0)\cdot d_yG(0) - d_yG(0)\cdot d_yU(0, 0) + d(d_yU)(0, 0)\cdot\begin{bmatrix} G(0) \\ \Lambda \end{bmatrix} \tag{6.377}$$

We recall from Section 6.3 that any matrix of the form

$$B = d_yU(0, 0)\cdot d_yG(0) - d_yG(0)\cdot d_yU(0, 0) \tag{6.378}$$

is in the tangent space of $\mathrm{Orb}\big(d_yU(0, 0)\big)$ at the origin; that is,

$$d_yU(0, 0)\cdot d_yG(0) - d_yG(0)\cdot d_yU(0, 0) \in T_o\mathrm{Orb}\big(d_yU(0, 0)\big) \tag{6.379}$$

Consequently, using (6.377) and (6.379), we see that Γ and $d(d_yU)(0, 0)\cdot\begin{bmatrix} G(0) \\ \Lambda \end{bmatrix}$ represent the same element in the quotient space Q_A, in which

$$A = d_yU(0, 0) \tag{6.380}$$

We have thus shown that the map

$$d(d_yU)(0, 0)\colon\ T_oS_U \to Q_A \tag{6.381}$$

is onto; by definition, then, $d_yU\colon S_U \to M_n$ is transverse to $\mathrm{Orb}(A)$ in M_n. To complete the argument proving Proposition 6.9, we must check that $U(y, \mu)$ is versal. But first-order Versality trivially implies versality, and Proposition 6.9 is proved.

Proposition 6.9 enables us to carry over from Section 6.4 an observation concerning first-order versality to one concerning first-order Versality: For the first-order Versal unfolding $U\colon R^n \times R^p \to R^n$, the stationary set S_U and the neutral set $D_U \subset S_U$ will be smooth submanifolds of $R^n \times R^p$, and they will be preserved in the sense of Proposition 6.2 by any first-order contact transformation. Moreover, if

$$P \subset M_n \tag{6.382}$$

is any invariant submanifold of M_n containing $d_yU(0, 0)$, then the corresponding secondary set in S_U

$$D_U(P) = \{(y, \mu) \in S_U \mid d_yU(y, \mu) \in P\} \tag{6.383}$$

will be a smooth submanifold of S_U preserved in the sense of Proposition 6.2 by any first-order contact transformation.

However, we still would like one further property: first-order Versality should be preserved by first-order contact transformations. And indeed, a straight-forward calculation proves the following proposition.

Proposition 6.10 (Invariance Lemma). A first-order contact transformation of a first-order Versal unfolding is again first-order Versal.

Our next proposition plays a major role in simplifying the computations that arise in applications of first-order Mather Theory.

Proposition 6.11 (Twisting Lemma). Suppose that near the origin the $n \times p$ matrix $N(x)$ satisfies the First-Order Transversality Condition with respect to $f(x)$ and suppose that the $n \times n$ matrix $M(x)$ is invertible. Then the matrix $M(x) \cdot N(x)$ satisfies the First-Order Transversality Condition with respect to $M(x) \cdot f(x)$.

Thus the Twisting Lemma holds for first-order Versality, although it does not hold for first-order versality. However, a regrettable peculiarity of first-order Versality is that no simple Extension Lemma (cf. Proposition 6.5) is available. That is, if the matrix $N(v)$ satisfies the First-Order Transversality Condition with respect to $g(v)$, then it is never true that the matrix

$$N_e(v) = \begin{bmatrix} 0 \\ N(v) \end{bmatrix} \tag{6.384}$$

satisfies the First-Order Transversality Condition with respect to

$$F(u, v) = \begin{bmatrix} u \\ g(v) \end{bmatrix} \tag{6.385}$$

To see this fact, we examine a function of the form

$$Y(u, v) = \begin{bmatrix} 1 & 0 \\ 0 & dg(v) \end{bmatrix} \cdot G(u, v) + H(u, v) \cdot \begin{bmatrix} u \\ g(v) \end{bmatrix} + \begin{bmatrix} 0 \\ N(v) \cdot \lambda \end{bmatrix} \tag{6.386}$$

with

$$H + dG = O(2) \tag{6.387}$$

We may write Y, G, and H in block form,

$$Y = \begin{bmatrix} Y_1 \\ Y_2 \end{bmatrix}, \quad G = \begin{bmatrix} G_1 \\ G_2 \end{bmatrix}, \quad H = \begin{bmatrix} H_{11} & H_{12} \\ H_{21} & H_{22} \end{bmatrix} \tag{6.388}$$

so that (6.387) implies that

$$\left.\begin{aligned} H_{11}(u, v) + d_u G_1(u, v) &= O(2) \\ H_{12}(u, v) + d_v G_1(u, v) &= O(2) \end{aligned}\right\} \tag{6.389}$$

Then we see from (6.386) that

$$Y_1(u, v) = G_1(u, v) - d_u G_1(u, v)\cdot u - d_v G_1(u, v)\cdot g(v) + O(3) \tag{6.390}$$

and finally that

$$Y_1(u, 0) = G_1(u, 0) - d_u G_1(u, 0)\cdot u + O(3) \tag{6.391}$$

because $g(0) = 0$. Writing the Taylor expansion of $G_1(u, 0)$ to the first order, we obtain

$$G_1(u, 0) = G_1(0, 0) + d_u G_1(0, 0)\cdot u + O(2) \tag{6.392}$$

Similarly, we calculate the Taylor expansion to the first order of $d_u G_1(u, 0)$ to be

$$d_u G_1(u, 0) = d_u G_1(0, 0) + d_u^2 G_1(0, 0)\cdot u + O(2) \tag{6.393}$$

Consequently, we find that

$$\begin{aligned} Y_1(u, 0) &= G_1(u, 0) - d_u G_1(u, 0)\cdot u + O(3) \\ &= G_1(0, 0) + \left(d_u G_1(0, 0) - d_u G_1(u, 0)\right)\cdot u + O(2) \\ &= G_1(0, 0) + O(2) \end{aligned} \tag{6.394}$$

Thus $Y_1(u, 0)$ cannot have a linear term, but may have constant, quadratic, and higher-order terms. A consequence of this peculiarity is that if $u = (u_1, \ldots, u_n)$, then a first-order Versal unfolding of $F(u, v)$ in (6.385) will require at least n^3 control parameters to describe the $n \times n$ matrix $d_u Y_1(0, 0)$.

To counterbalance this difficulty, we describe the simple device for obtaining a non-minimal, but certainly first-order Versal, unfolding of $F(u, v)$ in (6.385). We know from Chapter 2 that any column $Y(u, v)$ may be written in the form (6.386) if we do not require that G and H satisfy (6.387). Then we write

$$C = \text{the first-degree Taylor polynomial of } H + dG \tag{6.395}$$

Then, with

$$H_o = H - C \tag{6.396}$$

we have

$$Y = \begin{bmatrix} 1 & 0 \\ 0 & dg(v) \end{bmatrix} \cdot G + H_o \cdot \begin{bmatrix} u \\ g(v) \end{bmatrix} + c_o + C \cdot \begin{bmatrix} u \\ g(v) \end{bmatrix} + \begin{bmatrix} 0 \\ N(v) \cdot \lambda \end{bmatrix} \tag{6.397}$$

where C is a constant-plus-linear $n \times n$ matrix function of (u, v). Regarding the coefficients of C as unfolding parameters, we have obtained a first-order Versal unfolding of $\begin{bmatrix} u \\ g(v) \end{bmatrix}$ given by

$$F_e(u, v, \lambda, C) = \begin{bmatrix} u \\ g(v) \end{bmatrix} + \begin{bmatrix} 0 \\ N(v) \cdot \lambda \end{bmatrix} + C \cdot \begin{bmatrix} u \\ g(v) \end{bmatrix} \tag{6.398}$$

where, on the left side, we have abused notation slightly by identifing C with the list of its coefficients. The unfolding $F_e(u, v, \lambda, C)$ is first-order Versal because H_o and G satisfy, via (6.395) and (6.397),

$$H_o + dG = O(2) \tag{6.399}$$

With Proposition 6.11, we have finished describing our first-order point of view, and now we turn to the central substantial result that justifies taking that point of view. This result is the following theorem, and it is the precise analogue of Mather's Theorem I (Section 2.3), but our statement of it is not precisely parallel to that given in Chapter 2 because we have used a definition of first-order Versality that is not parallel to that of ordinary versality. However, by virtue of the following theorem, the meaning of first-order Versality becomes precisely parallel to that of ordinary versality.

The First-Order Versality Theorem. The unfolding $U(y, \mu)$ is first-order Versal if and only if any unfolding $F(x, \lambda)$ satisfying

$$F(z, 0) = U(z, 0) \tag{6.400}$$

near the origin is a pull-back of $U(y, \mu)$ by some first-order contact map

$$T = \big(M(x, \lambda), y(x, \lambda), \mu(\lambda)\big) \tag{6.401}$$

The immediate content of this theorem, analogous to that of Mather's Theorem I, is that the stability phase portrait of any unfolding $F(x, \lambda)$ satisfying (6.400) is generated by the stability phase portrait of $U(y, \mu)$; in a reasonable sense, the stability phase portrait of $U(y, \mu)$ already contains all the stability information of any unfolding $F(x, \lambda)$ satisfying (6.400).

Because a proof of the First-Order Versality Theorem is not available in the literature, we indicate very briefly how to construct one following Mather: Most of the steps are straightforward, if sometimes laborious, adaptations of the corresponding steps in the standard proof of Mather's Theorem I, given, for instance, in Mather (1968). However, where in the proof of that theorem there occur modules over a ring of germs of smooth functions, in the proof of this theorem there occur corresponding sets that no longer are such modules. The device for by-passing this difficulty is to observe that those sets are modules over a subring which is a sum, with one summand being the subring of functions in one of the variables and the other summand, the maximal ideal in the remaining variables, cubed. However, to apply the second form of the Mather Preparation Theorem, we need the base ring to be a whole ring of germs of smooth functions rather than the subring we have indicated. Accordingly, by means of a cubing map on the "remaining variables", we turn our subring into a finitely generated algebra over the whole ring of germs of smooth functions. Now the appropriate modules are finitely generated, we may apply the Second Form of the Mather Preparation Theorem as desired, and the theorem is proved.

To be able to apply the First-Order Versality Theorem as we applied Mather's Theorem I in Chapters 3 to 5, we need a major corollary to the First-Order Versality Theorem. In those chapters we arrived at a physically interpretable versal unfolding in three steps. In the first step we used the Lyapunov-Schmidt process to reduce the system to a system of very small dimension. In the second step, we found a versal unfolding of the small-dimensional system; the Lyapunov-Schmidt process already utilized, automatically extended that unfolding to a versal unfolding of the original system. Finally, in the third step we carried out alterations based on Mather's Theorem I to transform that versal unfolding to a physically interpretable one.

In Section 6.5 we already have noted that in the first-order case, the Lyapunov-Schmidt procedure involves a "twist" Γ (6.256) which might change drastically the first-order contact type of an unfolding.

In carrying out the first, or Lyapunov-Schmidt, step of our procedure, we encounter the difficulty mentioned after Proposition 6.11. This difficulty is that the simple version of the Extension Lemma does not hold; however, a first-order Versal unfolding of the function

$$F(u, v) = \begin{bmatrix} u \\ g(v) \end{bmatrix} \tag{6.402}$$

may still be constructed from a first-order Versal unfolding

$$G(v, \lambda) = g(v) + N(v)\cdot\lambda \tag{6.403}$$

of g(v) by setting

$$U(u, v, \lambda, C) = \begin{bmatrix} u \\ g(v) + N(v)\cdot\lambda \end{bmatrix} + C(u, v)\cdot\begin{bmatrix} u \\ g(v) \end{bmatrix} \qquad (6.405)$$

where, on the right side, C(u, v) is the general constant-plus-linear function of (u, v), and , on the left side, C is the list of the coefficients of C. The constant term of C(u, v) is a square matrix, and, as we have mentioned before, it is not possible to eliminate its block corresponding to u. However, it may be possible to eliminate some other entries of C. When this elimination has been fully carried out in an ad hoc manner, we will have a minimal first-order Versal unfolding of F(u, v). By applying the Twisting Lemma, Proposition 6.11, to the resulting unfolding, we finally arrive at a first-order Versal unfolding of our original system.

The second step is to be carried out in an essentially ad hoc manner, but it is greatly facilitated by having available a stock of first-order Versally unfolded small-dimensional systems; we will present the first-order Versal unfoldings of all one-dimensional systems in Example 21 below.

Finally, to make the third step of our procedure effective, we need a justification for carrying out elementary alterations. That justification is given by the following corollary to the First-Order Versality Theorem.

Corollary. Any two first-order Versal unfoldings of the same n-column g(x) that have the same number of control parameters are first-order contact equivalent; that is, a first-order contact transformation exists which carries one to the other.

The corollary, together with the results of Section 6.5, states that any alteration that we may make, carrying one minimal first-order Versal unfolding to another, is realized by a first-order contact transformation, which of course preserves the stability phase portrait. In particular, if N(x) and L(x) are $n \times p$ matrices, both satisfying the First-Order Transversality Condition with respect to the n-column f(x), then the two first-order Versal unfoldings

$$U(x, \lambda) = f(x) + N(x)\cdot\lambda \qquad (6.406)$$

and

$$V(x, \lambda) = f(x) + L(x)\cdot\lambda \qquad (6.407)$$

are first-order contact equivalent. That is, replacing N(x) by L(x) generates a first-order contact transformation from (6.406) to (6.407). The problem is to construct L(x) from N(x) in a reasonably versatile way so as to guarantee that L(x) satisfies the First-Order Transversality Condition if N(x) does. The simplest such construction, which we call a first-order elementary alteration is carried out as follows: Pick an index i and let $\ell(x)$ be an n-column such that the ith column $N_i(x)$ of N(x) may be written as

$$N_i(x) = df(x)\cdot G(x) + H(x)\cdot f(x) + L(x)\cdot\lambda \tag{6.408}$$

where L(x) is obtained from N(x) by replacing its ith column $N_i(x)$ with $\ell(x)$, and where

$$H(x) + dG(x) = O(2) \tag{6.409}$$

To guarantee that every possible replacement of N(x) by some L(x) may be obtained by repeating this construction, we state the following proposition.

Proposition 6.13 (Change of Basis Lemma). If N(x) and L(x) are both $n \times p$ matrices satisfying near the origin the First-Order Transversality Condition with respect to f(x), then L(x) may be obtained from N(x) by carrying out a finite sequence of first-order elementary alterations.

Now we turn to some examples in order to illustrate the ideas above. The first example may be used to facilitate the second step of the three-step procedure outlined above.

Example 21. The first-order Versal unfolding of x^n.
We let $f:R^1 \to R^1$ be defined by

$$f(x) = x^n \quad , \quad n > 1 \tag{6.410}$$

We seek a $1 \times p$ matrix

$$N(x) = [\alpha_1(x), \ldots, \alpha_p(x)] \tag{6.411}$$

satisfying the First-Order Transversality Condition, with p minimal, with respect to x^n. We need

$$df(x) = f'(x) = n\,x^{n-1} \tag{6.412}$$

We already know that the matrix $[x^{n-2}, x^{n-3}, \ldots, 1]$ satisfies the ordinary Transversality Condition with respect to x^n (see Section 2.3). That is, for any function y(x), there exist functions h(x) and g(x) such that

$$y(x) = h(x)\,x^n + n\,x^{n-1}\,g(x) + \lambda_{n-2}\,x^{n-2} + \cdots + \lambda_o \tag{6.413}$$

but, in general,

$$h(x) + dg(x) \neq O(2) \tag{6.414}$$

Instead, we may write

$$h(x) + dg(x) = \lambda_n + \lambda_{n+1}\, x + O(2) \tag{6.415}$$

and set

$$h_o(x) = h(x) - \lambda_n - \lambda_{n+1}\, x \tag{6.416}$$

Then we have

$$y(x) = h_o(x)\, x^n + n\, x^{n-1}\, g(x) + \lambda_{n+1}\, x^{n+1} + \lambda_n\, x^n + \lambda_{n-2}\, x^{n-2} + \cdots + \lambda_o \tag{6.417}$$

in which

$$h_o(x) + dg(x) = O(2) \tag{6.418}$$

Thus, the matrix $[x^{n+1}, x^n, x^{n-2}, x^{n-3}, \ldots, 1]$ satisfies the First-Order Transversality Condition, and we conclude that

$$U(x, \lambda_{n+1}, \lambda_n, \lambda_{n-2}, \lambda_{n-3}, \ldots, \lambda_o) = x^n + \lambda_{n+1}\, x^{n+1} + \lambda_n\, x^n + \lambda_{n-2}\, x^{n-2} + \cdots + \lambda_o \tag{6.419}$$

is a first-order Versal unfolding of x^n. This unfolding is not minimal; for $n \neq 2$ we may write

$$x^{n+1} = h(x)\, x^n + n\, x^{n-1}\, g(x) \tag{6.420}$$

with

$$g(x) = \frac{1}{n-2}\, x^2 \tag{6.421}$$

and

$$h(x) + g'(x) = 0 \tag{6.422}$$

Consequently, the term $\lambda_{n+1}\, x^{n+1}$ is superfluous when $n \neq 2$. Similarly, because $n > 1$ we may write

$$x^n = h(x)\, x^n + n\, x^{n-1}\, g(x) \tag{6.423}$$

with

$$g(x) = \frac{1}{n-1} x^n \tag{6.424}$$

and

$$h(x) + g'(x) = 0 \tag{6.425}$$

Thus, the term $\lambda_n x^n$ is superfluous for all $n > 1$. Both are superfluous for $n > 2$, and we obtain as minimal first-order Versal unfoldings of x^n the unfoldings

$$\begin{aligned} F_2(x, \lambda) &= x^2 + \lambda_3 x^3 + \lambda_o && \text{for } n = 2 \\ F_n(x, \lambda) &= x^n + \lambda_{n-2} x^{n-2} + \cdots + \lambda_o && \text{for } n > 2 \end{aligned} \tag{6.426}$$

#

In the unfolding $F_2(x, \lambda)$ of Example 21, we see a rather mysterious term $\lambda_3 x^3$; such terms do not appear in the remaining unfoldings $F_n(x, \lambda)$ for $n > 2$, but unfortunately, in general, the appearance of such terms is the rule rather than the exception. As we have indicated earlier, first-order Versal unfoldings involve a vast number of parameters, far too many to interpret or understand. Consequently, it is of crucial importance for us to find some way to reduce their number while preserving a good part of the pull-back property of first-order Versal unfoldings. We will suggest one means for doing so after the next example, which lays the groundwork for our suggestion.

In the following example, we examine the interplay between the neutral sets defined by the two submanifolds N_n^i and N_n^r. We will unfold a function $f(x)$ with a singular point at the origin and $df(0)$ in the intersection of the closures of the two sets. In particular, we note that the unfolding will necessarily contain Hopf bifurcations.

Example 22. First-order Versal unfolding of a fold

For an illustrative example, we will consider a family of fairly simple 3-column functions, one of which we define by setting

$$f(x) = \begin{bmatrix} x_1 \\ x_2 \\ x_3^{\,2} \end{bmatrix} \tag{6.427}$$

The rest of the family we will define later. Using the methods of Chapter 2, we find an ordinary versal unfolding of $f(x)$ to be given by adding a single parameter,

$$f(x, \lambda) = \begin{bmatrix} x_1 \\ x_2 \\ x_3^2 + \lambda \end{bmatrix} \tag{6.428}$$

It follows from Mather's Theorem I that any 3-column Y(x) is given near the origin by an expression of the form

$$Y(x) = df(x)\cdot G(x) + H(x)\cdot f(x) + \begin{bmatrix} 0 \\ 0 \\ \lambda \end{bmatrix} \tag{6.429}$$

where, as usual, G(x) is an n-column and H(x) an $n \times n$ matrix. Now we write

$$C(x) = \text{the first-order Taylor expansion of } H(x) + dG(x) \text{ at the origin} \tag{6.430}$$

and

$$H_o(x) = H(x) - C(x) \tag{6.431}$$

Regarding the $3^2 + 3^3 = 36$ coefficients of C(x) as unfolding parameters, we may use (6.397) to write an arbitrary function Y(x) as

$$Y(x) = df(x)\cdot G(x) + H_o(x)\cdot f(x) + \begin{bmatrix} 0 \\ 0 \\ \lambda \end{bmatrix} + C(x)\cdot f(x) \tag{6.432}$$

in which

$$dG(x) + H_o(x) = O(2) \tag{6.433}$$

Thus, a first-order Versal unfolding of f(x) is given by

$$U(x, \lambda, C) = \begin{bmatrix} x_1 \\ x_2 \\ x_3^2 \end{bmatrix} + \begin{bmatrix} 0 \\ 0 \\ \lambda \end{bmatrix} + C(x)\cdot \begin{bmatrix} x_1 \\ x_2 \\ x_3^2 \end{bmatrix} \tag{6.434}$$

where we have used the letter C to denote the list of 36 coefficients of the 3×3 matrix function C(x).

These 36 coefficients may be reduced to 16 by carrying out algebraic manipulations. These manipulations begin with the observation that the third term in (6.434) contributes 3-column vectors whose entries are arbitrary linear combinations of the monomials given in the list

$$\text{monomial} = x_1,\ x_2,\ x_1^2,\ x_1x_2,\ x_2^2,\ x_1x_3,\ x_2x_3,\ x_3^2,\ x_1x_3^2,\ x_2x_3^2,\ x_3^3 \tag{6.435}$$

Because of the six appearances of x_1x_2 in (6.434), three are redundant and so may be deleted.

By finding repeatedly 3-columns Z(x) which are simultaneously of the two forms

$$\left.\begin{aligned} &C(x)\cdot\begin{bmatrix} x_1 \\ x_2 \\ x_3^2 \end{bmatrix} = Z(x) = df(x)\cdot G(x) + H(x)\cdot f(x) \\ &dG(x) + H(x) = O(2) \end{aligned}\right\} \tag{6.436}$$

we may reduce the number of parameters each time by one. An easy way to find such columns is to calculate a succession of simple polynomial columns for $H(x) = -\,dG(x)$ and G(x). Thus, for example, we arrive at the following table with the aid of (6.427) and (6.436)

$$G(x) =$$

$$\begin{bmatrix} x_1 \\ 0 \\ 0 \end{bmatrix} \quad \begin{bmatrix} x_2 \\ 0 \\ 0 \end{bmatrix} \quad \begin{bmatrix} x_3 \\ 0 \\ 0 \end{bmatrix} \quad \begin{bmatrix} x_1^2 \\ 0 \\ 0 \end{bmatrix} \quad \begin{bmatrix} x_1x_2 \\ 0 \\ 0 \end{bmatrix} \quad \begin{bmatrix} x_2^2 \\ 0 \\ 0 \end{bmatrix}$$

$$df(x)\cdot G(x) - dG(x)\cdot f(x) = \tag{6.437}$$

$$\begin{bmatrix} 0 \\ 0 \\ 0 \end{bmatrix} \quad \begin{bmatrix} 0 \\ 0 \\ 0 \end{bmatrix} \quad \begin{bmatrix} x_3^2 - x_3 \\ 0 \\ 0 \end{bmatrix} \quad \begin{bmatrix} x_1^2 \\ 0 \\ 0 \end{bmatrix} \quad \begin{bmatrix} x_1x_2 \\ 0 \\ 0 \end{bmatrix} \quad \begin{bmatrix} 2x_1x_2 - x_2^2 \\ 0 \\ 0 \end{bmatrix}$$

The third entry of (6.437) tells us, for example, that if $\begin{bmatrix} \nu_1x_3^2 \\ 0 \\ 0 \end{bmatrix}$ appears in our unfolding with ν_1 an unfolding parameter, then $\begin{bmatrix} \nu_2 x_3 \\ 0 \\ 0 \end{bmatrix}$ may be deleted. By completing this table and using it to eliminate coefficients, we finally arrive at the first-order Versal unfolding of f(x).

$$U(x, \lambda, \alpha, \beta, \gamma) = \begin{bmatrix} x_1 \\ x_2 \\ x_3^2 \end{bmatrix} + \begin{bmatrix} \alpha_1 x_1 + \alpha_2 x_2 + \alpha_3 x_3^2 + \alpha_4 x_1 x_3 + \alpha_5 x_2 x_3^2 \\ \beta_1 x_1 + \beta_2 x_2 + \beta_3 x_3^2 + \beta_4 x_1 x_3 + \beta_5 x_2 x_3^2 \\ \lambda + \gamma_1 x_1 + \gamma_2 x_2 + \gamma_3 x_1^2 + \gamma_4 x_2^2 + \gamma_5 x_1 x_2 + \gamma_6 x_3^3 \end{bmatrix} \tag{6.438}$$

which contains 17 coefficients instead of 37. From this unfolding, using the Twisting Lemma (Proposition 6.11), we obtain another first-order Versal unfolding $U_\Gamma(x, \lambda, \alpha, \beta, \gamma)$ of $\Gamma(x)\cdot f(x)$, where $\Gamma(x)$ is an invertible 3×3 matrix, by setting

$$U_\Gamma(x, \lambda, \alpha, \beta, \gamma) = \Gamma(x)\cdot U(x, \lambda, \alpha, \beta, \gamma) \tag{6.439}$$

From this first-order Versal unfolding, we seek to obtain a family of minimal first-order versal unfoldings, using the De-Capitalization Lemma (Proposition 6.9). We do so because first-order versal unfoldings are easy to interpret geometrically and because they may contain many fewer parameters. Proposition 6.9 tells us, first of all, that U_Γ is first-order versal. Then we proceed to find minimal first-order versal unfoldings by reducing the set of control parameters to a set exactly large enough to imply that the differential map from the stationary set to M_3 is transverse to $\mathrm{Orb}\big(df(0)\big)$. To simplify this calculation, we restrict ourselves to the one member of the family given by

$$\Gamma = \begin{bmatrix} 0 & 1 & 0 \\ -1 & 0 & 0 \\ 0 & 0 & 1 \end{bmatrix} \tag{6.440}$$

so that we are unfolding $g(x) = \Gamma\cdot f(x)$. Obviously, this case will be interesting because it will lead to a first-order contact-invariant relationship between the set of fold points and the set of Hopf bifurcation points. For simplicity of notation, with Γ given by (6.440), we write U_Γ in (6.439) as

$$V = U_\Gamma \tag{6.441}$$

We begin by calculating $T_A\mathrm{Orb}(A)$ for $A = dg(0)$ as we did in (6.280), with the result that this tangent space is given by

$$T_A \mathrm{Orb}(A) = \left\{ \begin{bmatrix} u & v & p \\ v & -u & q \\ r & s & 0 \end{bmatrix} \;\middle|\; u, v, p, q, r, s \in R \right\} \qquad (6.442)$$

Next, using (6.438)-(6.441), we calculate the two differentials

$$d_x V(0) \begin{bmatrix} \xi_1 \\ \xi_2 \\ \xi_3 \end{bmatrix} = \begin{bmatrix} \xi_2 \\ -\xi_1 \\ 0 \end{bmatrix} \qquad (6.443)$$

and writing $\varepsilon = (\alpha, \beta, \lambda, \gamma)$ and $e = (a, b, \ell, c)$ we have

$$d_\varepsilon V(0)e = \begin{bmatrix} 0 \\ 0 \\ \ell \end{bmatrix} \qquad (6.444)$$

From these we find the tangent space at the origin to the stationary set,

$$T_0 S_V = \{(\xi, e) \mid d_x V(0)\xi + d_\varepsilon V(0)e = 0\} \qquad (6.445)$$

$$= \{(0, 0, \xi_3, 0, a, b, c) \mid \xi_3 \in R;\ a, b \in R^5 ;\ c \in R^6\}$$

where we have written everything as a row in the last version with the aid of (6.438).

Finally, we calculate the differential map

$$d(d_x V)(0):\ T_0 S_V \to M_n \qquad (6.446)$$

and we see that

$$d\big(d_x V(0)\big)\ (0, 0, \xi_3, 0, a, b, c) = \begin{bmatrix} b_1 & b_2 & 0 \\ -a_1 & -a_2 & 0 \\ c_1 & c_2 & 2\xi_3 \end{bmatrix} \qquad (6.447)$$

The condition for transversality to Orb(A) of the differential map $d_x V: S_V \to M_3$ is given by

$$T_A \mathrm{Orb}(A) + d(d_x V)(0) T_0 S_V = M_3 \qquad (6.448)$$

Consequently, we may define a minimal first-order versal unfolding of $g(x) = \Gamma \cdot f(x)$ by writing

$$v(x, \alpha_1, \alpha_2) = \begin{bmatrix} x_2 \\ -x_1 \\ x_3^2 \end{bmatrix} + \begin{bmatrix} \alpha_1 x_1 + \alpha_2 x_2 \\ 0 \\ \lambda \end{bmatrix} \tag{6.449}$$

Our next goal is to work out the stability phase portrait of (6.449). To begin with, we see that near the origin the stationary set of v is given by

$$S_v = \{(x, \lambda, \alpha) \mid x_3^2 + \lambda = 0\} \tag{6.450}$$

with $\dim(S_V) = 5$. Near dg(0), the real parts of the eigenvalues of a matrix may change sign in one of two ways:

i) A real eigenvalue may vanish

ii) A complex conjugate pair may become imaginary

Thus, we need to consider the two invariant subsets of M_3,

$$P_o = \{B \in M_3 \mid \text{exactly one eigenvalue vanishes}\} \tag{6.451}$$

and

$$P_1 = \{B \in M_3 \mid \text{exactly one conjugate pair of eigenvalues is pure imaginary}\} \tag{6.452}$$

Near dg(0), both are smooth submanifolds of M_3, with

$$\dim(P_o) = 8 \tag{6.453}$$

and

$$\dim(P_1) = 8 \tag{6.454}$$

Clearly $A = dg(0) \in P_o \cap P_1$. A straightforward calculation using Milnor's trick from Example 20 shows us that

$$T_A P_o = \{X = [x_{ij}] \in M_3 \mid x_{33} = 0\} \tag{6.455}$$

On the other hand, the path

$$X(t) = \begin{bmatrix} 0 & 1 & 0 \\ -1 & 0 & 0 \\ 0 & 0 & t \end{bmatrix} \tag{6.456}$$

is in P_1 with $X(0) = A$ so that its velocity vector $X'(0)$ is in T_AP_1. Thus

$$\begin{bmatrix} 0 & 0 & 0 \\ 0 & 0 & 0 \\ 0 & 0 & 1 \end{bmatrix} \varepsilon\ T_AP_1 \tag{6.457}$$

and

$$T_AP_0 + T_AP_1 = M_3 \tag{6.458}$$

That is, P_0 and P_1 are transverse at A and consequently must meet in a smooth invariant 7-dimensional submanifold of M_3. Because v is first-order versal, then near the origin, the differential map $d_xv:S_v \to M_3$ must be transverse to all three smooth submanifolds P_0, P_1 and $P_0 \cap P_1$. Consequently we have the neutral sets

$$D_v(P_0) = \{z \ \varepsilon\ S_v \mid d_xv(z) \ \varepsilon\ P_0\} \tag{6.459}$$

$$D_v(P_1) = \{z \ \varepsilon\ S_v \mid d_xv(z) \ \varepsilon\ P_1\} \tag{6.460}$$

$$D_v(P_0 \cap P_1) = \{z \ \varepsilon\ S_v \mid d_xv(z) \ \varepsilon\ P_0 \cap P_1\} \tag{6.461}$$

from which we conclude that

$$\dim(D_v(P_0)) = 4 \tag{6.462}$$

$$\dim(D_v(P_1)) = 4 \tag{6.463}$$

$$\dim(D_v(P_0 \cap P_1)) = 3 \tag{6.464}$$

and that

$$0 \ \varepsilon\ D_v(P_0 \cap P_1) = D_v(P_0) \cap D_v(P_1) \tag{6.465}$$

We see (Fig. 6.17) that $D_v(P_0)$ and $D_v(P_1)$ each divide S_v near the origin into two pieces, so that a real eigenvalue changes sign as we cross $D_v(P_0)$ and the real part of a complex conjugate pair of eigenvalues changes sign as we cross $D_v(P_1)$. Furthermore, from (6.458), we may conclude after a little calculating that

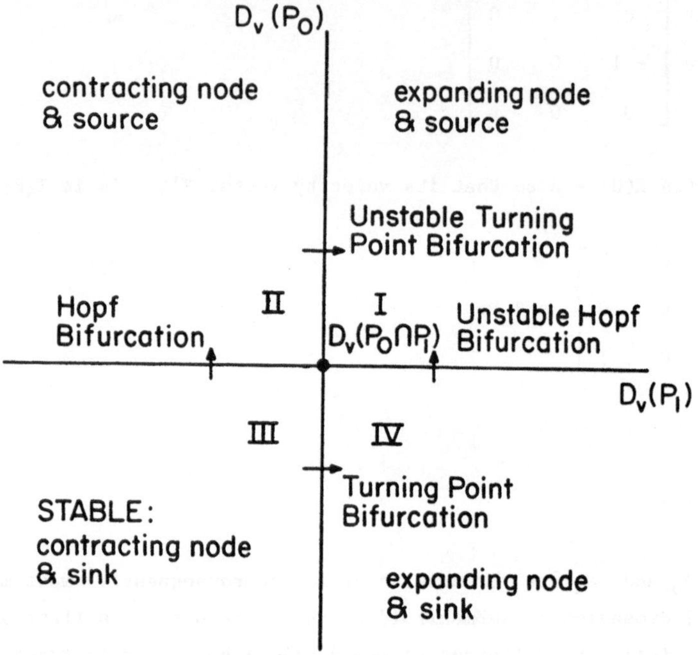

Fig. 6.17 The schematic arrangement of the sets I, II, III, IV, (6.467)-(6.470), in S_v.

$$T_o D_v(P_o) + T_o D_v(P_1) = T_o S_v \tag{6.466}$$

so that $D_v(P_o)$ and $D_v(P_1)$ meet <u>transversally</u> along $D_v(P_o \cap P_1)$ in S_v. It follows that together $D_v(P_o)$ and $D_v(P_1)$ divide S_v near the origin into four connected regions I, II, III, IV, with

$$\text{I} = \{z \mid \text{all eigenvalues have positive real part}\} \tag{6.467}$$

$$\text{II} = \{z \mid \text{the real eigenvalue is negative; the other two have positive real part}\} \tag{6.468}$$

$$\text{III} = \{z \mid \text{all eigenvalues have negative real part}\} \tag{6.469}$$

$$\text{IV} = \{z \mid \text{the real eigenvalue is positive; the other two have negative real part}\} \tag{6.470}$$

Schematically, these sets are arranged within S_v as pictured in Figure 6.17.

Two further interesting sets are the singularity sets in control parameter space. These are defined by setting

$$B_v(P) = \{(\lambda, \alpha) \mid \text{there is some } x \text{ such that } (x, \lambda, \alpha) \in D_v(P)\} \tag{6.471}$$

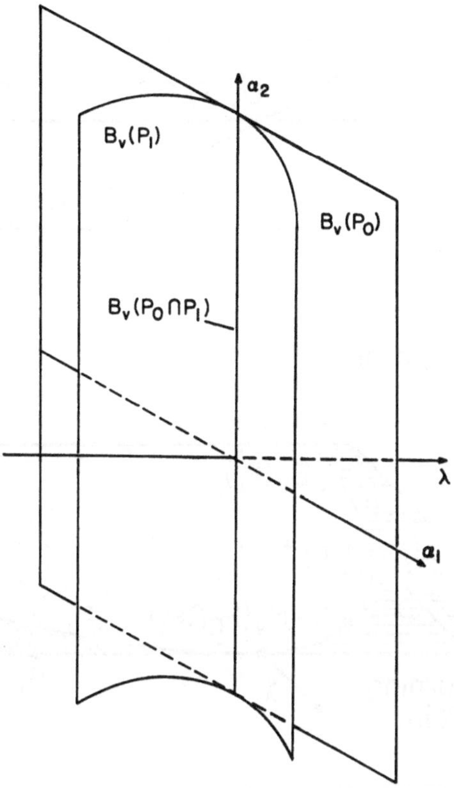

Fig. 6.18 The singularity sets of $v(x, \lambda, \alpha)$ in (6.476). The surface $B_v(P_0)$ is the set of fold points of $v(x, \lambda, \alpha)$. When the parameter (λ, α) crosses $B_v(P_0)$ by increasing λ past 0, the system loses two stationary solutions at a turning point of S_v. The surface $B_v(P_1)$ is the set of Hopf bifurcation parameters for the system; when the parameter (λ, α) crosses $B_v(P_1)$, the system undergoes a Hopf bifurcation, provided that the stationary solution was originally on the right sheet of S_v. Compare with Fig. 6.19.

in general. To find $B_v(P_0)$, we find $D_v(P_0)$ first; the calculation is very easy, and similar to the one in Example 19, resulting in the fact that $D_v(P_0)$ is the set of fold points, and that

$$D_v(P_0) = \{(x_1, x_2, 0, 0, \alpha_1, \alpha_2) \mid x_1, x_2, \alpha_1, \alpha_2 \in R\} \tag{6.472}$$

Thus

$$B_v(P_0) = \{(0, \alpha_1, \alpha_2) \mid \alpha_1, \alpha_2 \in R\} \tag{6.473}$$

is the (α_1, α_2)-plane in $(\lambda, \alpha_1, \alpha_2)$-space.

Again, in the same way, we may find

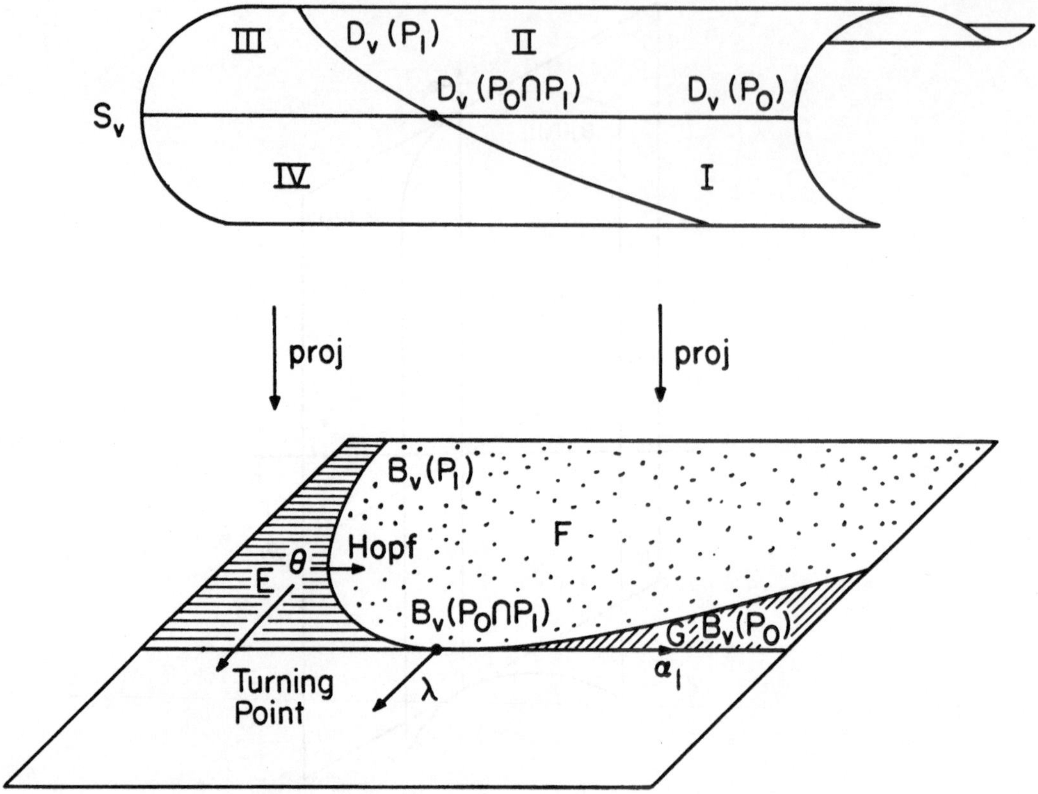

Fig. 6.19 Schematic representation of the first-order versal unfolding $v(x, \lambda, \alpha)$ in (6.476). Region I projects to region G; region II projects to $F \cup G$; region III, to E; and IV, to $E \cup F$. If the control parameter θ is in E with the state stable, then moving θ across $B_v(P_1)$ will bring about a Hopf bifurcation, and moving θ across $B_v(P_0)$ will bring about a catastrophic change to another (perhaps unsteady) state.

$$D_v(P_0 \cap P_1) = \{(x_1, x_2, 0, 0, 0, \alpha_2) \mid x_1, x_2, \alpha_2 \in R\} \tag{6.474}$$

$$B_v(P_0 \cap P_1) = \{(0, 0, \alpha_2) \mid \alpha_2 \in R\} \tag{6.475}$$

That is, $B_v(P_0 \cap P_1)$ is the α_2-axis. Without further calculation, we may see that the transversal intersection of $D_v(P_0)$ and $D_v(P_1)$ in S_v is turned by the folding projection

$$\text{proj}: \ S_v \to R^3, \quad \text{proj}(x, \lambda, \alpha) = (\lambda, \alpha) \tag{6.476}$$

into a quadratic tangency, resulting in Figure 6.18. Finally, Figure 6.19 suggests schematically what is going on, and our pictorial description of the unfolding of $g(x) = \Gamma \cdot f(x)$ is complete. #

In Yost and Shirer (1982), Figs. 6-8, the situation of Example 22 occurs; that is, the neutral sets $D(P_0)$ and $D(P_1)$ meet at a singular point. Because $D(P_0)$ coincides with the set of fold points in the stationary set, the singularity set $B(P_0)$ coincides with the set of fold points in the control parameter space. In the case of Yost and Shirer (1982), the dimensions are all reduced by one from that shown in Fig. 6.18; apparently that case is a section of a more general first-order versal unfolding. Because the dimensions are all reduced by one, the control parameter space is the plane and $B(P_0)$ and $B(P_1)$ are curves meeting tangentially as in Fig. 6.19. Finally, in that case only Hopf bifurcations from stable solutions are of interest, so that Yost and Shirer show only one branch of the curve $B(P_1)$.

We arrive now at a new element in our procedure, borrowed, as were invariant sets, tangent spaces, differentials, transversality and versality, from the realm of smooth topology. In that discipline, the various notions of homotopy are of central importance, especially those of structure-preserving homotopy. Here we will define only the particular version of homotopy that we need, and we will call it a deformation in order to suggest its intuitive meaning. More particularly, we will say that a deformation of first-order versal unfoldings, or simply a first-order deformation, is a piecewise smooth family $u_s(x, \mu)$ of first-order versal unfoldings with $0 \leq s \leq 1$; that is, $u_s(x, \mu)$ is continuous, and smooth for $s_i \leq s \leq s_{i+1}$ where $0 = s_0 < s_1 < \cdots < s_r = 1$. If $u_s(x, \mu)$ is a first-order deformation, then each map

$$u_s\colon\ R^n \times R^q \to R^n \tag{6.477}$$

is transverse to $\{0\}$ in R^n. To see a consequence of this, we introduce a lemma.

Proposition 6.14 (Deformation Lemma) If

$$f_s\colon\ Q \to R^k\ ,\quad 0 \leq s \leq 1 \tag{6.478}$$

is a piecewise smooth family of maps, each transverse to $P_1, \ldots, P_r$ in R^k near q_0 in Q, then there is a piecewise smooth family

$$\phi_s\colon\ Q \to Q \tag{6.479}$$

of diffeomorphisms near q_0, such that

$$\phi_s(q_0) = q_0 \tag{6.480}$$

and such that near q_0

$$\phi_s\colon\ f_0^{-1}(P_i) \to f_s^{-1}(P_i) \tag{6.481}$$

is a diffeomorphism for $i = 1, \ldots, r$.

To see the intuitive content of this lemma, we think of the parameter s as a "time" during which the map f_0 is deformed to f_1. If, during this deformation, transversality is preserved, then the only effect on the smooth submanifolds $f_0^{-1}(P_1), \ldots, f_0^{-1}(P_r)$ of Q will be to warp and rearrange them by a motion of the ambient space Q. A very good picture is that Q is filled with a fluid that flows, via ϕ_s, to a new position, carrying with it the submanifolds of Q $f_0^{-1}(P_1), \ldots, f_0^{-1}(P_r)$ to $f_1^{-1}(P_1), \ldots, f_1^{-1}(P_r)$. Thus, the smooth submanifolds are neither torn, nor glued together, nor passed through one another.

Applying the Deformation Lemma to a first-order deformation $u_s(x, \mu)$, we obtain near the origin a piecewise smooth family ϕ_s of diffeomorphisms

$$\phi_s: \; R^n \times R^q \to R^n \times R^q \tag{6.482}$$

that preserve the origin, such that

$$\phi_s: \; S_{u_0} \to S_{u_s} \tag{6.483}$$

is a diffeomorphism. Now, because each differential map

$$d_x u_s: \; S_{u_s} \to M_n \tag{6.484}$$

is transverse near the origin to the invariant smooth submanifolds $P_1, \ldots, P_r$ in M_n, so is the composition

$$(d_x u_s) \circ \phi_s: \; S_{u_0} \to M_n \tag{6.485}$$

Applying the Deformation Lemma (Proposition 6.14) to this smooth family, we obtain a piecewise smooth family ϕ_s of diffeomorphisms near the origin

$$\psi_s: \; S_{u_0} \to S_{u_0} \tag{6.486}$$

each preserving the origin, such that near the origin

$$\psi_s: \; \phi_0^{-1}\big(D_{u_0}(P_i)\big) \to \phi_s^{-1}\big(D_{u_s}(P_i)\big) \tag{6.487}$$

is a diffeomorphism. But then $\phi_s \circ \psi_s \circ \phi_0^{-1}$ near the origin is a family of diffeomorphisms

$$\phi_s \circ \psi_s \circ \phi_0^{-1}: \; S_{u_0} \to S_{u_s} \tag{6.488}$$

such that

$$\phi_s \circ \psi_s \circ \phi_o^{-1}: \; D_{u_o}(P_i) \to D_{u_s}(P_i) \tag{6.489}$$

is a diffeomorphism near the origin. Consequently, the entire two configurations $(S_{u_o}; D_{u_o}(P_1), \ldots, D_{u_o}(P_r))$ and $(S_{u_1}; D_{u_1}(P_1), \ldots, D_{u_1}(P_r))$ are diffeomorphic in the sense that there is a diffeomorphism carrying the ambient set S_{u_o} reversibly onto the ambient set S_{u_1}, so that the distinguished subsets $D_{u_o}(P_i)$ and $D_{u_1}(P_i)$ correspond. To put it more succinctly, if there is a first-order deformation connecting $u_o(x, \mu)$ and $u_1(x, \mu)$, then $u_o(x, \mu)$ and $u_1(x, \mu)$ have the same stability phase portrait.

Now we return to the unfolding of Example 22.

<u>Example 23. The stability phase portrait of a general first-order versal unfolding of $g(x) = [x_2, -x_1, x_3^2]^T$</u>

In Example 22 we found a certain first-order versal unfolding

$$v(x, \lambda, \alpha) = \begin{bmatrix} \alpha_1 x_1 + (1 + \alpha_2)x_2 \\ -x_1 \\ x_3^2 + \lambda \end{bmatrix} \tag{6.490}$$

of

$$g(x) = \begin{bmatrix} x_2 \\ -x_1 \\ x_3^2 \end{bmatrix} \tag{6.491}$$

and we found the two most important of the smooth submanifolds $D_v(P)$ of S_v in its stability phase portrait. In this example, we wish to show that any minimal first-order versal unfolding of $g(x)$ may be first-order deformed to a first-order versal unfolding of $g(x)$, which, in turn, is first-order equivalent to a canonical first-order versal unfolding. Then we will be able to conclude that any two minimal first-order versal unfoldings of $g(x)$ have the same stability phase portrait. In particular, the stability phase portrait of $v(x, \lambda, \alpha)$, given by (6.490), is the stability phase portrait of any other minimal first-order versal unfolding of $g(x)$.

We begin with any minimal first-order versal unfolding $u(x, \mu)$ of $g(x)$. For example, $u(x, \mu) = v(x, \lambda, \alpha)$, given by (6.490) is such an unfolding. Because $V(x, \lambda, \alpha, \beta, \gamma)$, given in (6.441) is first-order Versal, there exists a first-order contact map

$$T = (M(x, \mu); y(x, \mu); \lambda(\mu), \alpha(\mu), \gamma(\mu)) \tag{6.492}$$

such that u pulls back from V; that is,

$$u(x, \mu) = M(x, \mu)\cdot V\big(y(x, \mu), \lambda(\mu), \alpha(\mu), \beta(\mu), \gamma(\mu)\big) \tag{6.493}$$

To a certain extent, this notation, which was natural for the purposes of Example 22, is not natural for our present purposes. To avoid obscuring the siutation in the present example we must introduce some supplementary notation. Consequently, we set

$$\left.\begin{aligned} \theta(\mu) &= \big(\lambda(\mu), \alpha(\mu), \beta(\mu), \gamma(\mu)\big) \\ \phi(x, \mu) &= \big(y(x, \mu), \theta(\mu)\big) \end{aligned}\right\} \tag{6.494}$$

To supplement this notation, with $D = d_x y(0)$, we introduce a conjugation map

$$\mathrm{conj}_D: \ M_3 \to M_3 \tag{6.495}$$

defined by

$$\mathrm{conj}_D(B) = D\ B\ D^{-1} \tag{6.496}$$

Notice that conj_D carries Orb(A) into itself. Finally, for $A = dg(0)$, we denote the natural vector space projection from M_3 to the quotient space $M_3/T_A\mathrm{Orb}(A) = Q_A$ by π_1,

$$\pi_1: \ M_3 \to Q_A \tag{6.497}$$

The notation just introduced facilitates the following calculations, in which we seek a criterion for $d_x u: S_u \to M_3$ to be transverse to Orb(A); this criterion finally emerges in (6.518). We start with

$$u = M\cdot V \tag{6.498}$$

and apply d_x and d to u. Using the chain rule, we see that

$$d_x u = d_x M\cdot V + M\cdot dV\cdot d_x\phi \tag{6.499}$$

$$du = dM\cdot V + M\cdot dV\cdot d\phi \tag{6.500}$$

But $d_x\theta = 0$, and so

$$dV\cdot d_x\phi = d_y V\cdot d_x y \tag{6.501}$$

and then

$$d_x u = d_x M \cdot V + M \cdot d_y V \cdot d_x y \tag{6.502}$$

Next we apply d to $d_x u$ and obtain, with the last two terms interchanged,

$$d(d_x u) = d_x^2 M \cdot V + d_x M \cdot dV \cdot d\phi + dM \cdot d_y V \cdot d_x y + M \cdot d_y V \cdot d(d_x y) \tag{6.503}$$
$$+ M \cdot d(d_y V) \cdot (d\phi,\ d_x y)$$

The peculiar notation in the last term takes care of the double sum that appears.

Now we evaluate (6.500) and (6.503) at the origin. To evaluate (6.500), we recall that $V(0) = 0$ and obtain

$$du(0) = M(0) \cdot dV(0) \cdot d\phi(0) \tag{6.504}$$

In addition, we recall that

$$T_o S_u = \{\xi \mid du(0) \cdot \xi = 0\} \tag{6.505}$$

and then, using the fact that M(0) is invertible, we see from (6.504) that

$$d\phi(0):\quad T_o S_u \to T_o S_V \tag{6.506}$$

Having verified (6.506), we turn to evaluation of (6.503) at the origin. To evaluate (6.503), we recall that the condition defining a first-order contact transformation

$$M(x, \mu) \cdot d_x y(x, \mu) = 1 + O(2) \tag{6.507}$$

implies

$$M(0)^{-1} = d_x y(0) = D \tag{6.508}$$

and also implies both

$$dM(0) = -D^{-1} \cdot d(d_x y)(0) \cdot D^{-1} \tag{6.509}$$

and

$$d_x M(0) = -D^{-1} \cdot d_x^2 y(0) \cdot D^{-1} \tag{6.510}$$

Now, we evaluate (6.503) at the origin and on a vector $\xi \in T_oS_u$. We use (6.502) - (6.510) and recall that $A = dg(0) = d_xu(0) = d_yV(0)$ to obtain

$$d(d_xu)(0)\cdot\xi = -D^{-1}\big(d(d_xy)(0)\cdot\xi\big)D^{-1}AD + D^{-1}A\big(d(d_xy(0))\cdot\xi\big) + D^{-1}\big(d(d_yV)(0)\cdot d\phi(0)\cdot\xi\big)D \tag{6.511}$$

$$d(d_xu)(0)\cdot\xi = \big[D^{-1}AD\big]\cdot\big[D^{-1}\big(d(d_xy)(0)\cdot\xi\big)\big] - \big[D^{-1}\big(d(d_xy)(0)\cdot\xi\big)\big]\cdot\big[D^{-1}AD\big] + D^{-1}\big(d(d_yV)(0)\cdot d\phi(0)\cdot\xi\big)D \tag{6.512}$$

Having carried out these evaluations, we observe now that the sum of the first two terms in (6.512) is in $T_A\mathrm{Orb}(A)$, and consequently that sum is carried to zero by the map π_1 (6.497). Applying π_1 then to both sides of (6.512), we see that

$$\pi_1\big[D\big(d(d_xu)(0)\cdot\xi\big)D^{-1}\big] = \pi_1\big[d(d_yV)(0)\cdot d\phi(0)\cdot\xi\big] \tag{6.513}$$

for all $\xi \in T_oS_u$. In other words, when both sides are regarded as maps from T_oS_u to Q_A, we have the equation

$$\pi_1 \circ \mathrm{conj}_D \circ d(d_xu)(0) = \pi_1 \circ d(d_yV)(0) \circ d\phi(0) \tag{6.514}$$

This equation expresses the first fact that we need in order to find our first-order deformation. Equation (6.447) in Example 22 expresses the second fact that we need; we recall that, in the present terms, it is given by

$$d\big(d_yV(0)\big)\ (0,\ 0,\ \eta_3,\ 0,\ a,\ b,\ c) = \begin{bmatrix} b_1 & b_2 & 0 \\ -a_1 & -a_2 & 0 \\ c_1 & c_2 & 2\eta_3 \end{bmatrix} \tag{6.515}$$

Finally, equation (6.442) expresses the third fact that we need; we recall that it is given by

$$T_A\mathrm{Orb}(A) = \left\{ \begin{bmatrix} u & v & p \\ v & -u & q \\ r & s & 0 \end{bmatrix} \,\middle|\, u,\ v,\ p,\ q,\ r,\ s \in R \right\} \tag{6.516}$$

From these three facts, it follows that

$$d(d_xu)(0):\ T_oS_u \to Q_A \tag{6.517}$$

is onto if and only if the matrix

$$X_1 = \begin{bmatrix} d_\mu\beta_1(0) - d_\mu\alpha_2(0) \\ d_\mu\beta_2(0) + d_\mu\alpha_1(0) \end{bmatrix} \tag{6.518}$$

has rank two. Because (6.517) is equivalent to the transversality of $d_x u: S_u \to M_3$ to Orb(A), (6.518) is the criterion for transversality that we sought.

Our next goal is to define a first-order deformation $u_s(x, \mu)$ from $u(x, \mu)$ to an unfolding which pulls back from a canonical unfolding; the canonical unfolding itself will result from our construction of the deformation. To define the smooth family $u_s(x, \mu)$ of unfoldings, for $0 \leq s \leq 1$, we define λ_s, $\alpha_s = (\alpha_{1,s}, \ldots, \alpha_{5,s})$, β_s, and γ_s as follows,

$$\left.\begin{aligned} &\lambda_s = \lambda \\ &\alpha_{i,s} = (1 - s)\,\alpha_i && \text{for } i = 3, 4, 5 \\ &\beta_{i,s} = (1 - s)\,\beta_i && \text{for } i = 3, 4, 5 \\ &\gamma_{i,s} = (1 - s)\,\gamma_i && \text{for } i = 1, \ldots, 6 \end{aligned}\right\} \tag{6.519}$$

and

$$\left.\begin{aligned} &\alpha_{1,s} = \alpha_1 - \frac{s}{2}(\alpha_1 - \beta_2) \\ &\alpha_{2,s} = \alpha_2 - \frac{s}{2}(\alpha_2 + \beta_1) \\ &\beta_{1,s} = \beta_1 - \frac{s}{2}(\alpha_2 + \beta_1) \\ &\beta_{2,s} = \beta_2 + \frac{s}{2}(\alpha_1 - \beta_2) \end{aligned}\right\} \tag{6.520}$$

and

$$\theta_s(\mu) = \big(\lambda_s(\mu), \alpha_s(\mu), \beta_s(\mu), \gamma_s(\mu)\big) \tag{6.521}$$

Then

$$T_s = \big(M(x, \mu), y(x, \mu), \theta_s(\mu)\big) \tag{6.522}$$

is a contact map. We define

$$u_s(x, \mu) = M(x, \mu)\cdot V\big(y(x, \mu), \theta_s(\mu)\big) \tag{6.523}$$

so that $u_0(x, \mu) = u(x, \mu)$ and $u_1(x, \mu) = M(x, \mu)\cdot V\big(y(x, \mu), \theta_1(\mu)\big)$, where $\theta_1(\mu) = \big(\lambda(\mu), [\alpha_1(\mu) + \beta_2(\mu)]/2, [\alpha_2(\mu) - \beta_1(\mu)]/2, 0, 0, 0, [\beta_1(\mu) - \alpha_2(\mu)]/2, [\alpha_1(\mu) + \beta_2(\mu)]/2, 0, \ldots, 0\big)$.

Having obtained the smooth family in (6.523), we have to see that it is first-order versal. It is clearly versal so that it is the transversality of the differential map that we must check. We begin by checking that

$$T_o S_{u_s} = T_o S_u \qquad \text{for } 0 \leq s \leq 1 \tag{6.524}$$

Then we repeat the argument leading to equation (6.514) to obtain the corresponding equation for $u_s(x, \mu)$,

$$\pi_1 \circ \mathrm{conj}_D \circ d(d_x u_s)(0) = \pi_1 \circ d(d_y V) \circ d\phi_s(0) \tag{6.525}$$

as maps with domain $T_o S_u = T_o S_{u_s}$. Now, it is very easy to check, using (6.515), that the right side of (6.525) is onto. Finally, we check that, because

$$D\ A = A\ D \tag{6.526}$$

and

$$\mathrm{conj}_D\colon\ T_A \mathrm{Orb}(A) \to T_A \mathrm{Orb}(A) \tag{6.527}$$

there is a vector space isomorphism

$$C_D\colon\ Q_A \to Q_A \tag{6.528}$$

such that

$$\pi_1 \circ \mathrm{conj}_D = C_D \circ \pi_1 \tag{6.529}$$

It follows now that

$$\pi_1 \circ d(d_x u_s)(0)\colon\ T_o S_{u_s} \to Q_A \tag{6.530}$$

is onto. Thus the family $u_s(x, \mu)$ is first-order versal.

We also omit the easy argument that μ has three entries; that is, $\mu = (\mu_1, \mu_2, \mu_3)$. Then the first-order contact map T_1 may be regarded as a first-order contact <u>transformation</u> from $u_1(x, \mu)$ to

$$V_1(y,\ \lambda,\ \zeta_1,\ \zeta_2) = \begin{bmatrix} x_2 - \zeta_2 x_1 + \zeta_1 x_2 \\ -x_1 - \zeta_1 x_1 - \zeta_2 x_2 \\ x_3^2 + \lambda \end{bmatrix} \qquad (6.531)$$

by setting $T_1 = \big(M(x, \mu), y(x, \mu), \zeta_1(\mu), \zeta_2(\mu)\big)$ with $\zeta_1(\mu) = [\alpha_1(\mu) + \beta_2(\mu)]/2$ and $\zeta_2(\mu) = [\alpha_2(\mu) - \beta_1(\mu)]/2$.

Finally, we conclude via Proposition 6.14 that our arbitrary first-order versal unfolding $u(x, \mu)$ of $g(x)$, where $g(x)$ is given by (6.491), has the same stability phase portrait as $u_1(x, \mu)$ because there exists a first-order deformation $u_s(x, \mu)$ from $u(x, \mu) = u_0(x, \mu)$ to $u_1(x, \mu)$. And we conclude that $u_1(x, \mu)$ has the same stability phase portrait as $V_1(y, \lambda, \zeta_1, \zeta_2)$ because there exists a first-order contact transformation carrying $V_1(y, \lambda, \zeta_1, \zeta_2)$ into $u_1(x, \mu)$. That is, the stability phase portrait of <u>any</u> minimal first-order versal unfolding is that of V_1, which we may find once and for all by making routine computations beginning with (6.531). But because any two minimal versal unfoldings of $g(x)$ have the same stability phase portrait, we may compute it for any other such unfolding. This computation we have carried out already in Example 22, and we conclude that we now know the stability phase portrait for any minimal first-order versal unfolding of $g(x)$. We notice however that the parameters in V_1 are pleasantly related to the stability phase portrait: λ alone determines the stationary set via $x_3^2 + \lambda = 0$, and ζ_2 determines the Hopf bifurcation set via $x_3^2 + \lambda = 0$ and $\zeta_2 = 0$. #

6.7 Conclusion

We note first of all that the last example is actually a theorem thinly disguised as an illustrative example. We state it independently as follows.

<u>Theorem</u> Two first-order versal unfoldings with the same number of control parameters have the same stability phase portrait.

Of course, now we must finally define the stability phase portrait, and we do so by saying that the <u>stability phase portrait</u> is the configuration in $R^n \times R^p$ of the stationary set and its neutral subsets defined by invariant submanifolds of M_n, together with a specification of the way eigenvalues change near the neutral sets. We define the <u>strong stability phase portrait</u> to include the configuration of the projections of these sets to the control parameter space, and we note that the theorem above does not hold for the strong stability phase portrait.

To pass from one strong stability phase portrait to another within the same stability phase portrait, we must compute explicit first-order deformations. This is an interesting and enlightening task which here we must defer. Instead, we will note that now we have enough information to find the canonical first-order versal unfoldings and their stability phase portraits.

To find a physically interpretable first-order versal unfolding of a given physical dynamical system

$$\dot{x} = g(x) \tag{6.532}$$

we may proceed as follows:

i) We perform a first-order Lyapunov-Schmidt splitting of g(x). That is, we find an ordinary contact transformation $T = \big(M(x);\ u(x),\ v(x)\big)$ such that

$$g(x) = M(x)\cdot\begin{bmatrix} u(x) \\ h\big(v(x)\big) \end{bmatrix} \tag{6.533}$$

with $h(0) = dh(0) = 0$. The map h(v) must contain only a few variables so that we may unfold h(v) in an <u>ad hoc</u> manner.

ii) We find, in an <u>ad hoc</u> manner, a first-order Versal unfolding

$$H(v, \lambda) = h(v) + N(v)\cdot\lambda \tag{6.534}$$

of h(v).

iii) As in (6.405), we extend this unfolding to a first-order Versal unfolding

$$F(u, v, \lambda, C) = \begin{bmatrix} u \\ h(v) + N(v)\cdot\lambda \end{bmatrix} + C(u, v, \lambda)\cdot\begin{bmatrix} u \\ h(v) \end{bmatrix} \tag{6.535}$$

iv) We note that successive application of Proposition 6.10 and Proposition 6.11 shows that

$$G(x, \lambda, C) = M(x)\cdot F\big(u(x), v(x), \lambda, C\big) \tag{6.536}$$

is a first-order Versal unfolding of g(x). Using equation (6.514), we may calculate

$$d(d_xG)(0):\quad T_oS_G \to Q_A \tag{6.537}$$

without finding dM(0), which is somewhat difficult to find. The calculation (6.537) enables us to see which coefficients of C we will need in order to achieve first-order versality.

v) We make first-order alterations, using the result of (6.537) as a guide, so that the coefficients c of C that remain achieve first-order versality and are physically interpretable. Of course, this is the step that requires ingenuity -- this is not a routine step. The whole point of our theoretical machinery is to make it possible for us to begin this step.

vi) Finally we delete all the other coefficients of C. The theorem above guarantees that we lose no information about the stability phase portrait by doing so.

The result of the above procedure is a physically interpretable, minimal first-order versal unfolding $v(x, \lambda, c)$ of $g(x)$. To find its stability phase portrait, we proceed as follows:

i) We find S_v using the methods of Chapter 2.

ii) We find, much more easily by direct calculation, all terms in the following map,

$$d(d_x v)(0): \quad T_o S_v \to Q_A \tag{6.538}$$

iii) For each invariant smooth submanifold P of M_n, we find

$$T_o^A P = d(\text{proj})(0)\cdot(T_o P) \subset Q_A \tag{6.539}$$

where proj: $M_n \to Q_A$ is the vector space projection. We note that the vector spaces $T_o^A P$ are exactly the invariant sub-vector spaces of M_n.

iv) We find for each such P, using (6.538)

$$T_o D_v(P) = d(d_x v(0))^{-1}\,(T_o^A(P)) \tag{6.540}$$

Now any smooth submanifold of S_v through the origin and tangent to $T_o D_v(P)$ may be used for our first-order approximation to $D_v(P)$, and our process is complete.

APPENDIX

SUMMARY OF SPECTRAL MODELS

In this appendix we present brief reviews of the equations used to develop the spectral models that we studied in this monograph, and we give a summary of the principal results of application of the contact catastrophe procedure (Section 2.8). For convenience we list in tables in each section the dimensional and nondimensional variables and parameters used in each model.

A.1 The Lorenz Model

Tables A.1 and A.2 give the variables and parameters used to describe this model. The governing partial differential system is

$$\frac{\partial}{\partial t^*} \tilde{\nabla}^2 \psi^* = - K(\psi^*, \tilde{\nabla}^2 \psi^*) + \sigma(1 + a^2)^{-1} \tilde{\nabla}^4 \psi^* + \sigma(1 + a^2) \frac{\partial \theta^*}{\partial x^*} + \sigma(1 + a^2) h \tag{A.1}$$

$$\frac{\partial \theta^*}{\partial t^*} = - K(\psi^*, \theta^*) + r \frac{\partial \psi^*}{\partial x^*} + h \frac{\partial \psi^*}{\partial z^*} + (1 + a^2)^{-1} \tilde{\nabla}^2 \theta^* \tag{A.2}$$

The spectral expansions for ψ^* and θ^* are

$$\psi^* = \sqrt{2}\, x_1 \sin x^* \sin z^* \tag{A.3}$$

$$\theta^* = \sqrt{2}\, x_2 \cos x^* \sin z^* - x_3 \sin 2z^* \tag{A.4}$$

and the spectral model is

$$\dot{x}_1 = - \sigma x_1 + \sigma x_2 - 8\sqrt{2}\, \sigma h \pi^{-2} \tag{A.5}$$

$$\dot{x}_2 = - x_1 x_3 + r x_1 - x_2 \tag{A.6}$$

$$\dot{x}_3 = x_1 x_2 - b x_3 - 16\sqrt{2}\, h(3\pi^2)^{-1} x_1 \tag{A.7}$$

The singular points are cusp points, given by

$$\left.\begin{aligned} r_s &= 1 \\ h_s &= 0 \\ x_1 &= x_2 = x_3 = 0 \end{aligned}\right\} \tag{A.8}$$

and

Table A.1

Dimensional Variables: Lorenz Model

Symbol	Name
g	acceleration of gravity
H	domain height
L	domain width
t	time
T	temperature
T_0	value of T at (x, z) = (0, 0) when θ = 0
$\Delta_x T$	horizontal temperature difference
$\Delta_z T$	vertical temperature difference
x	horizontal distance
z	elevation
θ	perturbation temperature
κ	eddy thermometric conductivity
ν	eddy viscosity
ψ	stream function

$$\left.\begin{aligned} r_S &= -26 \\ h_S &= \mp 27\pi^2\sqrt{b}\,(8\sqrt{2})^{-1} \\ x_1 &= \pm 3\sqrt{b} \\ x_2 &= \mp 24\sqrt{b} \\ x_3 &= -18 \end{aligned}\right\} \tag{A.9}$$

The canonical unfolding parameters d_o and d_1 for the constant and linear terms of the cubic polynomial are related near (r_s, h_s) to r and h by

$$d_o = -8\sqrt{2}\,h\,\pi^{-2} \tag{A.10}$$

$$d_1 = r - r_s = r - 1 \tag{A.11}$$

Table A.2

Nondimensional Variables & Parameters: Lorenz Model

Symbol	Name	Definition
a	aspect ratio	H/L
b	shape parameter	$4(1 + a^2)^{-1}$
d_0, d_1	canonical unfolding parameters	
h	Hadley number	$- \Delta_x T\ r\ (\Delta_z T)^{-1}$
h_S	critical value of h	$\mp 27\ \pi^2 \sqrt{b}\ (8\sqrt{2})^{-1}$
$K(f,g)$	Jacobian operator	$\frac{\partial f}{\partial x^*} \frac{\partial g}{\partial z^*} - \frac{\partial f}{\partial z^*} \frac{\partial g}{\partial x^*}$
n	integer horizontal wavenumber	
r	normalized Rayleigh number	R/R_s
r_s	critical value of r: cusp point	1
r_S	critical value of r: cusp point	- 26
R	Rayleigh number	$- g\ H^3\ \Delta_z T(T_0\ \nu\ \kappa)^{-1}$
R_s	critical Rayleigh number	$(1 + a^2)^3\ \pi^4\ a^{-2}$
t^*	time	$\pi^2(1 + a^2)\kappa t\ H^{-2}$
u^*	horizontal velocity component	$- \partial\psi^*/\partial z^*$
w^*	vertical velocity component	$\partial\psi^*/\partial x^*$
x_1	Fourier coefficient for ψ^*	

Table A.2 (Con't)

Symbol	Name	Definition
x_2, x_3	Fourier coefficients for θ^*	
x^*	horizontal distance	$\pi\ x\ L^{-1}$; $0 \leq x^* \leq \pi$
X	value of x_1 at cusp point	
z^*	elevation	$\pi\ z\ H^{-1}$; $0 \leq z^* \leq \pi$
$\underset{\sim}{\alpha}$	parameter vector	$[r, \sigma, b]$
$\underset{\sim}{\alpha}_s$	critical value of α: cusp point	$[1, \sigma, b]$
θ^*	perturbation temperature	$- a^2\ R\ \theta[\pi^3(1 + a^2)^3 \Delta_z T]^{-1}$
σ	Prandtl number	ν/κ
ψ^*	stream function	$a\ \psi(1 + a^2)^{-1}\ \kappa^{-1}$
$\tilde{\nabla}^2$	Laplacian operator	$a^2 \dfrac{\partial^2}{\partial x^{*2}} + \dfrac{\partial^2}{\partial z^{*2}}$
$\tilde{\nabla}^4$		$\tilde{\nabla}^2(\tilde{\nabla}^2\ )$

A.2. The Vickroy and Dutton Model

The nondimensional parameters and variables are given in Table A.3. The governing quasi-geostrophic equation is

$$\frac{\partial}{\partial t} \nabla_H^2 \psi' + J(\psi', \nabla_H^2 \psi') - \frac{\partial \psi'}{\partial x} \frac{\partial^2 U}{\partial y^2} + U \frac{\partial}{\partial x} \nabla_H^2 \psi' + \beta \frac{\partial \psi'}{\partial x} - \nu \nabla_H^4 \psi' \tag{A.12}$$

$$= H'(x, y)$$

The Fourier expansion for ψ is

$$\psi = a_1\ \pi^{-1} \sin y + a_2 \sqrt{2}\ \pi^{-1} \cos y \sin \ell x + a_3 \sqrt{2}\ \pi^{-1} \cos 3y \cos \ell x \tag{A.13}$$

and the spectral model is

$$\dot{a}_1 = \Lambda_1 a_2 a_3 - \nu \lambda_1 a_1 - H_1/\lambda_1 \tag{A.14}$$

$$\dot{a}_2 = \Lambda_2 a_1 a_3 - \nu \lambda_2 a_2 + \Gamma_1 a_3 - H_2/\lambda_2 \tag{A.15}$$

$$\dot{a}_3 = \Lambda_3 a_1 a_2 - \nu \lambda_3 a_3 - \Gamma_2 a_2 - H_3/\lambda_3 \tag{A.16}$$

The singular point is a cusp point when U = 0 and it is given by

$$\left.\begin{aligned} a_1 &= 0 \\ a_2 &= -\nu[\lambda_1 \lambda_3(\Lambda_1 \Lambda_3)^{-1}]^{1/2} \\ a_3 &= 0 \end{aligned}\right\} \tag{A.17}$$

In this case the canonical unfolding parameters μ_1 and μ_2 for the constant and linear terms of the quintic polynomial are related to H_1 and H_2 by

$$\mu_1 = -H_1/\lambda_1 \tag{A.18}$$

$$\mu_2 = (H_2/\hat{H}_2 - 1)2\nu\lambda_1 \tag{A.19}$$

or to H_2 and H_3 by

$$\mu_1 = \Lambda_1 \hat{H}_2(\nu^2 \lambda_2^2 \lambda_3^2)^{-1} H_3 \tag{A.20}$$

$$\mu_2 = (H_2/\hat{H}_2 - 1)2\nu\lambda_1 \tag{A.21}$$

The singular point is a butterfly point when $U \neq 0$; for the quadratic wind profile

$$U(y) = |U|_1 (\pi^2/4 - y^2) \tag{A.22}$$

the point is specified by

$$\left.\begin{aligned} |\hat{U}|_1 &= 16 \lambda_2 \lambda_3 d[3\ell(\lambda_3^2 \Lambda_3 + \lambda_2^2 \Lambda_2)]^{-1} \\ a_1 &= -d(\lambda_3^2 \Lambda_3 - \lambda_2^2 \Lambda_2)[\Lambda_2 \Lambda_3(\lambda_3^2 \Lambda_3 + \lambda_2^2 \Lambda_2)]^{-1} \\ a_2 &= 0 \\ a_3 &= 0 \end{aligned}\right\} \tag{A.23}$$

The canonical unfolding parameters γ_1, γ_2, γ_3, and γ_4 for the constant, linear, quadratic and cubic terms of the quintic polynomial are related to H_1, H_2, H_3 and $|U|_1$ by

$$\gamma_1 = - H_3 \lambda_3^{-1} + H_2 e_6(\lambda_2 e_5)^{-1} \quad (A.24)$$

$$\gamma_2 = 3\ell \lambda_3(|U|_1 - |\hat{U}|_1) (8 \lambda_2 e_2)^{-1} \quad (A.25)$$

$$\gamma_3 = - H_2(\lambda_2 e_5)^{-1} \quad (A.26)$$

$$\gamma_4 = - (H_1 - \hat{H}_1) (\lambda_1 e_4)^{-1} - 3\ell \lambda_3 e_1(|U|_1 - |\hat{U}|_1) (8\lambda_2 e_2 e_4)^{-1} \quad (A.27)$$

Table A.3

Nondimensional Variables & Parameters: Vickroy and Dutton Model

Symbol	Name	Definition
a_1, a_2, a_3	Fourier coefficients for ψ	
A_1, A_2, A_3	Fourier coefficients for basic steady solution	
$\hat{A}_1$	critical value of A_1: butterfly point	$\dfrac{- d(\lambda_3^2 \Lambda_3 - \lambda_2^2 \Lambda_2)}{\Lambda_2\Lambda_3(\lambda_3^2 \Lambda_3 + \lambda_2^2 \Lambda_2)}$
$\hat{A}_2$	critical value of A_2: cusp point	$- \nu[\lambda_1 \lambda_3(\Lambda_1 \Lambda_3)^{-1}]$
c	coefficient in $\|\hat{U}\|_2$	$\ell[\frac{9}{2} (\lambda_3\Lambda_3 + \lambda_2\Lambda_2) +(\frac{3}{16} \pi^2 - \frac{45}{32})(\lambda_3^2\Lambda_3 + \lambda_2^2\Lambda_2)] \times(\lambda_2\lambda_3)^{-1}$
d	coefficient in $\|\hat{U}\|_1$ or $\|\hat{U}\|_2$	$\pm \nu(- \Lambda_2 \Lambda_3 \lambda_2 \lambda_3)^{1/2}$
D		$- 8 \ell(15 \pi^2)^{-1}$
e_1		$\dfrac{\lambda_1 d(\lambda_2^2 \Lambda_2 - \lambda_3^2 \Lambda_3)}{[2 \Lambda_2 \Lambda_3 \lambda_3(\lambda_3^2 \Lambda_3 + \lambda_2^2 \Lambda_2)]}$
e_2		$d \lambda_3[\nu(\lambda_3^2 \Lambda_3 + \lambda_2^2 \Lambda_2)]^{-1}$

Table A.3 (Con't)

Symbol	Name	Definition
e_3		$-d\,\lambda_2^{\,2}[\nu\lambda_3(\lambda_3^{\,2}\Lambda_3 + \lambda_2^{\,2}\Lambda_2)]^{-1}$
e_4		$-d\,\nu^2\lambda_1^{\,2}\lambda_2[2\,\Lambda_1\Lambda_2^{\,2}\Lambda_3\lambda_3]^{-1}$
e_5		$-\nu\,\lambda_1\lambda_2\,d(\lambda_3\Lambda_1\Lambda_2\Lambda_3)^{-1}$
e_6		$\nu^2\,\lambda_1\,\lambda_2(\Lambda_1\,\Lambda_2)^{-1}$
$H(x,y)$	Newtonian heating rate	
$H'(x,y)$	perturbation heating rate	$H(x,y) - H_o(y)$
$H_o(y)$	basic Newtonian heating rate	$\nu\,\partial^3U/\partial y^3$
H_1	Newtonian heating coefficient	$-\pi^{-1}\int\int_A H(x,y)\sin y\,dA$
$\hat{H}_1$	critical value of H_1: butterfly point	$\dfrac{d\,\nu\lambda_1^{\,2}(\lambda_3^{\,2}\,\Lambda_3 - \lambda_2^{\,2}\,\Lambda_2)}{\Lambda_2\,\Lambda_3(\lambda_3^{\,2}\,\Lambda_3 + \lambda_2^{\,2}\,\Lambda_2)}$
H_2	Newtonian heating coefficient	$-2\sqrt{2}\,\pi^{-1}\int\int_A H(x,y)\sin \ell x \times \cos y\,dA$
$\hat{H}_2$	critical value of H_2: cusp point	$\nu^2\,\lambda_2^{\,2}[\lambda_1\,\lambda_3(\Lambda_1\,\Lambda_3)^{-1}]^{1/2}$
H_3	Newtonian heating coefficient	$-2\sqrt{2}\,\pi^{-1}\int\int_A H(x,y)\cos \ell x \times \cos 3y\,dA$
ΔH	differential heating gradient for quartic profile	$24\nu\,\lvert U\rvert_2$
$J(f,g)$	Jacobian operator	$\dfrac{\partial f}{\partial x}\dfrac{\partial g}{\partial y} - \dfrac{\partial f}{\partial y}\dfrac{\partial g}{\partial x}$
ℓ	integer wavenumber	

Table A.3 (Con't)

Symbol	Name	Definition
t	time	
$U(y)$	basic current:	
	quadratic profile	$\lvert U\rvert_1(\pi^2/4 - y^2)$
	quartic profile	$\lvert U\rvert_2(\pi^4/16 - y^4)$
$\lvert U\rvert_1$	amplitude of quadratic profile	
$\lvert\hat{U}\rvert_1$	critical amplitude of quadratic profile: butterfly point	$\dfrac{16\,\lambda_2\,\lambda_3\,d}{3\ell(\lambda_3^{\,2}\,\Lambda_3 + \lambda_2^{\,2}\,\Lambda_2)}$
$\lvert U\rvert_2$	amplitude of quartic profile	
$\lvert\hat{U}\rvert_2$	critical amplitude of quartic profile: butterfly point	$2\,d/c$
x	length	$0 \leq x \leq 2\pi$
y	width	$-\pi/2 \leq y \leq \pi/2$
$\alpha_1, \alpha_2, \alpha_3$	Fourier coefficients for perturbation solution	
β	latitudinal variation of Coriolis parameter	
$\gamma_1, \gamma_2, \gamma_3, \gamma_4$	canonical unfolding parameters for butterfly point	
Γ_1	Fourier coefficient of $U(y)$:	
	quadratic profile	$3\ell\,\lambda_3\lvert U\rvert_1(8\,\lambda_2)^{-1}$
	quartic profile	$[(\frac{3}{16}\pi^2 - \frac{45}{32})\,\lambda_3 + \frac{9}{2}]\,\lvert U\rvert_2\,\ell\lambda_2^{\,-1}$

Table A.3 (Con't)

Symbol	Name	Definition
Γ_2	Fourier coefficient of U(y):	
	quadratic profile	$3\ell\ \lambda_2 \lvert U \rvert_1 (8\ \lambda_3)^{-1}$
	quartic profile	$[(\frac{3}{16}\pi^2 - \frac{45}{32})\ \lambda_2 + \frac{9}{2}] \lvert U \rvert_2\ \ell \lambda_3^{-1}$
λ_1	eigenvalue	1
λ_2	eigenvalue	$1 + \ell^2$
λ_3	eigenvalue	$9 + \ell^2$
μ_1, μ_2	canonical unfolding parameters for cusp point	
Λ_1	interaction coefficient	$(\lambda_2 - \lambda_3)D/\lambda_1$
Λ_2	interaction coefficient	$(\lambda_3 - \lambda_1)D/\lambda_2$
Λ_3	interaction coefficient	$(\lambda_1 - \lambda_2)D/\lambda_3$
ν	eddy viscosity	
ψ	stream function	
ψ'	perturbation stream function	$\psi - \Psi$
$\Psi(y)$	basic stream function	
ω	characteristic exponent	
∇_H^2	Laplacian operator	$\frac{\partial^2}{\partial x^2} + \frac{\partial^2}{\partial y^2}$
∇_H^4		$\nabla_H^2 (\nabla_H^2)$

A.3 The Charney and DeVore Model

In Tables A.4 and A.5 we give the dimensional and nondimensional variables and parameters. The governing quasi-geostrophic equation in this case is

$$\frac{\partial}{\partial t}(\nabla_H^2\psi - \psi\lambda^{-2}) + J(\psi, \nabla_H^2\psi + h) + \beta\frac{\partial\psi}{\partial x} = -k[\nabla_H^2(\psi - \psi^*)] \tag{A.28}$$

The expansion for ψ is

$$\psi = \psi_C\sqrt{2}\cos\frac{2y}{L} + \psi_K\, 2\cos\frac{nx}{L}\sin\frac{y}{L} + \psi_N\, 2\sin\frac{nx}{L}\sin\frac{2y}{L} \tag{A.29}$$

and the spectral system is

$$\dot{\psi}_K = -\delta_{n1}\,\psi_C\,\psi_N - k\,\psi_K + k\,\psi_K^* \tag{A.30}$$

$$\dot{\psi}_C = \varepsilon_n\,\psi_K\,\psi_N - k\,\psi_C + h_{02}\,\psi_N + k\,\psi_C^* \tag{A.31}$$

$$\dot{\psi}_N = \delta_{n2}\,\psi_C\,\psi_K - k\,\psi_N - h_{n2}\,\psi_C + k\,\psi_N^* \tag{A.32}$$

Table A.4

Dimensional Variables: Charney and DeVore Model

Symbol	Name
D_E	Ekman depth $(2\,\nu_E/f_o)^{1/2}$
f_o	Coriolis parameter
g	acceleration of gravity
h_o	lower boundary elevation
H	domain height
L	domain width
ν_E	eddy viscosity

The singular point is a butterfly point only when n = 1; it is given by

$$\left.\begin{aligned}(h_o/H)_b &= \pm \frac{2k(-\varepsilon_n \delta_{n2})^{1/2}}{\varepsilon_n c_2 + \delta_{n2} c_1} \\ \psi_K &= \hat{\psi}_K{}^* = \frac{(\varepsilon_n c_2 - \delta_{n2} c_1)}{2\varepsilon_n \delta_{n2}} (h_o/H)_b \\ \psi_C &= 0 \\ \psi_N &= 0 \end{aligned}\right\} \quad (A.33)$$

and the four unfolding parameters are $(h_o/H)-(h_o/H)_b$, $\psi_K{}^* - \hat{\psi}_K{}^*$, $\psi_C{}^*$, and $\psi_N{}^*$.

Table A.5

Nondimensional Variables & Parameters: Charney and DeVore Model

Symbol	Name	Definition
c_1	coefficient in h_{02}	$8\sqrt{2}\, n(15\,\pi)^{-1}$
c_2	coefficient in h_{n2}	$32\sqrt{2}\, n[15\,\pi(n^2 + 4)]^{-1}$
h	lower boundary elevation	$(h_o/H)\cos(n\, x\, L^{-1})\sin(y\, L^{-1})$
h_{02}	Fourier coefficient of h	$c_1(h_o/H)$
h_{n2}	Fourier coefficient of h	$c_2(h_o/H)$
$(h_o/H)_b$	critical value of amplitude of lower boundary: butterfly point	$\pm\left[\frac{2\,k(-\varepsilon_n \delta_{n2})^{1/2}}{\varepsilon_n c_2 + \delta_{n2} c_1}\right]$
J(f,g)	Jacobian operator	$\frac{\partial f}{\partial x}\frac{\partial g}{\partial y} - \frac{\partial f}{\partial y}\frac{\partial g}{\partial x}$
k	dissipation rate	$D_E(2\,H)^{-1}$
n	integer wavenumber	
x	length	

Table A.5 (Con't)

Symbol	Name	Definition
y	width	
β	latitudinal variation of Coriolis parameter	
δ_{n1}	interaction coefficient	$64\sqrt{2}\, n^3 [15\,\pi(n^2 + 1)]^{-1}$
δ_{n2}	interaction coefficient	$64\sqrt{2}\, n(n^2 - 3)\,[15\,\pi(n^2 + 4)]^{-1}$
ε_n	interaction coefficient	$16\sqrt{2}\, n(5\,\pi)^{-1}$
λ^2		$g\, H\, f_o^{-2}\, L^{-2}$
ψ	stream function	
ψ_C, ψ_K, ψ_N	Fourier Components of ψ	
$\hat{\Psi}_K$	critical value of ψ_K: butterfly point	$\dfrac{(\varepsilon_n\, c_2 - \delta_{n2}\, c_1)}{2\varepsilon_n\, \delta_{n2}}\,(h_o/H)_b$
ψ^*	momentum source	
ψ_C^*, ψ_K^*, ψ_N^*	Fourier components of ψ^*	
Ψ_K^*	critical value of ψ_K^*: butterfly point	$\hat{\Psi}_K$
∇_H^2	Laplacian operator	$\dfrac{\partial^2}{\partial x^2} + \dfrac{\partial^2}{\partial y^2}$

A.4 The Veronis Model

The dimensional and nondimensional variables are summarized in Tables A.6 and A.7. The Boussinesq system is

$$\frac{\partial}{\partial t} \tilde{\nabla}^2 \psi^* = - K(\psi^*, \tilde{\nabla}^2 \psi^*) - f^* \frac{\partial v^*}{\partial z^*} + \sigma(1 + a^2)^{-1} \tilde{\nabla}^4 \psi^* \tag{A.34}$$

$$+ \sigma(1 + a^2) \frac{\partial \theta^*}{\partial x^*} - \sigma\, \alpha(1 + a^2)\, a^{-1} \frac{\partial \theta^*}{\partial z^*} + \sigma(1 + a^2)\,(h + \alpha\, r\, a^{-1})$$

$$\frac{\partial v^*}{\partial t^*} = - K(\psi^*, v^*) + f^* \frac{\partial \psi^*}{\partial z^*} + \sigma(1 + a^2)^{-1} \tilde{\nabla}^2 v^* \tag{A.35}$$

$$\frac{\partial \theta^*}{\partial t^*} = - K(\psi^*, \theta^*) + r \frac{\partial \psi^*}{\partial x^*} + h \frac{\partial \psi^*}{\partial z^*} + (1 + a^2)^{-1} \tilde{\nabla}^2 \theta^* + Q(x^*) \tag{A.36}$$

The Fourier expansions for the variables are

$$\psi^* = \sqrt{2}\, x_1 \sin x^* \sin z^* \tag{A.37}$$

$$\theta^* = \sqrt{2}\, x_2 \cos x^* \sin z^* - x_3 \sin 2z^* \tag{A.38}$$

$$v^* = - \sqrt{2}\, x_4 \sin x^* \cos z^* + x_5 \sin 2x^* \tag{A.39}$$

and the spectral model is

$$\dot{x}_1 = - \sigma\, x_1 + \sigma\, x_2 + 16\sqrt{2}\, \sigma\, \alpha(3\pi^2 a)^{-1} x_3 + f^*(1 + a^2)^{-1} x_4 \tag{A.40}$$

$$- 8\sqrt{2}\, \sigma(h + \alpha\, r\, a^{-1})\pi^{-2}$$

$$\dot{x}_2 = - x_1 x_3 + r\, x_1 - x_2 + 2\sqrt{2}\, q\pi^{-1} \tag{A.41}$$

$$\dot{x}_3 = x_1 x_2 - b\, x_3 - 16\sqrt{2}\, h(3\pi^2)^{-1} x_1 \tag{A.42}$$

$$\dot{x}_4 = - x_1 x_5 - f^* x_1 - \sigma\, x_4 \tag{A.43}$$

$$\dot{x}_5 = x_1 x_4 - \sigma\, b\, a^2 x_5 \tag{A.44}$$

Table A.6

Dimensional Variables: Veronis Model

Symbol	Name
f	Coriolis parameter
g	acceleration of gravity
H	domain height
L	domain width
p_o	basic state pressure
p_{oo}	value of $p_o(0,0)$
t	time
T	temperature
T_o	value of T at $(x,z) = (0,0)$ when $\theta = 0$
$\Delta_x T$	horizontal temperature difference
$\Delta_z T$	vertical temperature difference
v	meridional velocity component
x	horizontal distance
z	elevation
θ	perturbation temperature
κ	eddy thermometric conductivity
ν	eddy viscosity
ρ_o	(constant) basic density
ψ	stream function

Table A.7

Nondimensional Variables & Parameters: Veronis Model

Symbol	Name	Definition
a	aspect ratio	H/L
b	shape parameter	$4(1 + a^2)^{-1}$
f^*	Coriolis parameter	$(1 + a^2)\ \pi^2\ \kappa\ f^{-1}\ H^{-2}$
f_c^*	critical value of f*: butterfly point	$\sigma^2\ a(1 + a^2)^{1/2}(1 - \sigma^2\ a^2)^{1/2}$
h	Hadley number	$-\ \Delta_x T\ r\ (\Delta_z T)^{-1}$
$K(f,g)$	Jacobian operator	$\frac{\partial f}{\partial x^*}\frac{\partial g}{\partial z^*} - \frac{\partial f}{\partial z^*}\frac{\partial g}{\partial x^*}$
q	Fourier coefficient of Q(x*)	$\frac{2}{\pi}\int_0^\pi Q(x^*)\cos x^*\,dx^*$
$Q(x^*)$	Newtonian heating rate	
r	normalized Rayleigh number	R/R_s
r_c	critical value of r: butterfly point	$-\ (\sigma^2\ a^2 - 1)^{-1}$
r_s	critical value of r: cusp point	$1 + f^{*2}\ \sigma^{-2}(1 + a^2)^{-1}$
R	Rayleigh number	$-\ g\ \Delta_z T\ H^3(T_0\ \nu\ \kappa)^{-1}$
R_s	critical Rayleigh number (f* = 0)	$(1 + a^2)^3\ \pi^4\ a^{-2}$
t^*	time	$\pi^2(1 + a^2)\ \kappa\ t\ H^{-2}$
u^*	latitudinal velocity component	$-\ \partial\psi^*/\partial z^*$
v^*	meridional velocity component	$a\ H\ v(1 + a^2)^{-1}\ \kappa^{-1}\ \pi^{-1}$
w^*	vertical velocity component	$\partial\psi^*/\partial x^*$

Table A.7 (con't)

Symbol	Name	Definition
x_1	Fourier coefficient for ψ^*	
x_2, x_3	Fourier coefficients for θ^*	
x_4, x_5	Fourier coefficients for v^*	
x^*	horizontal distance	$\pi \; x \; L^{-1}; \quad 0 \leq x^* \leq \pi$
y	value of x_1 at butterfly point	
z^*	elevation	$\pi \; z \; H^{-1}; \quad 0 \leq z^* \leq \pi$
α	tilt angle of domain	
θ^*	perturbation temperature	$- a^2 \; R \; \theta[\pi^3(1 + a^2)^3 \; (\Delta_z T)]^{-1}$
μ_1, μ_2, μ_3, μ_4	canonical unfolding parameters	
σ	Prandtl number	ν/κ
ψ^*	stream function	$a \; \psi(1 + a^2)^{-1} \; \kappa^{-1}$
$\tilde{\nabla}^2$	Laplacian operator	$a^2 \dfrac{\partial^2}{\partial x^{*2}} + \dfrac{\partial^2}{\partial z^{*2}}$
$\tilde{\nabla}^4$		$\tilde{\nabla}^2(\tilde{\nabla}^2 \;)$

There are several butterfly points in this model, and they are listed in Table A.8. For the butterfly point $r = -(\sigma^2 a^2 - 1)^{-1}$, the canonical parameters μ_1, μ_2, μ_3, and μ_4 for the constant, linear, quadratic, and cubic terms of the quintic polynomial are related to the parameters r, h, f*, and q via

$$f^* = f_c^*[1 + (\mu_2 + b\,\mu_4)\,(2\sigma)^{-1}] \tag{A.45}$$

$$r = r_c[1 + (\mu_2 + b\,\sigma^2\,a^2\,\mu_4)\sigma^{-1}] \tag{A.46}$$

$$h = -\,3\pi^2(\mu_1 + b\,\mu_3)\,(8\sqrt{2}\,\sigma)^{-1} \tag{A.47}$$

$$q = -\,\pi(2\,\mu_1 + 3\,b\,\mu_3)\,(2\sqrt{2}\,\sigma)^{-1} \tag{A.48}$$

or to the parameters r, α, f*, and q via

$$f^* = f_c^*[1 + (\mu_2 + b\,\mu_4)\,(2\sigma)^{-1}] \tag{A.49}$$

$$r = r_c[1 + (\mu_2 + b\,\sigma^2\,a^2\,\mu_4)\sigma^{-1}] \tag{A.50}$$

$$\alpha = 3\pi^2\,a\,(\mu_1 + b\,\mu_3)\,(\sigma^2\,a^2 - 1)\,(8\sqrt{2}\,\sigma)^{-1} \tag{A.51}$$

$$q = -\,\pi(2\,\mu_1 + 3\,b\,\mu_3)\,(2\sqrt{2}\,\sigma)^{-1} \tag{A.52}$$

Table A.8

Butterfly Points in the Veronis Model

$(q = \alpha = 0)$

r	h	f^{*2}	σ	x_1	x_2	x_3	x_4	x_5
-2.76	$\frac{-32.93}{(1+a^2)^{1/2}}$	$\frac{11.91(1+a^2)}{a^2}$	$\frac{0.409}{a}$	$\frac{2.52}{(1+a^2)^{1/2}}$	$\frac{-18.13}{(1+a^2)^{1/2}}$	4.43	-2.02	$\frac{-3.12(1+a^2)^{1/2}}{a}$
-2.76	$\frac{32.93}{(1+a^2)^{1/2}}$	$\frac{11.91(1+a^2)}{a^2}$	$\frac{0.409}{a}$	$\frac{-2.52}{(1+a^2)^{1/2}}$	$\frac{18.13}{(1+a^2)^{1/2}}$	4.43	2.02	$\frac{-3.12(1+a^2)^{1/2}}{a}$
-71.06	$\frac{-139.5}{(1+a^2)^{1/2}}$	$\frac{158.3(1+a^2)}{a^2}$	$\frac{7.34}{a}$	$\frac{10.66}{(1+a^2)^{1/2}}$	$\frac{-128.7}{(1+a^2)^{1/2}}$	-58.98	-11.96	$\frac{-4.34(1+a^2)^{1/2}}{a}$
-71.06	$\frac{139.5}{(1+a^2)^{1/2}}$	$\frac{158.3(1+a^2)}{a^2}$	$\frac{7.34}{a}$	$\frac{-10.66}{(1+a^2)^{1/2}}$	$\frac{128.7}{(1+a^2)^{1/2}}$	-58.98	11.96	$\frac{-4.34(1+a^2)^{1/2}}{a}$
$\frac{-1}{(\sigma^2 a^2-1)}$	0	$\frac{-\sigma^4 a^2(1+a^2)}{(\sigma^2 a^2-1)}$	—	0	0	0	0	0

REFERENCES

Ahlers, G. and R. P. Behringer, 1978a: Evolution of turbulence from the Rayleigh-Bénard instability. Phys. Rev. Letters, 40, 712-716.

Ahlers, G. and R. P. Behringer, 1978b: The Rayleigh-Bénard instability and the evolution of turbulence. Sup. Prog. Theor. Phys., No. 64, 186-201.

Boldrighini, C. and V. Franceschini, 1979: A five-dimensional truncation of the plane incompressible Navier-Stokes equations. Comm. Math Phys., 64, 159-170.

Chandrasekhar, S., 1961: Hydrodynamic and Hydromagnetic Stability. Clarendon Press, 652 pp.

Charney, J. G. and J. G. DeVore, 1979: Multiple flow equilibria in the atmosphere and blocking. J. Atmos. Sci., 36, 1205-1216.

Chillingworth, D. R. J. and P. J. Holmes, 1980: Dynamical systems and models for reversals of the earth's magnetic field. Math. Geology, 12, 41-59.

Clark, J. H. E., 1983: The effect of topography on the evolution of unstable disturbances in a baroclinic atmosphere. Submitted to J. Atmos. Sci. for review.

Curry, J. H., 1978: A generalized Lorenz system. Comm. Math Phys., 60, 193-204.

Curry, J. H., 1979: Chaotic response to periodic modulation of model of a convecting fluid. Phys. Rev. Letters, 43, 1013-1016.

Dutton, J. A., 1976a: The nonlinear quasi-geostrophic equation. Part II: Predictability, recurrence and limit properties of thermally-forced and unforced flows. J. Atmos. Sci., 33, 1431-1453.

Dutton, J. A., 1976b: The Ceaseless Wind: An Introduction to the Theory of Atmospheric Motion. McGraw-Hill, 579 pp.

Dutton, J. A., 1982: Fundamental theorems of climate theory--some proved, some conjectured. SIAM Review, 24, 1-33.

Fenstermacher, P. R., H. L. Swinney, S. V. Benson, and J. P. Gollub, 1979: Bifurcations to periodic, quasiperiodic, and chaotic regimes in rotating and convecting fluids. Bifurcation Theory and Applications in Scientific Disciplines. O. Gurel and O. E. Rossler, Eds., N.Y. Acad. Sci., 708 pp.

Fultz, D., R. R. Long, G. B. Owens, W. Boham, R. Kaylor and J. Weil, 1959: Studies of thermal convection in a rotating cylinder with some implications for large-scale atmospheric motions. Meteor. Monographs, 4, No. 21, American Meteorological Society, 104 pp.

Gollub, J. P. and S. V. Benson, 1978: Chaotic response to periodic perturbation of a convecting fluid. Phys. Rev. Letters, 41, 948-951.

Golubitsky, M. and D. G. Schaeffer, 1979: A theory for imperfect bifurcation via singularity theory. Comm. Pure Appl. Math, 32, 21-98.

Gromoll, D. and W. Meyer, 1969: On differentiable functions with isolated critical points. Topology, 8, 361-369.

Hirsch, M. and S. Smale, 1974: Differentiable Equations, Dynamical Systems, and Linear Algebra. Academic Press, 358 pp.

Iooss, G. and D. Joseph, 1980: Elementary Stability and Bifurcation Theory. Springer-Verlag, 286 pp.

Krishnamurti, R., 1970a: On the transition to turbulent convection. Part 1. The transition from two- to three-dimensional flow. J. Fluid Mech., 42, 295-307.

Krishnamurti, R., 1970b: On the transition to turbulent convection. Part 2. The transition to time-dependent flow. J. Fluid Mech., 42, 309-320.

Krishnamurti, R., 1973: Some further studies on the transition to turbulent convection. J. Fluid Mech., 60, 285-303.

REFERENCES (Con't)

Levine, H., 1971: Singularities of differentiable mappings. Proceedings of the Liverpool Singularities Symposium, Lecture Notes in Mathematics, 192. C.T.C. Wall, Ed., Springer-Verlag, 1-89.

Lorenz, E. N., 1962: Simplified dynamic equations applied to the rotating basin experiments. J. Atmos. Sci., 19, 39-51.

Lorenz, E. N., 1963: Deterministic nonperiodic flow. J. Atmos. Sci., 20, 130-141.

Lorenz, E. N., 1980: Attractor sets and quasi-geostrophic equilibrium. J. Atmos. Sci., 37, 1685-1699.

McLaughlin, J. B. and P. C. Martin, 1975: Transition to turbulence in a statically stressed fluid system. Phys. Rev. A, 12, 186-203.

Marcus, P. S., 1981: Effects of truncation in modal representations of thermal convection. J. Fluid Mech., 103, 241-255.

Marsden, J. E. and M. McCracken, 1976: The Hopf Bifurcation and Its Applications. Applied Mathematical Sciences, 19, Springer-Verlag, 408 pp.

Mather, J., 1968: Stability of C^∞ mappings III: Finitely determined map germs. Publ. Math I.H.E.S., 35, 127-156.

Maurer, J. and A. Libchaber, 1979: Rayleigh-Bénard experiment in liquid helium; frequency locking and the onset of turbulence. Le Journal of Physique-Lettres, 40, 419-423.

Milnor, J., 1965: Topology from the Differentiable Viewpoint, University Press of Virginia, 61 pp.

Mitchell, K. E. and J. A. Dutton, 1981: Bifurcations from stationary to periodic solutions in a low-order model of forced, dissipative, barotropic flow. J. Atmos. Sci., 38, 690-716.

Ogura, Y. and A. Yagihashi, 1970: A numerical study of finite-amplitude time-dependent convection induced by time-dependent internal heating: truncated systems. J. Meteo. Soc. Japan, 48, 1-17.

Poston, T. and I. Stewart, 1978: Catastrophe Theory and Its Applications, Pitman, 491 pp.

Richards, P. I., 1959: Manual of Mathematical Physics. Pergamon Press, 486 pp.

Saltzman, B., 1962: Finite amplitude free convection as an initial value problem--I. J. Atmos. Sci., 19, 329-341.

Shirer, H. N., 1980: Bifurcation and stability in a model of moist convection in a shearing environment. J. Atmos. Sci., 37, 1586-1602.

Shirer, H. N., 1982: Toward a unified theory of atmospheric convective instability. Cloud Dynamics, E.M. Agee and T. Asai, Eds., D. Reidel Publishing Co., 163-177.

Shirer, H. N. and J. A. Dutton, 1979: The branching hierarchy of multiple solutions in a model of moist convection. J. Atmos. Sci., 36, 1705-1721.

Shirer, H. N. and R. Wells, 1982: Improving spectral models by unfolding their singularities. J. Atmos. Sci., 39, 610-621.

Sommeria, G. and M. LeMone, 1978: Direct testing of a three-dimensional model of the planetary boundary layer against experimental data. J. Atmos. Sci., 35, 25-39.

Swinney, H. L., 1978: Hydrodynamic instabilities and the transition to turbulence. Sup. Prog. Theor. Phys., No. 64, 164-175.

Swinney, H. L. and J. P. Gollub, 1981: Hydrodynamic Instabilities and the Transition to Turbulence. Springer-Verlag, 292 pp.

Tavantzis, J., E. L. Reiss and B. J. Matkowsky, 1978: On the smooth transition to convection. SIAM J. Appl. Math, 34, 322-337.

REFERENCES (Con't)

Thom, R., 1964: Local properties of differentiable mappings. Differential Analysis, Bombay Colloquium. Oxford University Press, 191-202.

Thom, R., 1976: Structural Stability and Morphogenesis. W. A. Benjamin, 339 pp.

Veronis, G., 1966: Motions at subcritical values of the Rayleigh number in a rotating fluid. J. Fluid Mech., 24, 545-554.

Vickroy, J. G. and J. A. Dutton, 1979: Bifurcation and catastrophe in a simple, forced, dissipative, quasi-geostrophic flow. J. Atmos. Sci., 36, 42-52.

Wiin-Nielsen, A., 1979: Steady states and stability properties of a low-order barotropic system with forcing and dissipation. Tellus, 31, 375-386.

Yost, D. A. and H. N. Shirer, 1982: Bifurcation and stability of low-order steady flows in horizontally and vertically forced convection. J. Atmos. Sci., 39, 114-125.

Vol. 144: Topics in Nuclear Physics I. A Comprehensive Review of Recent Developments. Edited by T.T.S. Kuo and S.S.M. Wong. XX, 567 pages. 1981.

Vol. 145: Topics in Nuclear Physics II. A Comprehensive Review of Recent Developments. Proceedings 1980/81. Edited by T. T. S. Kuo and S. S. M. Wong. VIII, 571-1.082 pages. 1981.

Vol. 146: B. J. West, On the Simpler Aspects of Nonlinear Fluctuating. Deep Gravity Waves. VI, 341 pages. 1981.

Vol. 147: J. Messer, Temperature Dependent Thomas-Fermi Theory. IX, 131 pages. 1981.

Vol. 148: Advances in Fluid Mechanics. Proceedings, 1980. Edited by E. Krause. VII, 361 pages. 1981.

Vol. 149: Disordered Systems and Localization. Proceedings, 1981. Edited by C. Castellani, C. Castro, and L. Peliti. XII, 308 pages. 1981.

Vol. 150: N. Straumann, Allgemeine Relativitätstheorie und relativistische Astrophysik. VII, 418 Seiten. 1981.

Vol. 151: Integrable Quantum Field Theory. Proceedings, 1981. Edited by J. Hietarinta and C. Montonen. V, 251 pages. 1982.

Vol. 152: Physics of Narrow Gap Semiconductors. Proceedings, 1981. Edited by E. Gornik, H. Heinrich and L. Palmetshofer. XIII, 485 pages. 1982.

Vol. 153: Mathematical Problems in Theoretical Physics. Proceedings, 1981. Edited by R. Schrader, R. Seiler, and D.A. Uhlenbrock. XII, 429 pages. 1982.

Vol. 154: Macroscopic Properties of Disordered Media. Proceedings, 1981. Edited by R. Burridge, S. Childress, and G. Papanicolaou. VII, 307 pages. 1982.

Vol. 155: Quantum Optics. Proceedings, 1981. Edited by C.A. Engelbrecht. VIII, 329 pages. 1982.

Vol. 156: Resonances in Heavy Ion Reactions. Proceedings, 1981. Edited by K.A. Eberhard. XII, 448 pages. 1982.

Vol. 157: P. Niyogi, Integral Equation Method in Transonic Flow. XI, 189 pages. 1982.

Vol. 158: Dynamics of Nuclear Fission and Related Collective Phenomena. Proceedings, 1981. Edited by P. David, T. Mayer-Kuckuk, and A. van der Woude. X, 462 pages. 1982.

Vol. 159: E. Seiler, Gauge Theories as a Problem of Constructive Quantum Field Theory and Statistical Mechanics. V, 192 pages. 1982.

Vol. 160: Unified Theories of Elementary Particles. Critical Assessment and Prospects. Proceedings, 1981. Edited by P. Breitenlohner and H.P. Dürr. VI, 217 pages. 1982.

Vol. 161: Interacting Bosons in Nuclei. Proceedings, 1981. Edited by J.S. Dehesa, J.M.G. Gomez, and J. Ros. V, 209 pages. 1982.

Vol. 162: Relativistic Action at a Distance: Classical and Quantum Aspects. Proceedings, 1981. Edited by J. Llosa. X, 263 pages. 1982.

Vol. 163: J. S. Darrozes, C. Francois, Mécanique des Fluides Incompressibles. XIX, 459 pages. 1982.

Vol. 164: Stability of Thermodynamic Systems. Proceedings, 1981. Edited by J. Casas-Vázquez and G. Lebon. VII, 321 pages. 1982.

Vol. 165: N. Mukunda, H. van Dam, L.C. Biedenharn, Relativistic Models of Extended Hadrons Obeying a Mass-Spin Trajectory Constraint. Edited by A. Böhm and J.D. Dollard. VI, 163 pages. 1982.

Vol. 166: Computer Simulation of Solids. Edited by C.R.A. Catlow and W.C. Mackrodt. XII, 320 pages. 1982.

Vol. 167: G. Fieck, Symmetry of Polycentric Systems. VI, 137 pages, 1982.

Vol. 168: Heavy-Ion Collisions. Proceedings, 1982. Edited by G. Madurga and M. Lozano. VI, 429 pages. 1982.

Vol. 169: K. Sundermeyer, Constrained Dynamics. IV, 318 pages. 1982.

Vol. 170: Eighth International Conference on Numerical Methods in Fluid Dynamics. Proceedings, 1982. Edited by E. Krause. X, 569 pages. 1982.

Vol. 171: Time-Dependent Hartree-Fock and Beyond. Proceedings, 1982. Edited by K. Goeke and P.-G. Reinhard. VIII, 426 pages. 1982.

Vol. 172: Ionic Liquids, Molten Salts and Polyelectrolytes. Proceedings, 1982. Edited by K.-H. Bennemann, F. Brouers, and D. Quitmann. VII, 253 pages. 1982.

Vol. 173: Stochastic Processes in Quantum Theory and Statistical Physics. Proceedings, 1981. Edited by S. Albeverio, Ph. Combe, and M. Sirugue-Collin. VIII, 337 pages. 1982.

Vol. 174: A. Kadić, D.G.B. Edelen, A Gauge Theory of Dislocations and Disclinations. VII, 290 pages. 1983.

Vol. 175: Defect Complexes in Semiconductor Structures. Proceedings, 1982. Edited by J. Giber, F. Beleznay, J. C. Szép, and J. László. VI, 308 pages. 1983.

Vol. 176: Gauge Theory and Gravitation. Proceedings, 1982. Edited by K. Kikkawa, N. Nakanishi, and H. Nariai. X, 316 pages. 1983.

Vol. 177: Application of High Magnetic Fields in Semiconductor Physics. Proceedings, 1982. Edited by G. Landwehr. XII, 552 pages. 1983.

Vol. 178: Detectors in Heavy-Ion Reactions. Proceedings, 1982. Edited by W. von Oertzen. VIII, 258 pages. 1983.

Vol. 179: Dynamical Systems and Chaos. Proceedings, 1982. Edited by L. Garrido. XIV, 298 pages. 1983.

Vol. 180: Group Theoretical Methods in Physics. Proceedings, 1982. Edited by M. Serdaroğlu and E. İnönü. XI, 569 pages. 1983.

Vol. 181: Gauge Theories of the Eighties. Proceedings, 1982. Edited by R. Raitio and J. Lindfors. V, 644 pages. 1983.

Vol. 182: Laser Physics. Proceedings, 1983. Edited by J. D. Harvey and D. F. Walls. V, 263 pages. 1983.

Vol. 183: J. D. Gunton, M. Droz, Introduction to the Theory of Metastable and Unstable States. VI, 140 pages. 1983.

Vol. 184: Stochastic Processes – Formalism and Applications. Proceedings, 1982. Edited by G. S. Agarwal and S. Dattagupta. VI, 324 pages. 1983.

Vol. 185: H. N. Shirer, R. Wells, Mathematical Structure of the Singularities at the Transitions between Steady States in Hydrodynamic Systems. XI, 276 pages. 1983.

Vol. 186: Critical Phenomena. Proceedings, 1982. Edited by F.J.W. Hahne. VII, 353 pages. 1983.

Selected Issues from

Lecture Notes in Mathematics

Vol. 836: Differential Geometrical Methods in Mathematical Physics. Proceedings, 1979. Edited by P. L. García, A. Pérez-Rendón, and J. M. Souriau. XII, 538 pages. 1980.

Vol. 837: J. Meixner, F. W. Schäfke and G. Wolf, Mathieu Functions and Spheroidal Functions and their Mathematical Foundations Further Studies. VII, 126 pages. 1980.

Vol. 838: Global Differential Geometry and Global Analysis. Proceedings 1979. Edited by D. Ferus et al. XI, 299 pages. 1981.

Vol. 840: D. Henry, Geometric Theory of Semilinear Parabolic Equations. IV, 348 pages. 1981.

Vol. 841: A. Haraux, Nonlinear Evolution Equations- Global Behaviour of Solutions. XII, 313 pages. 1981.

Vol. 842: Séminaire Bourbaki vol. 1979/80. Exposés 543-560. IV, 317 pages. 1981.

Vol. 843: Functional Analysis, Holomorphy, and Approximation Theory. Proceedings. Edited by S. Machado. VI, 636 pages. 1981.

Vol. 845: A. Tannenbaum, Invariance and System Theory: Algebraic and Geometric Aspects. X, 161 pages. 1981.

Vol. 849: P. Major, Multiple Wiener-Itô Integrals. VII, 127 pages. 1981.

Vol. 851: Stochastic Integrals. Proceedings, 1980. Edited by D. Williams. IX, 540 pages. 1981.

Vol. 852: L. Schwartz, Geometry and Probability in Banach Spaces. X, 101 pages. 1981.

Vol. 856: R. Lascar, Propagation des Singularités des Solutions d'Equations Pseudo-Différentielles à Caractéristiques de Multiplicités Variables. VIII, 237 pages. 1981.

Vol. 858: E. A. Coddington, H. S. V. de Snoo: Regular Boundary Value Problems Associated with Pairs of Ordinary Differential Expressions. V, 225 pages. 1981.

Vol. 861: Analytical Methods in Probability Theory. Proceedings 1980. Edited by D. Dugué, E. Lukacs, V. K. Rohatgi. X, 183 pages. 1981.

Vol. 866: J.-M. Bismut, Mécanique Aléatoire. XVI, 563 pages. 1981.

Vol. 878: Numerical Solution of Nonlinear Equations. Proceedings, 1980. Edited by E. L. Allgower, K. Glashoff, and H.-O. Peitgen. XIV, 440 pages. 1981.

Vol. 881: R. Lutz, M. Goze, Nonstandard Analysis. XIV, 261 pages. 1981.

Vol. 888: Padé Approximation and its Applications. Proceedings, 1980. Edited by M. G. de Bruin and H. van Rossum. VI, 383 pages. 1981.

Vol. 898: Dynamical Systems and Turbulence, Warwick, 1980. Proceedings. Edited by D. Rand and L.-S. Young. VI, 390 pages. 1981.

Vol. 901: Séminaire Bourbaki vol. 1980/81 Exposés 561-578. III, 299 pages. 1981.

Vol. 904: K. Donner, Extension of Positive Operators and Korovkin Theorems. XII, 182 pages. 1982.

Vol. 905: Differential Geometric Methods in Mathematical Physics. Proceedings, 1980. Edited by S.J. Andersson, H.-D. Doebner, and. H.R. Petry. VI, 309 pages. 1982.

Vol. 909: Numerical Analysis. Proceedings, 1981. Edited by J.P. Hennart. VII, 247 pages. 1982.

Vol. 912: Numerical Analysis. Proceedings, 1981. Edited by G. A. Watson. XIII, 245 pages. 1982.

Vol. 920: Séminaire de Probabilités XVI, 1980/81. Proceedings. Edité par J. Azéma et M. Yor. V, 622 pages. 1982.

Vol. 921: Séminaire de Probabilités XVI, 1980-81 Supplément: Géométrie Différentielle Stochastique. Proceedings. Edité par J. Azéma et M. Yor. III, 285 pages. 1982.

Vol. 922: B. Dacorogna, Weak Continuity and Weak Lower Semicontinuity of Non-Linear Functionals. V, 120 pages. 1982.

Vol. 923: Functional Analysis in Markov Processes. Proceedings, 1981. Edited by M. Fukushima. V, 307 pages. 1982.

Vol. 926: Geometric Techniques in Gauge Theories. Proceedings, 1981. Edited by R. Martini and E.M.de Jager. IX 219 pages. 1982.

Vol. 927: Y. Z. Flicker, The Trace Formula and Base Change for GL (3). XII, 204 pages. 1982.

Vol. 928: Probability Measures on Groups. Proceedings 1981. Edited by H. Heyer. X, 477 pages. 1982.

Vol. 929: Ecole d'Eté de Probabilités de Saint-Flour X - 1980. Proceedings, 1980. Edited by P.L. Hennequin. X, 313 pages. 1982.

Vol. 930: P. Berthelot, L. Breen, et W. Messing, Théorie de Dieudonné Cristalline II. XI, 261 pages. 1982.

Vol. 931: D.M. Arnold, Finite Rank Torsion Free Abelian Groups and Rings. VII, 191 pages. 1982.

Vol. 932: Analytic Theory of Continued Fractions. Proceedings, 1981. Edited by W.B. Jones, W.J. Thron, and H. Waadeland. VI, 240 pages. 1982.

Vol. 934: M. Sakai, Quadrature Domains. IV, 133 pages. 1982.

Vol. 935: R. Sot, Simple Morphisms in Algebraic Geometry. IV, 146 pages. 1982.

Vol. 936: S.M. Khaleelulla, Counterexamples in Topological Vector Spaces. XXI, 179 pages. 1982.

Vol. 937: E. Combet, Intégrales Exponentielles. VIII, 114 pages. 1982.

Vol. 938: Number Theory. Proceedings, 1981. Edited by K. Alladi. IX, 177 pages. 1982.

Vol. 942: Theory and Applications of Singular Perturbations. Proceedings, 1981. Edited by W. Eckhaus and E.M. de Jager. V, 363 pages. 1982.

Vol. 953: Iterative Solution of Nonlinear Systems of Equations. Proceedings, 1982. Edited by R. Ansorge, Th. Meis, and W. Törnig. VII, 202 pages. 1982.

Vol. 956: Group Actions and Vector Fields. Proceedings, 1981. Edited by J.B. Carrell. V, 144 pages. 1982.

Vol. 957: Differential Equations. Proceedings, 1981. Edited by D.G. de Figueiredo. VIII, 301 pages. 1982.

Vol. 963: R. Nottrot, Optimal Processes on Manifolds. VI, 124 pages. 1982.

Vol. 964: Ordinary and Partial Differential Equations. Proceedings, 1982. Edited by W.N. Everitt and B.D. Sleeman. XVIII, 726 pages. 1982.

Vol. 968: Numerical Integration of Differential Equations and Large Linear Systems. Proceedings, 1980. Edited by J. Hinze. VI, 412 pages. 1982.

Vol. 970: Twistor Geometry and Non-Linear Systems. Proceedings, 1980. Edited by H.-D. Doebner and T.D. Palev. V, 216 pages. 1982.

Vol. 972: Nonlinear Filtering and Stochastic Control. Proceedings, 1981. Edited by S.K. Mitter and A. Moro. VIII, 297 pages. 1983.

Vol. 978: J. Ławrynowicz, J. Krzyż, Quasiconformal Mappings in the Plane. VI, 177 pages. 1983.

Vol. 979: Mathematical Theories of Optimization. Proceedings, 1981. Edited by J.P. Cecconi and T. Zolezzi. V, 268 pages. 1983.